THE ISLANDS AND THE STARS

The Islands and the Stars

A History of Japan's Space Programs

SUBODHANA WIJEYERATNE

STANFORD UNIVERSITY PRESS
Stanford, California

Stanford University Press
Stanford, California

Library of Congress Cataloging-in-Publication Data
Names: Wijeyeratne, Subodhana, author.
Title: The islands and the stars : a history of Japan's space programs / Subodhana Wijeyeratne.
Description: Stanford, California : Stanford University Press, 2026. | Includes bibliographical references and index.
Identifiers: LCCN 2025018710 (print) | LCCN 2025018711 (ebook) | ISBN 9781503644144 (cloth) | ISBN 9781503644786 (paperback) | ISBN 9781503644793 (ebook)
Subjects: LCSH: Uchū Kōkū Kenkyū Kaihatsu Kikō—History. | Astronautics and state—Japan—History. | Rocketry—Japan—History. | Outer space—Exploration—Japan—History.
Classification: LCC QB500.266.J3 W55 2026 (print) | LCC QB500.266.J3(ebook) | DDC 629.40952—dc23/eng/20250418
LC record available at https://lccn.loc.gov/2025018710
LC ebook record available at https://lccn.loc.gov/2025018711

Cover design: Daniel Benneworth-Gray
Cover art: based on *Sketches of Famous Places in Japan; Asakusa Kinryuzan Temple* by Tsuchiya Kōitsu, 1938

The authorized representative in the EU for product safety and compliance is: Mare Nostrum Group B.V. | Mauritskade 21D | 1091 GC Amsterdam | The Netherlands | Email address: gpsr@mare-nostrum.co.uk | KVK chamber of commerce number: 96249943

For Luna,
my favorite celestial object

CONTENTS

Acknowledgments ix

Notes on Naming Conventions xi

List of Abbreviations xiii

Introduction 1

PART I

Child of War, 1920–1960

ONE The Transwar Origins of Japanese Rocketry 27

TWO Demilitarizing Rocketry in the Postwar Period 53

PART II

The Institutionalization of Japanese Space Research, 1960–1980

THREE Manager-Specialists in Japan's Space Program 79

FOUR Influence of Commercial Interest 92

FIVE Welcome and Resistance to Japanese Space Facilities 108

PART III
The Challenges of Advanced Spacefaring, 1980–2003

SIX Disseminating and Debating Japanese Space Policy 125

SEVEN Growing Ambitions and Difficulties with the United States 143

EIGHT Success and Failure in the 1990s 167

NINE Reform and the Creation of JAXA 182

Conclusion 196

Glossary 203

Notes 207

Bibliography 273

Index 319

ACKNOWLEDGMENTS

A man may write about exploring the vacuum, but he does not live in one. Accordingly, this book is the product of the input of far more people than just the name that appears on the cover.

My sincerest thanks go, first, to my PhD supervisor and mentor, Ian Miller. Thank you for your support and guidance of a neurotic grad student who snuck in promising to write about race, and then decided that it was all too depressing, and he'd rather write about giant rockets instead. My thanks also to Janet Browne, Matthew Hersch, and Andrew Gordon for their invaluable advice. I would also like to thank my Japanese mentors, Yasushi Sato and Takehiko Hashimoto, for all their insights into not only the technical aspects of this work, but also the culture and lived experience of the people this book is primarily about.

I am also very grateful for the intellectual guidance and help of Yulia Frumer, Charla Griffy-Brown, Nakao Gyo, Inaba Hajime, David Howell, Dong-won Kim, Angelika Koch, Shi-lin Loh, Christine Luk, Kuniko McVey, Jürgen P. Melzer, Takashi Nishiyama, Michael Thornton, and Brett Walker. I am particularly grateful as well to Yasunori Matogawa for his kindness, and for his patience in answering my many emails and questions. My thanks also to Satoki Kurokawa and Tomoko Kitagawa at JAXA, as well as Mitsubishi Heavy Industries and Nippon Kayaku, for taking the time to provide me with invaluable information necessary for this work.

This research would not have been possible without the kind sponsorship of the Dong-won Kim Foundation, the Japan Foundation, the Reischauer Institute for Japanese Studies, Tokyo University, the History Department of Harvard University, the Harvard Graduate School of Arts and Sciences, and the Purdue University History Department. I'd also like to thank the staff and faculty of Purdue University, whose warm welcome catalyzed the completion of this work—in particular David Atkinson, Will Gray, Michael Smith, and Margaret Tillman. My deepest appreciation also goes to the staff and faculty at Tokyo Woman's Christian University, who rekindled my love of teaching and confidence after a difficult time.

Lastly, my love and thanks to Sara Gothard, Kathleen Jackson, Jonathan Liang, Jim Neeson, David Schneider, Mutsumi Takahata, Chris White, and Luke Willert, for their help and support, and for being skilled and committed guardians of my sanity these past nine years. My love also to Balto, Boicey, Buddy, Trigger, Buzz, Leela, Saskia, Wolf, Lady Turanga Leela, and Lord Nibbler, all of whom reminded me that, sometimes, a walk in the sun is more important than all the rockets in Japan.

My sincerest apologies to anyone I've left out.

NOTES ON NAMING CONVENTIONS

The naming of Japanese rockets has been inconsistent. Within the H-series, for example, the first vehicle is often represented in sources using Roman numerals (e.g., H-I) and sometimes with Arabic numbering (e.g., H-1). Furthermore, the hyphen is excluded in later names (e.g., H3). For consistency's sake, this book refers to all rockets with Arabic numerals, and with the hyphen (H-1, H-2, and H-3).

Unlike the European and American tradition, Japanese satellites are known by a technical designation, such as ETS-IV, until launch, and are named only after they are successfully orbited. Again, for consistency's sake, this book will refer to all satellites by their technical designation. Exceptions will be made for devices such as Ohsumi and Hayabusa, which are widely known by their given names.

Macrons are used for names, but not for those well-known in the English language, such as Tokyo.

ABBREVIATIONS

AAF	Army Air Force Scientific Advisory Group.
AS	Asahi Shimbun.
AVSA	Avionics and Supersonic Aerodynamics group of the University of Tokyo, active 1954–1964.
CNSA	China National Space Administration.
ESA	European Space Agency, founded in 1975.
ETS	Series of experimental satellites that helped advance Japanese technology in the area between 1975 and 1994.
GEO	Geostationary orbit, a stable orbit where a satellite orbits Earth at the same speed as the planet's rotation, maintaining a fixed position relative to the surface.
GTO	Transition orbit enabling satellites to reach geostationary orbit by first achieving the required altitude.
ICBM	Intercontinental Ballistic Missile, a rocket capable of hitting distant locations with warheads.
IHI	Ishikawajima-Harima Heavy Industries.
IJA	Imperial Japanese Army, abolished in 1945, and replaced with JSDF.

IJN	Imperial Japanese Navy, abolished in 1945, and replaced with JSDF.
ISAS	Institute of Space and Aeronautical Science (1964–1984); Institute of Space and Astronautical Science (since 1984).
ISRO	Indian Space Research Organisation.
JAXA	Japan Aerospace Exploration Agency.
JEM	Japanese Experiment Module, a component of the International Space Station (ISS).
JSDF	Japan Self-Defense Force.
JSP	Japan Socialist Party.
KDD	Kokusai Denshin Denwa, Japanese communications corporation.
LDP	Liberal Democratic Party.
LFRs	Liquid-fuel rockets. Liquid fuel is advantageous in that engines using it can have their output varied ("throttling"), turned on, or turned off, allowing them to be used precisely. However, liquid fuel is unstable, combustible, and difficult to manufacture. Furthermore, LFRs can only be fueled a short time before launch and must be unfueled if launches are canceled.
MELCO	Mitsubishi Electric Corporation.
METI	Ministry of Economy, Trade, and Industry.
MHI	Mitsubishi Heavy Industries.
MITI	Ministry of International Trade and Industry, reformed into Ministry of Economy, Trade, and Industry (METI) in 2001.
MLV	Medium lift vehicle. These are rockets capable of lifting between 4,400 and 44,000 lb (2,000 and 20,000 kg).
NavTech	US Naval Technical Mission to Japan.
NASA	US National Aeronautics and Space Administration.
NASDA	National Space Development Agency of Japan, established 1969.
NDHC	National Diet of Japan House of Councilors.
NDHC Com.	National Diet of Japan House of Councilors Communications Committee.

NDHC Cab. National Diet of Japan House of Councilors Cabinet Committee.

NDHC E&C National Diet of Japan House of Councilors Committee on Education and Culture.

NDHC For. National Diet of Japan House of Councilors Foreign Affairs Committee.

NDHC S&T National Diet of Japan House of Councilors Special Committee on Science and Technology.

NDHC Set. National Diet of Japan House of Councilors Settlement Committee.

NDHC TIC National Diet of Japan House of Councilors Transportation, Information and Communication Committee.

NDHR National Diet of Japan House of Representatives.

NDHR Bud. National Diet of Japan House of Representatives Budget Committee.

NDHR Cab. National Diet of Japan House of Representatives Cabinet Committee.

NDHR Com. National Diet of Japan House of Representatives Communications Committee.

NDHR E&C National Diet of Japan House of Representatives Committee on Education and Culture.

NDHR For. National Diet of Japan House of Representatives Foreign Affairs Committee.

NDHR LIT National Diet of Japan House of Representatives Land, Infrastructure, Transport and Tourism Committee.

NDHR SST National Diet of Japan House of Representatives Special Committee on the Promotion of Science and Technology.

NDHR S&T National Diet of Japan House of Representatives Science and Technology Committee.

NDHR Set. National Diet of Japan House of Representatives Settlement Committee.

NDHR SOC National Diet of Japan House of Representatives Settlement Administration Oversight Committee.

NDHC S&T	National Diet Special Committee on Promotion of Science and Technology.
NEC	Multinational information technology and electronics company headquartered in Japan, founded in 1899.
NHK	Nippon Hōsō Kyōkai, Japan Broadcasting Corporation..
NICT	National Institute of Information and Communications, Japan's primary national research institute for information and communications.
NTT	Nippon Telegraph and Telephone. Japanese telecommunications company, state-owned but privatized in 1985.
OTU	Ortho-tolyl-urethane, a fuel stabilizer developed in Japan during World War II.
QZSS	Quasi-Zenith Satellite System, a satellite navigation system developed by JAXA that aims to provide highly accurate positioning information in Japan and surrounding regions, particularly in urban and mountainous areas where traditional GPS signals may be obstructed.
RRL	Radio Research Laboratory, a state-owned telecommunications research institute with branches and institutes across Japan.
SCAP	Supreme Commander for the Allied Powers, the title held by General Douglas MacArthur during the Allied occupation of Japan following World War II. The term is often used as a synecdoche for the Allied occupation authorities between 1945 and 1953 generally.
SFRs	Solid-fuel rockets. Once ignited, these engines cannot be turned off, limiting the accuracy of these rockets. However, solid fuel—actually a thick gunk or putty-like substance—is far more stable than liquid fuel, and can be stored within a rocket for a longer time.

THE ISLANDS AND THE STARS

Introduction

THE STORY OF ROCKETRY IN JAPAN has at least three distinct epochs. The country's first encounter with the technology came sometime between the Mongol invasions of the late thirteenth century, and warlord Toyotomi Hideyoshi's invasion of Korea in the 1590s.[1] The inaccuracy and unreliability of the primitive rocket weapons they encountered, like "fire lances," meant their battlefield impact was minimal—but firework rockets had, by the seventeenth century, become a mainstay of harvest festivals across the country.[2] Then, in the early twentieth century, came the second phase of Japan's engagement with rocketry, this time through the work of theorists such as Konstantin Tsiolkovsky (1857–1935) and Robert Goddard (1882–1945). Initially, dreams of space exploration were as much a part of this iteration of rocket mania as military usage. Yet given Japan's ongoing quest for empire, the emblematic vehicles of this period—the Shūsui rocket plane and the Ōka rocket bomb included—were primarily weapons.

The third phase of Japan's development of rocket technology began in the aftermath of World War II. It also had its roots in the work of the aforementioned theorists. Defeat and pacifism had rendered military rocketry anathema to the public, but this simply meant that spacefaring dreams inspired by Tsiolkovsky and Goddard now took the limelight.[3] So when the third, and final, birth of rocketry in Japan came, in 1953, it was with space exploration in mind. That year, Tokyo University professor Hideo Itokawa formed the Avionics and Supersonic Aerodynamics

(AVSA) research group in order to explore the basics of rocketry in the country.[4] Within sixteen years, Japan had become the fourth country to put one of its own satellites into orbit atop one of its own rockets.[5]

Since then, Japan's scientific and civilian space program has grown to become one of the six biggest space enterprises in the world, alongside China's CNSA, Europe's ESA, India's ISRO, the US's NASA, and Russia's Roscosmos. The Japan Aerospace Exploration Agency (JAXA), a descendant of sorts of the AVSA, had a budget in 2017 of ¥173 billion ($1.35 billion)—bigger than that of France or Germany individually, and more than that of Italy, India, Canada, and the United Kingdom put together.[6] The agency employs more than 1,600 specialists at various locations across the country, has nineteen offices and test facilities, and works with some 677 contractors and R&D partners, including Mitsubishi Heavy Industries (MHI), Ishikawajima-Harima Heavy Industries (IHI), Hitachi, NEC, and Toshiba.[7] By the turn of the millennium, these corporations had generated a space industry worth around $8.6 billion.[8] Together, corporations and the space agency continue to produce and manage globally significant probes and space infrastructure, including the Hayabusa probes (which recovered materials from asteroids for analysis on Earth) and Japan's own supplement to GPS, the Quasi-Zenith Satellite System (QZSS).[9] They have ambitious medium-term plans, including developing a cost-effective launch vehicle, the H-3; a mission to Martian moon Phobos; and the development of a pressurized rover for use on the US Artemis moon program.[10] By any measure, Japan is one of the twenty-first century's premier spacefaring nations.

All this technological accomplishment rests on a complex bed of power politics, discursive contestation, and social tension. Hence, in order to explore how the Japanese space enterprise evolved, and what drove this evolution, this book looks beyond the impressive physical products of Japanese space exploration, and proposes that in the period from 1920 to 2003 (but particularly from 1953 to 2003), Japan's space effort evolved through the interactions of several key stakeholding groups: politicians and lawmakers; bureaucrats working in Japan's ministries and government agencies; private corporations; engineers, academics, and specialist technicians; the United States; and locals living near Japan's space facilities—in particular, fishermen. The analysis presented here will show how in the first stage of the process, from the 1920s to the 1960s, specialists and engineers dominated the story. Historian Richard Samuels has observed that during this period, but particularly in the war-torn decades of the 1930s and 1940s, Japan's body politic felt a profound sense of "insecurity" and "anxiety" about Japan's status in the world, and concluded that the country and its people must "make sacrifices to enhance national security in a hostile world." In this context, rockets were understood primarily as weapons of war. Weapons in turn—and, indeed, technology and industry writ large—were also understood as manifestations of Japan's native genius, the innovativeness of its people, and the dynamism of its empire. Together, these notions constituted a body of thought historians have dubbed "technonationalism."[11] It was under this regime that Japan's first astronautical and aeronautical specialists acquired their skills, devised solutions, and mastered essential production processes.

Then, from the early 1960s to the late 1990s, these same specialists and managers carved influential political niches for themselves within a space enterprise that became institutionalized in a markedly decentralized manner. The Japanese state became far more involved in space research at this time, passing key pieces of legislation (the various Space Laws), and promoting key domestic industries (satellite manufacturing); politicians were keen to capitalize on this potential new source of income, but also prioritized the reputational rewards of becoming one of the world's pioneer spacefaring nations. However, the actual management and planning of space research was carried out through multiple different institutions, ranging from government ministries to small laboratories like the National Aerospace Laboratory (NAL). Not only did the manager-engineers who had dominated the earliest phase of Japan's postwar rocketry project find that they could ensconce themselves in powerful and influential positions, but other stakeholders, ranging from fishermen to CEOs, also found ways of making their voices heard in policy and planning discussions. Private corporations, keen to capitalize on the potential of satellites for telecommunications, began to funnel resources into space research as well, and campaigned to speed up the development of rocketry and satellite production in Japan. Locals organized to welcome the country's expanding space facilities and extract maximum benefit for their communities from them. Furthermore, the success that space programs enjoyed in this period dovetailed with a broader sense of the country having moved away from the legacy of World War II. As Japan's economy and global status surged, so too did the dreams of space engineers and researchers, who proposed ever larger and more elaborate projects that captured the imagination of the public, lawmakers, and private corporations alike. This confidence culminated in a series of ambitious projects including the Kibō module of the International Space Station, and the H-2 medium-lift rocket.

The early 1990s saw the discrediting of the decentralized system due to a series of expensive launch and probe failures. The economic travails Japan experienced in that decade made the space programs' political overseers even more unhappy with the country's decentralized approach. As the public began to lose interest, and corporations began to leave the space industry, Japan's flagship H-2 rocket exploded after launch in late 1999. It took with it any hope that Japan's space programs could continue in its decentralized form. The resulting formation of the Japan Aerospace Exploration Agency (JAXA) in 2003 presents a suitable endpoint for the story told here, for several reasons. The new institution was formed on the back of extensive organizational restructuring. A concurrent industrial realignment further changed the topography of the space enterprise: IHI acquired Nissan's rocketry assets, consolidating missile production, while MELCO emerged as a dominant satellite manufacturer at the expense of Toshiba and NEC. Last, and perhaps most significantly, Japan's space objectives expanded to integrate military applications, spurred by regional security concerns. The 2008 Basic Space Law legitimized dual-use space technology, and thus "reversed Japan's decades-long policy that space be utilized for nonmilitary purposes only."[12] The foundation of JAXA therefore represented an institutional, programmatic, and policy transformation that resulted in Japan's space program as it stands in the early twenty-first century being markedly different from the program that existed in the twentieth.

A Holistic Portrait of Japan in Space

Japan's space program is not the only massive infrastructural and technological project the country initiated after World War II, but it is certainly one of the most understudied in English. The first formal space agency, the Institute of Space and Aeronautical Science in the University of Tokyo, was founded in 1964, the same decade when the country began work on its famous Shinkansen high-speed railway network, and its first nuclear power facility. Yet while extensive coverage exists of both the latter topics, Japan's space program received relatively little attention in Western academia until the early 2000s. In 1962, an article in the *Wall Street Journal* characterized Japan's rocketry program as one of "the world's least known [such] projects."[13] Twenty-seven years later, Harvard graduate student George Eberstadt began his thesis on the subject with the made-up epigraph, " 'I didn't know Japan has a space program'—everyone"—surely, only half in jest.[14] What little attention space research in Japan *did* garner tended to promote narratives of unsophistication, emphasized its reliance on foreign technologies, foregrounded its impact on Western economies, or fretted over its implications for Western geostrategy.[15] It was not until the 1990s and 2000s that the successes of the Hayabusa probes and the Kibō module of the International Space Station began to impress overseas observers with their complexity and uniqueness, responses that tallied with Eberstadt's observation that "upon further study, [observers] will find that space projects have been conducted in Japan for decades, which will prompt accusations of not noticing such progress going on right under our noses."[16] Yet even then, the country's rocketry development during World War II, the 1950s, and the 1960s; its sale of rockets to Indonesia and Yugoslavia; and its pioneering contributions to, for example, the exploration of Halley's Comet in 1986, got almost no attention at all.[17] To be fair, projects like the Shinkansen are far larger than the space effort, but that does not mean the space program lacks intrinsic interest. Japan's space program has a social and cultural imprint at least as important as that of other major industrial, technological, and infrastructural projects in the country. It has an ubiquitous physical presence in the form of infrastructure from "rocket roads" to testing ranges; enjoys a high public profile; garners a great deal of political interest; is worth billions of dollars to industry; and is a significant factor in Japanese diplomacy and geostrategy. For these reasons alone, it is high time the Japanese space effort received a thorough historical analysis.

One of the key interventions of this book is to approach Japan's space exploration holistically—that is, to focus not on one particular element of space research in the country, but to cast a wide net that includes infrastructure, discourse, anecdote, and past analysis. This is intended to form an antidote to the tendency for observers, thus far, to consider only one element of Japanese activities in space in their studies. Analysts and historians have thus written about the civilian space programs in Japan's postwar economic resurgence, in US–Japan relations, and in the development of a military-industrial complex in the country, among other topics.[18] Still other works focus on individual figures, which benefits those like the pioneering rocketeer Hideo Itokawa, generally regarded as the founder of Japan's space program. Yet this approach, while insightful, does not allow for delineating the dynamics that dictated Japanese space policy and development as a whole. Take, for example, the involvement of private enterprise in the space program. On its own, the sourcing of hardware in Japan does not seem strikingly different from that of other nations, in that corporations compete for government contracts to produce launch vehicles and their components. However, in practice, different parts of Japan's space effort had enduring relations with certain companies, and ensured an equitable division of assignments where there were multiple bidders. Furthermore, these corporations actively influenced and shaped the space programs on a scale far greater than nearly anywhere else, excepting perhaps Europe. Coordinating through groups like the powerful Japanese Business Federation, Keidanren, they were able to influence lawmakers, and even engage in bureaucratic skulduggery to assert their interests. Take, for example, Japan's development of liquid fuel rocketry, a technology that enables launchers to precision-guide the payloads they launch into orbit. Japan's manager-engineers, led by Hideo Itokawa, had long resisted developing the technology for fear of having to involve the Americans in their work (and thus give up their independence) and because ICBMs were liquid-fuel rockets (developing only solid-fuel rockets had enabled them to avoid suspicion of stealthily working on a military project). By the mid-1960s, however, communications corporations had concluded that they simply could not compete on the international stage without the technology. Furthermore, they needed it quickly, so as not to lose market share. And so they began a counter-campaign, one that resulted in the foundation of what would be Japan's biggest prereform space agency, National Space Development Agency of Japan (NASDA), in 1969, specifically to import and indigenize liquid-fuel rocket technology. Thus the true significance of what seems like a bureaucratic reorganization is difficult to understand unless these developments are compared to similar situations overseas, and unless the objectives and encounters of the various stakeholders within the dynamic are fully delineated.

Similarly, explaining the pace and limitations of Japanese rocket testing is almost impossible without first understanding the geography, economics, and demographics of the areas in which these tests occurred. It was geography that pushed Japan's space institutions to build their facilities so close to inhabited areas in Japan's south; the presence of large numbers of fishermen who opposed their construction caused delays; and it was perceived threats to livelihood that provoked the fishermen's resistance. At the same time, the fishermen's responses were primarily organized by trade unions from outside their localities; the pressure the state felt to get rocket testing going again came not only from locals relying on tourism, but also from communications companies and rocket manufacturers desperate for the production of liquid-fuel rockets; the coverage of the protests in the press provoked widespread public commentary, which placed pressure on both sides; and the fact that Japan had *two* space facilities to protect was the result of bureaucratic competition.

The primary mode of interaction between these power centers was contestation, and studying the squabbles that determined the direction of the space programs is not only interesting in and of itself. Exploring confrontations and compromises within the space programs enables observers to identify who, precisely, the confronters and compromisers were. In other words, the outcomes of the encounters described in this book enable us to see what the main centers of influence in the Japanese space program were, how they changed, and what they each wanted. In a context where the formal space institutions competed with each other, expanded their authority into areas hitherto beyond their remit, and contested areas of overlapping authority, institutional formation does not always equal actual power. Groups that technically have no say at all—such as fishermen—could end up having more influence than powerful state bodies such as the Science and Technology Commission. Looking at the contestations in Japan's space programs will better enable this study to define who the contestants are, while also speaking to one of the central proposals of this work: that the objectives, products, and public justifications of Japan's space programs were determined by the balance of power between these centers of influence.

However, the fact that there was no centralized space exploration organization in Japan until 2003 poses a challenge for one seeking to define what, precisely, the term "civilian space programs" means. Where exactly does one draw the analytical line, institutionally and technologically? How can one talk about a space program, if there was no single "space program" to begin with? Luckily, despite their not being centralized until 2003, there was considerable institutional continuity in Japan's space efforts that facilitates identifying who and what to talk about. For example, one of Japan's two major space agencies in the period discussed, the Institute of Space and Astronautical Science, is a direct heir to Tokyo University's AVSA. The AVSA began as primarily a rocket project in 1954, became the Institute of Space and *Aero*nautical Science in 1964, and took pride in the fact that until 1984, when it changed its name, it had both payload developers and rocketry designers working together in-house.[19] Meanwhile, the far bigger National Space Development Agency of Japan (NASDA), founded in 1969 to oversee the importation of American liquid-fuel technology, also had the responsibility of overseeing the development of engineering and test satellites.[20] Beyond these two, various ministries and agencies had funding for sounding rockets, balloons, satellites, and orbital platforms together; budgets tended to be determined by the purpose of these devices (for example, weather forecasting) and not the type of technology used. Most of these institutions did not actually build their rockets and payloads themselves; this work was contracted out to corporations ranging from giants like Nissan to a slew of smaller, specialized entities. Private companies often shared management and personnel with state space entities as well, blurring cultural and structural lines between them. Thus, which institution developed what in Japan's space program from 1953 to 2003 varied, but often involved the same personnel. Even if there was no unitary space program, there was a large, but distinct, cluster of institutions and individuals involved in space development.

A People's History of Space Exploration

Animating the approach in this book is a focus on the lived experience of those who interacted with Japan's space efforts between 1920 and 2003. A great deal of work on space programs tends to foreground the technological aspects of the story—astronauts over administration, rockets over researchers.[21] And certainly, there is plenty that is unique about Japanese hardware, ranging from the success of their first satellite Ohsumi to the fact that the pioneering H-1 rocket had a configuration of nine smaller boosters in order to accommodate the winds around Japan's launch facilities.[22] Here, the remarkable work of analysts like Brian Harvey and John O'Sullivan already provides students all they could ask for, though it is work that suffers from two drawbacks.[23] The first is a reliance on English-language sources. The second is that their approach inherently leaves out critical pieces of the story; as their focus is on Japan *in space*—the missions, the astronauts, and the technologies—much of the policy, cultural, political, and social aspects of Japan's space development remain beyond readers' reach.

Hence the second objective of this work is to tell the story not only of Japan in space, but of *space in Japan*, guided by E. P. Thompson's injunctions regarding the "enormous condescension of posterity." Thompson chose to turn his eye to beleaguered workers trying to survive the obfuscating narratives of capitalism and industrialization. This book, in a similar vein, tries to tell the story of the people who were involved in a vast techno-industrial project, one that was arguably a distant aftershock of the tectonic changes that intrigued Thompson. This includes the engineers who dreamed of travel on spaceships, the people whose taxes funded their making, and the children who stood on bunds by the sea and watched rockets soar into the night sky. What motivated them to sustain such a vast project? What did they stand to gain, and how did they react to its travails? How did the specific anxieties, experiences, and aspirations of twentieth-century Japan—recovery from war, both material and psychological; economic stagnation; the country's fraught relations with its Asian neighbors—feed into what the Japanese tried to accomplish in space, how they went about doing so?

One method of addressing such questions lies in works like Carlo Ginzburg's *The Cheese and the Worms*. One strength of his "microhistorical" approach ("micro" in its focus on details) is its ability to illuminate broader systemic and historical trends without subsiding into dryness and mere statistics. Indeed, one of *The Cheese and the Worms'* signal accomplishments is putting the reader, as much as possible, in the shoes of its subjects, men and women facing the Inquisition in early modern Europe. In the case of Japan, it's one thing to say that Japan launched its first rocket in 1954, and put its first satellite into orbit in 1970. It's entirely another to understand what these accomplishments meant to people, how they changed their expectations of the future, and altered their understanding of the world around them. Similarly, the space program relied on their support; as Asif Siddiqi has observed, "Justifications for spaceflight have been historically contingent; different historical periods required different justifications to be accentuated."[24] To engineers who had participated in Japan's failed war effort, rockets represented redemption for a country laid low by its own geopolitical mistakes. For citizens keen to put the war behind them, space technology represented a new Japan, which engaged in space research "exclusively for peaceful purposes."[25] For people in Japan's far-flung villages, it represented an ongoing failure to improve the lives of the marginalized in favor of big, noisy projects that benefited the few. For children, it represented a mobile, high-tech future. The hardware is fascinating, but what it means to people is more fascinating still.

Such an approach necessitates an engagement with the stories and justifications—that is, the discourse—associated with the space programs. As Michel Callon puts it, "science and technology are dramatic 'stories' in which the identity of the actors is one of the issues at hand."[26] Peter Fritzsche's observation that "the precision and power of the engine, the sophisticated instrumentation in the cockpit, the durable yet lightweight streamlined metal frame" of aircraft form a sort of material shorthand for the experience of living in a modern industrial society can be extended to space programs.[27] Far more so than, say, an aircraft industry, high-profile space projects require public support in order to exist at all. James Webb, NASA's second administrator, pointed out that "any major public undertaking requires for success a working consensus amongst diverse individuals, groups, and interests. A decision to do a large, complex job cannot be simply reached at the top and then carried through. Only through an intricate process can a major undertaking be gotten underway, and only through a continuation of that process can it be kept going."[28]

Hence the analysis presented here will look at the lives of those who lived near launch facilities, technicians, local inn owners, children who played with homemade rockets, and a swathe of other people whose lives were touched in one way or another by the space programs—and who touched them back. This is not to say that significant figures will be ignored; the influence of Hideo Itokawa, for example, is hard to overestimate, and no story about Japan's space program would be complete without reference to him. Nor can we ignore his influential protégés, such as future associate executive director of JAXA, Yasunori Matogawa, and director general of ISAS, Ryōjirō Akiba. However, their stories will be complemented by attention to lesser known figures, whose stories will provide equally significant insights into the development of Japanese space technology. This will include figures such as Eiichi Iwaya and Kumao Hino, World War II–era technical specialists involved in rocket design; Tomifumi Godai, who headed up the ill-fated H-2 project; and satellite specialist Fugono Nobuyoshi, a long-term observer of the space program. Emblematic of this approach, perhaps, is the choice to focus on fishermen over astronauts. Anyone decrying a history of a space program without spacemen and -women may want to consider that for those who actually administered launches, the astronauts' job was primarily to sit down, shut up, and not touch anything.[29]

Understanding the symbols and stories that animate the space effort is also crucial to fully comprehending what they mean to the Japanese populace. Consider, for example, the extensive coverage of Hayabusa's 2003–2010 mission to the asteroid 25143 Itokawa, and its return with the first sample of asteroid dust from space to Earth. The probe was as much a testament to human ingenuity as to Japan's technological capacity. Hayabusa in fact suffered from multiple technical failures during its mission: elements of the device didn't function, power was lost, telemetry malfunctioned, and at one point communications died entirely. It was only through the heroic efforts of the team involved that the device returned its samples to Earth. In fact, the very framing and naming of the project referred to key narratives in the public story of Japan's space program. The asteroid was named "Itokawa" after the founder of the Japanese civilian space programs, the aforementioned Hideo Itokawa. During World War II, he had worked as an aircraft engineer on devices such as the Nakajima Ki-43, dubbed the "Army Zero" by US servicemen who encountered it over the Pacific. In Japan, it was known by another nickname: Hayabusa. Thus, human experience, past and present, is a critical element in any space program. Telling a space program's story without placing this at its heart is akin to telling the story of galactic formation without reference to dark matter.

Foregrounding the justifications for and stories about the space program is an approach that also touches on significant debates more broadly within the history of Japan. For example, the experience of locals living near Japan's launch facilities speaks directly to the work of historians such as Martin Dusinberre and Daniel Aldrich, who have pointed out the complex interactions of social status, economic pressures, and nationalist discourse that govern local responses to such projects. In the case of launch facilities like the Tanegashima Launch Center, demographic decline, government incentives, and the rehabilitation of rocketry as a technology by disassociating it from war eventually led to acceptance by most of the local community. As both Dusinberre and Aldrich have pointed out, a great deal of the effort in securing these resolutions was based on outside pressure. However, unlike the examples they foreground—nuclear facilities, waste disposal facilities, and dams—the space facilities eventually became a critical part of local identity. This stands in stark contrast to ongoing unease about nuclear facilities, for example, speaking to the power of particular local circumstances to govern the responses of citizens to large industrial projects.[30] A focus on discourse also helps chart the life of technonationalism in postwar Japan. The country's space program was inspired by nationalist goals; the earliest rocket launches in Japan, in the late 1950s, were not functional—that is, they were purely displays of technical ability, at that point mostly for the purpose of showing that Japan could achieve such things. Later, as Japan's space effort became more institutionalized, the primary mode of justifying it to the public remained an appeal to nationalism, framed in terms of increasing both quality of life and international status. At the same time, for a technology intimately tied up with warfare, space pioneers had to work hard to demilitarize it; few other technologies, apart from nuclear power, have had to deal with the same issue. Yet Japan's space agencies, led by manager-engineers, were committed to a program of carefully formulated outreach that was extremely sensitive to the sentiments of the public.

In fact, the attitudes of planners and communicators in Japan's space programs are strikingly reflective of the country's zeitgeist—hence, the excitement of the 1950s, the ambition of the 1960s, the growing economic concerns of the 1970s, the confidence of the 1980s, and the malaise of the 1990s. Thus, this work can also be read as a study in the development of technonationalism, showing how it recalibrated not only to account for the legacy of wartime, but also the experiences of an industrial power navigating the economics and politics of the late twentieth century. These efforts continued to collapse the distinction between a people and their technological products, helping construct a vision of Japan as a technological superpower—albeit a peaceful one.

A Global History of Space Exploration

From the perspective of space histories generally, there are several aspects of Japan's space program that, at very least, provide notable counterexamples to the experiences of the superpower space projects. In doing so, they compel us to amend our understanding of how and why countries went into space in the twentieth century. Perhaps the most striking is that in both the United States and the USSR/Russia, centralized control and government initiative provided the framework for expanding space programs. In the USSR, the rocketry and space programs essentially functioned as a wing of the defense industry; its disparate and sometimes conflicting elements—from the Ministry of General Machine Building to rocket engineers such as Sergei Korolev—were all employed within the framework of a nationalized military-industrial network.[31] Similarly, the United States had also, by the 1980s, brought rocket design and space exploration under centralized governmental purview. Nevertheless, the emergence of what Walter McDougall has dubbed a state-controlled "technocracy" was enough to provoke the ire of scientists such as the Jet Propulsion Laboratory's Frank Malina.[32] In contrast, as mentioned above, there was no centralized space agency in Japan between 1954 and 2003. In fact, it wasn't until 2003 that the country's bewildering array of space research entities was finally integrated under a single governing body, the Japan Aerospace Exploration Agency (JAXA). Until then, it was the contestation and cooperation of several centers of power that determined how the whole enterprise developed, what it accomplished, when, and why. The reasons for the emergence of this system, its functioning, and eventual collapse will form a central part of the story told in this book.

Another distinction is that one of Japan's primary goals in space, from the beginning, was saving and making money. In this it diverges significantly from the United States and the Soviet Union, but this preoccupation with moneymaking is also apparent in Russia and the European Space Agency (ESA). Rather than striving for dominance in space exploration, Japan focused on developing space technologies that had clear economic benefits. From the outset, its civilian space programs concentrated on telecommunications, weather forecasting, and Earth observation satellites, ensuring that space investment would yield tangible returns. Its space budget remained significantly lower than those of the US and Soviet programs, and even as ambitions grew in the 1970s and 1980s, funding remained limited. This fiscal restraint led Japan to pursue space endeavors with a clear eye on commercial viability. To maximize its capabilities without excessive spending, in the 1960s and 1970s the country actively sought international collaboration, particularly with the United States. By leveraging American technology while maintaining selective independence, Japan contributed high-tech components—such as the Kibō module for the International Space Station—without developing costly heavy-lift rockets of its own. Yet at the same time, the H-2 medium-lift rocket project of the 1980s emphasized affordability by *not* relying on American launches, as well as the potential profits Japan could accrue from providing competitively priced commercial launches. Thus, a preoccupation with moneymaking could be the root cause of major changes in Japanese space planning and foreign cooperation.

Japan's space program also had a markedly different local impact compared to those of its international counterparts. For starters, locals had far more say in what was developed, and when; whereas even Japanese citizens in remote Tanegashima influenced national space policy, the Russian population, for example, wasn't informed of the 1972 failure of the USSR's answer to Saturn V, the N-1, until 1989.[33] While space facilities in the Soviet Union, the United States, and French Guiana spurred economic and population growth, Japan's facilities, particularly in Kyushu and Tanegashima, could not prevent rural decline. In contrast to other nations where space programs (re)vitalized regions, Japan's experience underscores the persistent challenges of rural depopulation, and reveals the limitations of large space projects as local investments.

A further reason to pay attention to the Japanese space program is that it is, in many ways, closer to the story of *everyone else* in space. Hitherto, space histories have tended to look at the space age from the perspective of the superpowers, whose developments were partly but consistently motivated by a desire to compete with each other. Yet this was not the only dynamic in the space age. Arguably, the dynamics behind the superpower space program were unique to them, in that both were highly funded, ambitious, and understood primarily as vehicles for weapons and the enhancement of reputation and influence. This becomes particularly evident when comparing Japan's space efforts to those of new spacefaring nations. Like India's ISRO, for example, Japan has smaller budgets and focuses on launching satellites and probes.[34] Like the ESA, Japan's program from 1954 to 2003 was entirely civilian.[35] Like Brazil and South Korea, Japan also had to rely heavily on overseas launch and component providers.[36] While the Soviets and Americans competed to accomplish impressive things in space, and to draw other countries into their spacefaring orbit, smaller space programs sought to leverage tensions to their own advantage. It is reasonable to state that Japanese rocket technology is derived from American models, but the country also forged links with left-wing countries, cooperating on weather forecasting with China, and selling rockets to Suharto's Indonesia and Tito's Yugoslavia. At one point the Japanese even trial-ran cross-broadcasting with Soviet communications satellites.[37] Thus, as this work will explore, Cold War tensions manifested in very particular ways in the nonsuperpower space programs, and these did not always conform to a simple binary.

The story of Japan's space program is also notable in that it is the story of a civilian space program that long resisted military involvement, but ultimately gave in. While NASA emerged from Cold War competition, and the Soviet space effort was effectively an extension of its missile programs, Japan's space policy was guided by its 1969 commitment to "peaceful purposes," explicitly barring military involvement.[38] Whereas the United States and USSR both integrated their military and space efforts, Japan's space program initially prioritized commercial and scientific applications, such as communications satellites and weather observation. This approach meant that Japan did not invest in heavy-lift rockets capable of carrying military payloads, nor did it develop an independent space-based reconnaissance system as early as the superpowers. Japan's stance began to shift only in the late 1990s, and primarily in reaction to external threats. The 1998 North Korean missile test over Japan, known as the "Taepodong shock," was critical in forcing this reevaluation, leading to the deployment of reconnaissance satellites, even if these were presented as "information-gathering" rather than military assets. Eventually, the 2008 Basic Space Law allowed for national security applications to be formally folded into JAXA's remit, ending more than fifty years of commitment to an entirely peaceful space program.[39]

Sources

The sort of history this book attempts—which brings together individual experiences with large-scale systemic changes—requires a commensurately diverse dataset. As a result, the work presented here is based on a variety of materials. These encompass the period from the 1890s to the 2020s, though the focus of the work is on the decades from 1920 to 2003. To provide both systemic insight and individual experience, it draws on poems in newspapers, children's magazines, advertisements, squabbles in the House of Councilors, private letters, US military reports, biographies, autobiographies, and interviews in order to paint as full a picture as possible.

One of the great advantages of studying a topic such as Japan's space program is the country's political system. Similar to the United States, nonmilitary records are generally not classified, and are open to the public. This accessibility takes the form not only of extensive resources available at the National Diet Library in Tokyo, but also of a substantial collection of materials available online. JAXA, for example, maintains a publicly accessible archive, which includes documents ranging from schematics to interviews with engineers, as well as images, sound recordings, and video footage. Also invaluable is the fact that all Japanese parliamentary records—including discussions of important institutions like the Science and Technology Agency, which was intimately involved in Japanese space developments—are accessible online. While certainly there is a great deal of political posturing and stonewalling to be found here, there are also earnest expressions of opinion. Crucially, members of Japan's Diet, especially in the lower House of Representatives, often approached policy and infrastructural discussions from the perspective of their localities, providing a valuable insight into the interface between the two. Due to their close cooperation, Japan's various space institutions have regularly produced reports, digests, and technical materials for overseas agencies such as ESA and NASA. To the extent that these, too, are openly accessible, the institutional fingerprint of Japan's space programs is a fairly easily examined one. Lastly, basic and applied research carried out both on the topic of, and onboard, space vessels is generally available to the public via a variety of academic archives; without any military secrecy, it is possible to ascertain what was going on during any particular day of a Japanese space mission, as John O'Sullivan's *Japanese Missions to the International Space Station* shows.

Beyond this there lies a huge trove of material related to the space program in the private sector, and in the possession of individuals. In the 2010s and 2020s, Japan's major print newspapers underwent a decline in readership in the face of digital media, but print still remains Japan's dominant format, especially in the period discussed here. The Big Four national newspapers—the *Mainichi Shimbun, Asahi Shimbun, Yomiuri Shimbun,* and *Nihon Keizai Shimbun*—all had multiple weekly editions, local supplements, and extensive networks of reporters with ability to report with speed from even the remotest parts of Japan. The number of newspapers printed in Japan annually (including local publications) in 1956 came to around 24 million copies. By 1970, this had reached 36 million; by 1990, it was 51 million. Most families subscribed to one newspaper or another; circulation per household peaked in 1982, with an average of 1.3 publications arriving daily at homes around the country. Indeed, so extensive is the engagement of the Japanese with print media that the decline of traditional family households in favor of single residencies can be traced in the fact that though per household newspaper purchasing declined in the 1990s, *per capita* circulation actually peaked in 1997 (before the gradual encroachment of digital platforms on the medium).[40] The medium was particularly important in providing a venue for senior figures within Japan's space industry to voice their opinions, through interviews, articles, and, in some cases, weekly columns. Meanwhile, local supplements for areas such as Kyushu and Akita are particularly valuable in drawing attention to the local consequences of space technological research, such as the "rocket play" issue described in chapter 5. They also provided a venue for local voices to be heard, such as that of Mikiko Kanadome, a grade-school student who dreamed of one day being a stewardess on a rocket plane. Another great source of information are the localities themselves; Uchinoura, Tanegashima, Noshiro, Sagamihara, and Kokubunji all maintain memorials and sources related to the space activities that were undertaken there.

Trade publications such as *Space Japan News* provide insights into the experiences of a great many individuals working within Japan's space program. Perhaps most valuable of all, however, are the extensive writings produced by the individuals discussed here themselves. Eiichi Iwaya, Kumao Hino, Hideo Itokawa, Yasunori Matogawa, Hideo Shima, and Ryōjirō Akiba were all prolific writers, who brought their experience and knowledge to bear in quite literally hundreds of texts available for public consumption. These range from descriptions of their daily lives (which, in wartime, included bombing raids and submarine chases), as well as analyses of their work and organization. In the case of Hideo Itokawa, these writings extend far beyond the topic of space and rocketry to subjects as diverse as David Ben-Gurion and violins.[41] This extraordinary outpouring of materials provides vivid insights into how the space program developed from an insider's perspective. Japan's commitment to intellectual openness also means that few topics were beyond the remit of their work (something bolstered in Itokawa's case, perhaps a little, by his departure from the space program on a sour note in 1967). A good example here is Yasunori Matogawa's *Uchū ni ichiban chikai machi—Uchinoura no roketto hasshajō,* which goes into detail on the ins and outs of negotiations between ISAS, NASDA, and the fishermen of southern Japan.

In stark contrast to this, private corporations remain somewhat secretive, perhaps on account of the proprietary nature of some of the technologies they researched; alternatively, this may be attributable to the generally closed-off and inward-looking cultures of large conglomerates, many of which do not have public archives. When I approached Mitsubishi regarding information on the development of the H-2 rocket, for example, they stated that the company does not provide such information to the public. Luckily, former employees are often willing to talk or write. In contrast, JAXA's ongoing willingness to share information has also been extremely helpful in putting researchers in touch with senior members in order to answer questions. Of particular importance here has been Yasunori Matogawa himself, still sprightly and engaged despite having been retired from the field of space research and administration for some time. Indeed, with his multiple appearances on TV, at seminars, and at workshops—on top of his huge written output—Matogawa has helped ensure that any scholar seeking to work on Japan's space program has plenty of insight into the ins and outs of its development.

Overview of Contents

This book is divided into three time periods, each characterized by its own set of dynamics within Japan's civilian space programs. As the enterprise changed in scale, cost, and institutional formation, so too did the pressures and contestations within it. These processes in turn led to major changes in direction in terms of technology, policy, and perception. Part I, "Child of War," covers the period from 1920 to

1960, which was characterized, on one hand, by the experiences and activities of a relatively small group of engineering specialists who initiated and managed Japan's earliest space efforts. On the other hand, the impact of World War II also loomed large over Japanese rocketry, as it had been the crucible of both the material infrastructure and human skills that enabled this process to occur.

Part II, "The Institutionalization of Japanese Space Research," takes as its central dynamic the institutionalization and proliferation of Japan's space organizations in the 1960s, 1970s, and 1980s. This period is characterized not only by the emergence of Japan's two largest and most accomplished space organisations, ISAS and NASDA, but also by the importation of key US technologies, the expansion of facilities, the involvement of major corporations, and growing public awareness of—and support for—the space enterprise. Private enterprises made their desires for the space enterprise known through lobbying and industrial linkages with particular space organizations, while engineer-managers entrenched their power and influence. This period was marked by a slew of successful space projects, from Japan's first satellite to large-scale rocketry development. It was also characterized by growing ambitions, as Japan dreamed of its own spaceplane, space base, and journeys to the Moon.

Part III, "The Challenges of Advanced Spacefaring," describes how and why most of these dreams did not come to pass. This third period, from 1980 to 2003, witnessed both the greatest successes of Japan's multiple space programs and also a signal failure—the H-2 rocket. The central dynamic in this period was between the entrenched space programs, which pursued their own goals and promoted their own projects, and their political overseers, who fretted about cost and control. As long as the space program continued to be successful, it could rely on increasing budgets, and its managers could hold political control at bay. With the failure of the H-2, however, Japan's multiheaded postwar space exploratory system finally came to an end. In 2003, centralizing forces won out, and subsumed ISAS, NASDA, and other institutions into a single entity: the Japan Aerospace Exploration Agency, better known as JAXA.

Chapter 1, "The Transwar Origins of Japanese Rocketry," identifies how Japan's early twentieth-century experience of imperial expansion produced the physical infrastructure (in the form of research laboratories, industrial facilities, institutions of technical education) and trained engineers (such as rocket pioneer Hideo Itokawa, fuel specialist Kumao Hino, and military engineer Eiichi Iwaya) who would be the driving force behind Japan's modern space program. The war also provided the opportunity for both institutions and individuals to garner some experience of working on rocketry development, which enabled these engineers to rapidly pick up in the mid-1950s where they left off in the mid-1940s. The chapter shows how these specialists developed rocket engines and weapons like the Shūsui rocket fighter and the Ōka suicide rocket. The experience of the postwar era—in particular, being forced to find their own opportunities during the US occupation, when aeronautics was proscribed—forced engineers to find their own ways to participate in Japan's economic and technological resurgence in various capacities. As R. W. Home and Morris Low have pointed out, "While Japanese mobilization of science was a downright failure when compared to the efforts of the Allies, the wartime experience was significant, because it provided the immediate historical backdrop to Japan's remarkable postwar industrialization."[42] It also meant that when the time came, Japan had the personnel and infrastructure needed to jumpstart its civilian space programs.

Chapter 2, "Demilitarizing Rocketry in the Postwar Period," turns to how these figures—rocket scientists, theoreticians, and engineers—sought to change public and political perceptions of rocketry in the 1950s and 1960s. Given that aeronautics had become so closely connected with death, how could they persuade the public that developing rockets was a project worth pursuing in the new, pacifist Japan? Their response was to produce a body of writing that sought to whitewash aeronautics more broadly, and rocketry more specifically, of wartime connotations. Their stories of wartime technological development became less about what these devices were for, and more about how wondrous the people and devices they made were in and of themselves. Their discussions also focused on aviation as aspirational and futuristic. Ultimately, their work helped ensure that rocketry did not go the way of battleships and bombers in the public consciousness. Instead, rockets became more like the other notable postwar product of wartime Japanese engineers, the Shinkansen "bullet train." All this technology was to be deployed in service of the talismanic "bright new" (*akarui*) life—a life of material comfort and peace for all citizens, the achievement of which replaced the quest for empire.[43]

Chapter 3, "Manager-Specialists in Japan's Space Program," examines the role of aeronautical engineers and specialists in the period 1960–1980. As historian Nishiyama Takashi put it, for these (overwhelmingly male) engineers, "the year 1945 was neither the beginning nor the end of their engineering lives; it was merely a period of transition."[44] This is evident in the fact that engineering specialists and their protégés went on to wield considerable power within the space institutions that formed in the 1970s, even resisting state pressure; ISAS, for example, steadfastly refused to develop liquid-fuel rockets, or cooperate with American institutions, despite the Japanese government being keen for both. Conversely, they created their own networks by moving between space agencies and private corporations, a process that generated a common working culture. Beyond their institutional power, they also created their own space myths, which served to both explain and justify their vision for space exploration. The stories, dreams, and priorities embodied in the stories they told audiences outside the industry about their work enabled them to both improve the security of their work and defend their agendas within the increasingly institutionalized space program.

Chapter 4, "Influence of Commercial Interest," examines the impact of private corporations on space technological development. It demonstrates how the rise of satellites dovetailed with surging Japanese economic growth in the 1960s to create a huge demand for space-based communications networks. Particularly enamored of its potential were companies such as NEC and Mitsubishi, who organized with and through the powerful trade federation Keidanren to push for state support for their plans. These companies also understood that as satellites grew in complexity, they would grow in size and weight, requiring larger and more capable rockets to put them in orbit. They thus pressured the state to develop an entirely Japanese, LFR-based space program through the import and indigenization of its more sophisticated American counterpart. Their efforts finally paid off in 1969, with the signing of a US–Japan technology import agreement that would determine the direction of Japanese rocketry research for the next fifteen years. It would also result in a major institutional reshuffling, with a new entity, NASDA, quickly eclipsing the older ISAS in importance and scale. Thus, private corporate interest was responsible for drastically changing both the direction and policy of Japan's civilian space programs in the 1970s and 1980s.

Chapter 5, "Welcome and Resistance to Japanese Space Facilities," examines how people living near Japanese space infrastructure made their voices heard in the enterprise that birthed it. On one hand, locals in places such as Akita and Kyushu welcomed the arrival of space facilities, excited by the prospect of investment and tourism. On the other hand, these same people had seen industrial excess in other parts of postwar Japan cause disease, death, and denials of responsibility from private enterprise and state alike. Strong traditions of activist NIMBYism ("not in my back yard") developed to counter the building of controversial facilities such as nuclear power stations and airports. These two responses clashed over the arrival of launch facilities in southern Japan, where one group, fishermen, organized to resist their development. By interrupting launch tests, suing for access to tracts of land, and pressuring the government to provide them with restitution for the disruption of their work, they delayed the development of Japan's civilian space programs, and increased their cost. Their success in doing so, and the subsequent agreements that limited rocket launches to specific times of year, had a major impact on the pace of Japanese space development well into the twenty-first century.

Chapter 6, "Disseminating and Debating Japanese Space Policy," charts the emergence of a new set of narratives justifying the development of space technology in Japan in the 1980s by engineers and managers in the space industry. Whitewashing was no longer required; instead, supporters emphasized the benefits of the civilian space programs to the average Japanese citizen in terms of everyday life and national pride. Concurrently, Japan's growing confidence, a sense that the country had finally moved out of the shadow of World War II, and the geopolitical anxiety of the 1990s provoked a long-lasting debate around the issue of peaceful usage. In the 1990s, in particular, this was reflected not only in a greater willingness among senior figures in the space industry to discuss rocketry's military past and applications, but also in a sense that Japan should "normalize" its approach to military matters, in line with other countries such as China.

Whereas the various public discourses described in chapter 6 took some time to develop and mature, a much more rapid change of policy occurred in the 1980s when Japan decided to move away from reliance on US LFR imports. Chapter 7, "Growing Ambitions and Difficulties with the United States," shows how the impetus for this came from the expanding scope of Japan's civilian space programs. Emboldened by successes in the 1970s, these entities developed a host of ambitious plans, ranging from establishing a permanent presence in Earth orbit to the manned exploration of Mars. Yet at the same time, many specialists and engineers found themselves frustrated by the limitations the United States had begun placing on technology imports. Meanwhile, lawmakers and corporations became intrigued by the idea that rocketry could boost Japan's already spectacular economic growth through providing commercial launches. This alignment of success, frustration, and ambition among these stakeholding groups inspired the decision that Japan's civilian space programs would move away from reliance on LFR technology from the United States, and develop their own, "purely domestic" vehicles.

Chapter 8, "Success and Failure in the 1990s," shows how this decision did not pay off. Rather, the 1990s witnessed a series of high-profile failures in Japan's civilian space programs, ranging from the loss of expensive Mars probes to massive budgetary overruns in their rocketry development programs. Just as bad was the fact that plans to create a profitable space launch industry by developing an all-Japanese rocket failed in the face of temperamental technology and economic circumstances. The results negatively affected nearly every major stakeholding group in Japan's space enterprise. Engineers were frustrated and lost credibility in the eyes of the public and of lawmakers. Corporations came in for criticism and suffered tensions in their relationships with space agencies. Crucially, in the eyes of the public, these failures dovetailed with a general sense of malaise that gripped Japan in the 1990s. Reeling from the dual impact of the revaluation of the yen and an economic collapse, Japanese growth slowed, and a general sense that something had gone badly wrong in the country took hold. The civilian space programs' travails were seen as emblematic of this unease, which placed intense pressure on their defenders to accept some kind of reform.

Chapter 9, "Reform and the Creation of JAXA," argues that one of the main consequences of this malaise was a shift in balance between those who had long agitated for the unification of Japan's civilian space programs and those who had resisted it. Leading the former groups were Japan's politicians and lawmakers, who had the responsibility of overseeing state-funded Japanese space activities via an array of parliamentary committees. On one hand, these administrative entities had profound concerns about expenditure, the duplication of infrastructure, and the decentralization of projects in the civilian space programs, which escalated throughout the 1990s. On the other hand, from the very beginning in the 1950s, members of various Diet committees had pushed to integrate Japan's civilian space programs into a single, NASA-style entity—a process dubbed *ichigenka* or *ipponka*, "unification." For a generation, however, they had found their efforts largely sidelined by bureaucratic and managerial defenders of these institutions, as well as the successes of the institutions themselves. Many failed to understand the differences between these institutions, why these differences existed, and how, if their independence was so important, they could be overseen by the same committees.

Most of all, they were infuriated by the fact that the programs' independence from each other meant that even though they suffered from similar problems, none of their solutions could be universalized due to turf wars and minor institutional differences. With the failures of the 1990s, however, Japan's bureaucrats and space managers finally moved into alignment with the public and politicians over the issue of reform. This shift marked the end of the decentralization that had characterized Japan's civilian space programs since the 1960s, and culminated in the 2003 formation of JAXA.

HISTORIES OF THE FUTURE

Space—as a field of academic study, as an area of public interest, and as an industry—has come of age. In the last two decades, humans have flown probes through geysers on alien moons, collected high-definition images of the most distant known planet in the system, and discovered a veritable bounty of planets orbiting distant stars.[45] In the century to come, our species may yet reach Mars, build a Moon station, and launch a telescope capable of looking back to a (cosmically) scant 300,000 years after the creation of the universe. These accomplishments are no longer the domain of a few, elite space powers. China recently produced the most detailed map of the Moon ever created.[46] India successfully reached Mars on its very first try—and on a budget a fraction of similar US missions. And North Korea, despite even Chinese reluctance to provide it with technology, has put a satellite in orbit that can, at very least, change position as and when required.[47] Add to this the rise of corporations such as SpaceX and Blue Origin, and it is not difficult to see that we are living at a time of extraordinary space-exploratory proliferation. Yet space is also part of our lived experience in a way that goes far beyond the high-profile accomplishments of Cassini, of New Horizons, and of Starship. Nearly everyone who will read this book uses satellite technology daily, to phone their families, read the news, and find their way to meetings and lessons and restaurants. In the sixty years since the Soviet Union launched a 183 lb (83 kg), polished-metal sphere into orbit, humanity has cluttered the skies with tens of thousands of orbital objects. Observing, relaying, and broadcasting have become nonnegotiable elements of twenty-first-century life for billions of people.

There is hence no better time to turn our attention to the development and growth of this vast network of machinery, manpower, and imagination. Space exploration is not, and never was, merely a frivolous side-project driven primarily by national pride. It is a key political, economic, and social component of human modernity, and it would behoove us to better understand how this came to be. It is also one of the most aspirational of human endeavors—a fabulously expensive, nerve-racking, and rewarding effort at understanding the vast cosmos our little blue marble exists in. This book illuminates, as much as it can, how all of this looked from the perspective of the people of Japan.

PART I

Child of War, 1920–1960

ONE

The Transwar Origins of Japanese Rocketry

THE WAR IN ASIA WAS not going well for the Empire of Japan in 1944. For starters, it had lost control of the skies; in 1944 and 1945, a staggering 40 percent of noncombat Japanese ferrying flights resulted in losses, and B-29 raids leveled 178 sq mi (285 sq km) of the country's urban areas. Allied navies also stalked the waves in search of potential kills, and their land forces were hopping across the Pacific.[1] In fact, their submarines were already in Japanese imperial waters, too—which was how it came to be that in spring 1944, a group of US submarine-killers began tracking two IJN B1 type submarines en route from Singapore to Japan.[2] One of them, USS *Sawfish,* sank one of these, the I-29 *Matsu,* in the Luzon Strait near the Philippines. The crew could not have known at the time, but what they'd sunk was no ordinary boat: *Matsu* had previously shipped Indian nationalist Subhas Chandra Bose from Nazi Germany back to Asia.

On the day of her encounter with *Sawfish,* she was carrying a contingent of Japanese officers who had spent years in Europe gathering samples and advice from their (often reluctant) Nazi allies on topics related to German aviation. Among these officers was Eiichi Iwaya, senior case officer in the fighter technology research division of the Imperial Japanese Naval Aviation Headquarters.[3] Iwaya had spent 1940–1944 in Nazi Germany, alongside Takeo Mori, a fuel liquefaction specialist at Ōfuna Naval Arsenal. The Japanese military viewed Germany as an ally that "shared the same weakness of having limited access to petroleum resources"; German technologies, it was expected, would therefore fit well with Japan's material circumstances.[4] Thus Iwaya had spent time accumulating such things as radio components, Me-63 Komet rocket plane schematics, and even a sample of the pioneering Walther HWK 509A rocket engine. This device, several times more powerful than anything the Japanese possessed, was a particularly treasured piece of cargo, and it was imperative that it reach Japan intact. It fell to *Matsu* to undertake the 87-day, 9,300 mile (15,000 km) voyage homeward with both Iwaya and his stuff. Luckily for him personally, he disembarked *Matsu* in Singapore, with the intention of flying the rest of the way. Neither boat nor materials survived the encounter with *Sawfish*.[5] Iwaya didn't find out that *Matsu* had been lost until he returned to Tokyo. There was nothing to be done about the lost materials, but Iwaya still had a variety of schematics and reports in his possession. Working from these, he and his Imperial Japan Navy (IJN) colleagues began working on Japan's very first rocket-driven weapons of war. These would eventually include the infamous Ōka, a suicide torpedo bomb utilized by the *tokubetsu kōgekitai*—better known in the West as the *kamikaze*.[6]

After Japan's surrender in 1945, leading Allied military advisors, such as Karl Compton, president of Massachusetts Institute of Technology (MIT) and leader of the postwar Army Scientific Advisory Survey (ArmTech), wondered about how Japan "had . . . organised its scientific effort in the war" and "what technical ideas or developments of the Japanese could [the Allies] advantageously incorporate in our own future national security program?"[7] The Empire of Japan had excelled, for example, in the field of naval submarines; its I-400 series submarines were some 2,000 tons (1,814 tonnes) heavier, and had a range of nearly 20,000 mi (32,000 km), more than US *Nautilus* class submarines launched in the late 1950s. Large-scale battleships such as the *Yamato* and *Musashi,* were also technically impressive, even if largely obsolete due to air power.[8] Compton knew, however, that answering his questions would be a challenge. In addition to lax recordkeeping in Japanese laboratories, and extensive bomb damage, the two weeks between surrender and the arrival of occupation forces had seen authorities destroy documents and materials they wanted to keep out of Allied hands.[9] Thus "investigators had to enter Japan with the occupation forces, before manufacturing plants, equipment, materials and records could be destroyed and experienced personnel dispersed."[10] Organized into groups dubbed "technical missions," these investigators moved so swiftly that they sometimes arrived at their objectives ahead of occupation forces, leaving organizers concerned about "fanatics" and "guerilla bands" attacking them.[11]

FIGURE 1.1. Iwaya Eiichi.
Source: Wikimedia Commons.

The first technical mission—the US Army Scientific Intelligence Survey—featured 750 personnel sent to "ferret out . . . any new discoveries of techniques of significance to war activities."[12] The second, the Army Air Force Scientific Advisory Group (AAF), flew in on Army Air Force Commanding General H. H. "Hap" Arnold's personal C-54.[13] This group included such luminaries as (future) father of the Chinese Silkworm missile, Qian Xuesen; physicist Fritz Zwicky; future Jet Propulsion Laboratory head Bill Pickering; and noted mathematician and aerospace engineer Theodore von Kármán.[14] The third, and largest, technical mission was the US Naval Technical Mission to Japan (NavTech), which arrived in Kyushu on September 23, 1945.[15] In December 1945, to facilitate their activities, the Allies designated that "all tools and equipment" in "Army and Navy arsenals . . . should be made available for removal as soon as possible," as well as the entirety of Japan's "aircraft and aircraft engine" inventory. [16] Among the materials collected was "a complete collection of magnetrons and related high frequency tubes," samples of neocyanine dye-based medicines, and two 403,000 lb (183 tonnes) naval guns from Kure shipyard. Indeed, so much material was collected by NavTech that it had to establish collection centers in Sasebo, Yokosuka, Kure, and Kobe; in total, the US Navy alone dispatched some 15,000 items to the United States.[17]

Almost none of this material, however, was related to the work of Iwaya and his colleagues. Vehicles like Ōka had garnered plenty of attention, especially as it became clear that the Japanese had been experimenting with ramjets, and had also produced a variety of solid propellant motors and propellants earlier than the Allies had.[18] On the whole, however, they were unimpressed by Japanese rocketry at the end of World War II. One problem was that a great deal of advanced research had been destroyed before the Allies could get to it; Matogawa Yasunori, one of the most significant space administrators in twentieth-century Japan, observes that "almost all the invaluable materials and documents on solid and liquid propellant rocket engines during the war were destroyed by the relevant people for fear of being designated as war criminals."[19] This included all of the country's work on, for example, an advanced liquid-fuel rocket series dubbed Funryū, which utilized a radar homing system to hit targets with an accuracy of ±165 ft (50 m).[20] In short, the Allies were severely limited by what was left to find.

Beyond this, Japan's limited resources, and flaws in the design and manufacture of rockets, undermined the quality and sophistication of wartime Japanese rockets. The IJN's largest rockets, while "formidable," were deemed to have been "put into production too rapidly for much refinement." The design of rockets had also apparently involved a large degree of trial and error, as opposed to "new theories on optimum design rules."[21] Reliability was also undermined by the fact that "Japanese methods of manufacture and quality control were rather elementary compared to American methods." Apart from a 28-rocket launcher designed for warships, "no launchers of interest" to the Allies were produced by the IJN.[22] Homing and control systems on the I-go series of guided missiles, Japan's most sophisticated, were deemed "most elementary" and "inferior, in general, to American designs."[23] Where sophistication was found, as in the development of high-powered hydrogen peroxide fuel, there was a tendency to attribute this to "technical information received from Germany."[24] On the whole, while German rocketry impressed, Japanese weapons like Ōka served largely to confirm presumptions about Japanese fanaticism.[25] Whereas the V-2 was seen as era-defining because of its sophisticated pathfinding technology, the Ōka was dubbed the "baka bomb" (*baka* meaning "idiot" in Japanese), because instead of a tracker, it carried young men who were quite literally sealed within—the eponymous "idiots."[26] And whereas the creators of the V-2 were regarded as important enough to be spirited away to the United States as part of a campaign called Operation Paperclip, no Japanese rocketeers were deemed worthy of the same treatment.

There were, however, some who believed this evaluation to be misguided. These included dark matter theorist (and inventor of the designation "spherical bastard")[27] Fritz Zwicky, a member of the AAF mission. His report warned his colleagues against assuming that the United States was "superior to the vanquished in every respect, not only materially, but also intellectually, ethically, and culturally." To the contrary, Japanese wartime technology was "in many respects very good and might have sufficed to win the war for them had they not forfeited possible victory mostly through fatal strategic and political mistakes of their leaders." Their work was, in fact, "more impressive than is mostly believed," particularly in areas of rocket developments such as solid propellants and motors.[28]

At least from the perspective of Japan's postwar space programs, Zwicky was, in fact, the most prophetic of NavTech's observers. The relatively low impact of Japanese rocketry in battle, Allied preconceptions, and the aforementioned "mistakes" obscured the fact that Japan's prewar and wartime rocketry programs did, in fact, bequeath an impressive long-term inheritance to the country. First, the prewar period saw Japan construct an impressive network of physical infrastructure—from institutions of higher education, to fuel depots, to wind tunnels—that made sophisticated aeronautical research possible. Second, it produced the skilled individuals needed to design, maintain, and exploit these facilities. In fact, these two processes went hand in hand; nearly every major figure associated with Japan's postwar rocketry project had studied and secured their earliest jobs in the context of the country's century-long quest for empire.[29] Infrastructure and personnel would together produce the intellectual and industrial soil in which Japan's postwar space industry could grow. Lastly, Japan's wartime rocketry effort produced not just the people and the places, but the *experience* of developing high-temperature alloys, refined fuels, and other crucial technologies and processes. As such, the core elements of Japan's postwar space program—its infrastructure, its personnel, and its institutional knowledge—can all trace their roots back to Japan's doomed attempt at creating its own colonial empire.

Aviation, Rocketry, and R&D in Japan, 1920–1945

ENGINEERS AND TECHNOLOGICAL EXPECTATION IN PREWAR JAPAN

The context in which modern rocketry development began in Japan is inextricably tied up with the country's rapid industrialization in the early twentieth century. Between 1895 and 1915, for example, Japan's manufacturing sector alone increased in size by 250 percent, led by textile and coal production.[30] Despite this, Japan on the eve of World War I found itself profoundly reliant on overseas knowledge, technology, and processed materials. This concerned its burgeoning academic communities; in 1910, for example, astronomers complained that Japan had plenty of talented researchers, but their work was being stifled because the equipment they needed had to be imported at great cost, usually from Europe.[31] During this period, technology—"modernisation's most visible product," as historian Aaron Moore puts it—acquired talismanic importance in the imaginaries of both elite and common Japanese. Technology's importance lay not only in the creation of important infrastructure, but in what Moore has described as technology's "subjective, ethical, and visionary" aspects—that is, the impact that Marconi wireless radios, electric plants, and motor vehicles (among other things) had on what people expected of their lives in the present, and in the future.[32] Increasingly, lawmakers agreed that whatever the outcome was, it would behoove their country to have the ability to produce these life-changing products.

Advanced knowledge was also primarily sourced from abroad. Japan relied heavily on the German education system to add the finishing touches to specialist and technical education; in 1914, two of every three years spent abroad by Japanese scientists were spent in Germany. More than 75 percent of the 155 Japanese scholars studying abroad did so in the Kaiser's domain.[33] Japan's first student pilots, military officers Yoshitoshi Tokugawa and Kumazō Hino, studied in France and Germany, respectively. Hino, whose son Kumao would one day play a role in the development of rocketry in Japan, was responsible for the importation of a German Grade aircraft to Japan; alongside two French planes, they were the first such vehicles brought to the country.[34] Yet World War I abruptly revealed the downsides of this reliance. Beyond having to worry about the safety of their nationals abroad (Japan joined the war on the Allied side), the Japanese now had to face disruptions to supplies of alloys, components, and medicine. Clearly, the country's networks of supply and development needed an overhaul if it was to have the status and security its rulers craved.

The Japanese state thus prioritized the growth of domestic research and development (R&D) capacity in the early twentieth century. First came a wave of research institutions, including the founding of the famed physical sciences research institute Rikagaku Kenkyūjo (RIKEN)—home of Japan's first cyclotrons—in Tokyo in 1917.[35] These were then bolstered by the provision of scholarships and research funds.[36] Industrial production was also targeted with the creation of entities such as the Industrial Policy Association, formed to encourage researchers to serve national interests.[37] The disruption created by World War I was also a commercial opportunity for local producers, who prospered by filling gaps in the market left by stifled imports.[38] This in turn stimulated the creation of various private sector grants promoting basic scientific research—the company Mitsui, for example, partly funded RIKEN's cyclotrons.[39] The armed forces also began to establish their own R&D branches, beginning with IJA's Technical Department (Rikugun gijutsu honbu) and Army Institute of Scientific Research (Rikugun kagaku kenkyūjo) in 1919. The IJN established its own Technical Research Institute (Kaigun gijutsu kenkyūjo) in 1923.[40] Thus, with remarkable speed, a spread of institutions in Japan set out to develop their own homegrown, cutting-edge technologies—and rocketry was among these.

AVIATION AND ROCKETRY IN PREWAR JAPAN

Throughout this period, Japanese aviation enthusiasts had kept a close eye on developments abroad, and tried to replicate or outdo them at home. Just as the work of Wilbur Wright, Louis Bleriot, and Charles Lindbergh was fêted abroad, the accomplishments of pioneering Japanese aviators like Kumazō Hino and Yoshitoshi Tokugawa received extensive press coverage at home. Also covered were women like Tadashi Hyōdō (Japan's first qualified woman pilot) and the enterprising duo of Chōko Mabuchi and Kiku Nishizaki, the first women to fly across the Sea of Japan.[41]

Conversely, the successes of Western pilots such as Charles Lindbergh could breed an "irresistible sense of antagonism" amongst Japanese engineers, and fostered a powerful impulse to catch up technologically.[42] Flying machines and their masters became important parts of Japanese national self-image in the lead-up to Japan's invasion of China in 1937. The result, in terms of discourse, was what Hiromi Mizuno has dubbed "scientific nationalism" and the "techno-fascism" of Japan's interwar reform bureaucrats examined by Janis Mimura, both of which combined social engineering with technocratic knowledge in aid of creating a vigorous and respected Empire of Japan. Joining these bodies of thought was, of course, a healthy dose of technonationalism.[43]

Even at this early stage, aviation R&D had notable differences from overseas equivalents. Whereas in the United States, 70 percent of prewar R&D endeavors had been conducted by private enterprise, one German technologist visiting the country in 1938 observed that in Japan R&D was largely carried out by universities and state institutions, notably the military.[44] This was particularly the case in aviation. The Japanese were well aware of the sky's military potential—since the 1860s onward, the government had investigated manned surveillance balloons that would reach heights of up to 200 m; experiments in fixed-balloon reconnaissance continued during the Russo-Japanese War (1904–1905).[45] In 1909, future admiral Eisuke Yamamoto[46] argued for the creation of a naval air wing as soon as possible.[47] The direct result of this appeal was the establishment in July 1909 of a Research Commission on the Military Uses of Balloons (Rikugun gunyō kikyū kenkyū kai) at Tokyo Imperial University.[48] In the following decades, the militarizing Japanese state further stepped up a "national mobilisation of science," partly through the creation of more such research bureaus.[49]

Educational institutions were not, however, merely home to military-funded laboratories. Japan had a robust system of engineering schools; by the 1930s, 0.3 percent of the Japanese population studied such institutions, compared to 0.15 percent in Britain, and 0.18 percent in France.[50] Twenty-eight new research institutes attached to universities were established between 1939 and 1945, including new schools of engineering at Nagoya, Keio, and Tokyo Imperial.[51] Between February 1938 and August 1945, eighty more were established across Japan's imperial holdings.[52] Of these, the aeronautical institute at Tokyo Imperial University was almost certainly the most prestigious and significant of its kind. Many of the key figures involved in Japan's early postwar rocketry program, including Hideo Itokawa, studied there; Japan's space program would begin in 1953 at the same institution (by then renamed Tokyo University). Another important alumnus was Ichirō Tani (sometimes styled Itiro Tani in English), who traveled and studied in Europe in the 1930s before completing his dissertation during World War II. His work on laminar flow was so advanced that effectively capitalizing on it proved beyond the skills of Japan's aircraft industry; his later experimentation was critical in the successful wind-tunnel testing of Ōka.[53] His work proved to be of great interest to the incoming Allies at the end of the war, even if they mistakenly attributed some of it to someone else.[54] After the conflict, he too continued to work at Tokyo University, and won one of Japan's highest honors, the Order of the Rising Sun, in 1977.[55]

For those who didn't have the connections, couldn't afford the outlay to get into Tokyo University, or preferred to work with their hands, technical training offered careers ranging from commissioned officers in the IJN or IJA, to organizational staff, to engineers in Japan's burgeoning heavy industrial sector.[56] The expanding budgets, and ambitions, of the IJA's twenty-one research laboratories, and the IJN's eight, ensured there were plenty of research projects to go around.[57] Aeronautics was a particular beneficiary; the Japanese, for example, built a supersonic air tunnel long before anything similar existed in Allied lands.[58] This commitment only increased in wartime; in 1945, for example, $6 million in research funds was earmarked specifically for aeronautics.[59] Postwar Allied observers were impressed by "the tremendous importance . . . attached to research" by the Japanese throughout this period, even if, in the end, the empire had lost the war.[60] This system was fertile ground for career military engineers such as Misao Wada, the vice admiral who oversaw the development of Ōka, and naval officer Tadanao Miki, who worked with Ichirō Tani; both secured educations at Tokyo Imperial University, but pursued their aeronautical careers—until 1945, at any rate—within the IJN.[61] Their work was supported by a cadre of skilled technical specialists who could maintain the machinery, source the materials, and assist with management. Among these figures was a young man called Kumao Hino, a rocket fuel specialist who would go on to establish a successful postwar career as a trailblazer in the world of shaped charges.[62]

Rocketry was among the many technologies targeted for applied research in this period, particularly after Robert Goddard's successful work with LFRs in 1926.[63] Within five years, the Japanese armed forces were testing military rockets.[64] The IJA's research focused on vehicles that would complement its artillery.[65] Experiments went fairly well until 1934, when during a barrage rocket bomb test, one looped over the experimenters' heads and crashed in the forest *behind* them, instead of in the sea in front.[66] It thus became evident that Japan needed dedicated testing facilities, and by 1936, the IJA had become the first Asian armed force to build one.[67] Rocket and aviation facilities were often built in sparsely populated northern Japan; after the war, Japan's civilian space program would commence its own test launches there as well.[68] As tensions with Great Britain, the USSR, and the United States increased in the late 1930s, military figures redoubled their efforts, noting that "scientists in every country [were] competing" to develop effective military rockets.[69] Thus the development of key technologies such as spin-stabilization continued through the early years of the war, even as US bombing forced the dispersal of people and materials.[70]

The IJN also had its own aeronautical research system, which matched, if not outdid, the IJA's. By 1937, its Kasumigaura research facility boasted a 500-horsepower wind tunnel, the first in Asia powerful enough to model rockets.[71] Two years later it added a giant 4,000-horsepower apparatus, the most powerful in the world at the time. In fact, it was the IJN's Tsutomu Murata and his team who were responsible for what was probably Japan's first vertical rocket launch, in 1934.[72] By 1941, IJN aeronautical research facilities employed around 1,700 research staff working on projects ranging from rockets, to jet propulsion, to long-range bombing.[73] The navy also invested in huge facilities that supported research into, and the manufacturing of, fuels, alloys, and components. Among the biggest was the Ōfuna No. 1 Naval Fuel Depot near Kamakura, established in 1939, which required more than 3,200 employees at its peak.[74] Among its major products was hydrogen peroxide fuel for rockets. The site survived a major bombing raid in July 1945 largely intact, but was burned to the ground by the IJN itself after surrender.[75]

State authorities also deployed money to encourage Japanese corporations to cater to military needs. Where such corporations didn't exist, they were coaxed into existence: Japan's first domestic aircraft manufacturer, Nakajima, was founded by a former navy man in 1918. The industrial conglomerate (*zaibatsu*) Kawasaki joined the aviation fray that year, followed a decade later by Mitsubishi Heavy Industries. These three formed the foundational triad of Japan's wartime aeronautical industry. The IJN further bolstered local industry by favoring domestic production over imports.[76] Meanwhile, the same year as Nakajima's foundation, General Ikutarō Inoue, head of the Army Aeronautical Research Division, expressed his belief that a national Aviation Bureau should be founded to deal with "matters relating to the protection and encouragement of private aeronautical research and technology," "the supervision and qualification of private aviators," and "matters relating to private aircraft manufacturers." His recommendations were enacted by imperial edict on September 1, 1921.[77] Nor were subsidies and contracts the only way of encouraging local producers. Well before his turn as a star admiral in World War II, Isoroku Yamamoto, as head of the Naval Technology Department (*Kōhon gijutsubu*), initiated a series of competitions between 1931 and 1933 that provided prizes and incentives to companies making key contributions to aeronautics.[78] Thus, throughout the period leading up to World War II, money was made available to companies that proved themselves both innovative and responsive to military needs.[79] As a result, by 1938 Japan had a fairly robust domestic airplane industry.[80]

For corporations like MHI, entering aeronautics also meant entering the field of rocketry.[81] The materials Eiichi Iwaya had brought back from Germany had included schematics for a rocket plane, the Messerschmitt Me 163 Komet. When the Japanese decided to devise their own version, it was Mitsubishi that took the lead in this project. The eventual product, the Mitsubishi J8M, was deployed against the American B-29 bombers that had been laying waste to Japan's cities. Their mode of attack was using a built-in rocket to soar up past the bombers, strafing them at too high a speed to be countered, before gliding to ground at a distance. Mitsubishi's Nagasaki plant also specialized in producing and testing torpedoes, and conducted its own primary research into the high-energy fuels these required. A domestic rocket engine, the Ne series, was also developed there, though ultimately none saw manufacture due to American bombing.[82] Meanwhile, the Ōsone location became an aircraft-manufacturing center after the 1934 merger of Mitsubishi Aircraft and Mitsubishi Shipbuilding, and eventually turned to producing one of Japan's first rocket engines, the Tokurō-2 (a license-built version of the German engine that powered the Komet).[83] After bombing raids rendered the site nonfunctional, successor installations received supplies from as far afield as Fukuoka and Nagano.[84] The same area, in the postwar era, became the base for MHI's Nagoya Aeronautical and Space Systems Facility.

A host of smaller companies were also drawn into the effort through a system of subcontracting that would persist well into the next century. Wartime efforts to develop hydrogen peroxide rocket fuel, for example, drew on the work not only of companies in and near Tokyo, but of those in Kyoto, Nagasaki, and Kyushu as well. The relatively small Hodogaya Chemical Manufacturing Company's Yokohama plant produced approximately 27 tonnes of H_2O_2 fuel for the IJN between 1938 and 1945.[85] The US Strategic Bombing Survey also noted that a considerable percentage of airframe and engine manufacturing and subassembly was let out to subcontractors, and "shops scattered throughout the industrial areas supplied the thousands of bits and pieces that made up the finished aircraft."[86] Many of these corporations continued to use the same infrastructure in the postwar period for similar ends; Hodogaya Chemical Corporation still uses the same site in Tsurumi, Yokohama, that it did in 1945.[87]

LIMITATIONS AND FAILURES IN JAPAN'S WARTIME ROCKETRY PROGRAM

Rocketry came of age during World War II primarily as a weapons technology, and it was this objective that spawned the R&D networks described above.[88] Still, in the eyes of those it was intended to defeat—the Allies—Japanese wartime rocketry was far less formidable than the weapons of their Nazi allies. And, indeed, there were several circumstantial reasons for the failure of Japanese rocketry to make an impact on the battlefield. One was the disruption to supply chains. From 1941, the throttling of imports to the country by Allied navies meant that per capita consumption of rice was restricted to 12.5 oz (354 g) per day in cities, flour to 6.5 oz (186 g) *per month* per family of three, and vegetable oil to 0.4 quarts (0.36 liters) *per quarter* per person.[89] Even fish was in short supply.[90] In the absence of petroleum, 15 percent of Japan's forests were logged for fuel.[91] The situation only worsened with the onset of Allied bombing; Operation Meetinghouse in March 1945 leveled nearly 10 sq mi (16 sq sq km) of central Tokyo, and may well have killed more people than the atomic bombing of Nagasaki or Hiroshima.[92] One-quarter of Japan's factories and infrastructure were obliterated by war's end.[93] The Allies who landed in Japan in late 1945 were confronted with a nation where transport "was limited to crowded, creaky trains, hand-pulled two-wheel 'rear cars' designed to be attached to bicycles, and oxcarts"—a land in which a mere 41,000 operable motor cars survived, mostly driven by coal.[94]

Air raids also had a direct impact on aeronautical researchers, particularly in the Nagoya region, where much of Japan's aircraft production was located.[95] Researchers found supplies as basic as paper and electricity dwindling; despite producing promising research, the Glider Research Institute at Kyushu Imperial University published none of its findings on account of a shortage of paper, and because all of its printing presses had been requisitioned for the war effort.[96] Allied ships also cut off supply lines. In 1937, Japanese manufacturers imported 67,200 lb (30,000 kg) of the manganese required for temperature-resistant alloys from Malaysia, and 26,500 lb (12,000 kg) from Indonesia and the Philippines, respectively.[97] By 1945, both territories were cut off. And lastly, air raids and wartime exigencies rendered the lives of researchers and specialists miserable. Future Nobel Prize laureate Shin'ichirō Tomonaga would later recall how in just one day during the war, "one student's house was burned down, another [student] was drafted, and a third had his house burned down just before he was drafted."[98]

Despite these difficulties, Japan's bureaucrats and administrators continued to place some faith in the products of their country's R&D. Resources were still made available for projects with names like War Decider (Kessen), a "death ray" capable of murdering monkeys (but not much else), and performance-enhancing chemicals for soldiers and pilots.[99] Rockets also seemed worthwhile—as US military aviation pioneer H. H. "Hap" Arnold put it: "Why should bombers, each of which required thousands of man hours to construct, be the sole means of long-range aerial offense[,] when a V-2 rocket can be fabricated in a fraction of the time?"[100] For the resource-poor Japanese in 1945, jet engines and rockets also held the allure of obviating the difficulties they faced in the mass production of piston engines. Furthermore, both provided far more energy output than similarly sized piston engines (in technical parlance, a higher mass-to-weight ratio), and had relatively simple designs, making them lighter.

As a result, Japan's wartime rocketry R&D was shaped by material scarcity. The IJA had graduated to exploring LFRs in 1937, only to abandon the effort in 1939. As piston engine shortages escalated, however, the technology once more became a priority, particularly if the engines were to be used on airplanes. Thus, in 1942, the IJA resumed funding research at Mitsubishi's ordnance factory in Nagasaki.[101] Two years later, work commenced on a radio-controlled flying bomb, which by November 1944 resulted in the "the only flyable results of pilotless [i.e., remote-guided] missiles" used by Japan during World War II, the 1,765 lb (800 kg) Ichi-go 1A and the 660 lb (300 kg) Ichi-go 1B.[102] Again, despite some success in simplifying the production of these devices, and a great deal of fundamental manufacturing work, they were too slow, and of too short a range, to be effective battlefield weapons.[103] Reliability also remained an issue: a remotely guided rocket from 1945, for example, was not extensively used by the army because it was too dangerous to their own soldiers.[104]

Shortages and strategic defeats led to a focus on low-cost, easily produced armaments designed to hit specific Allied targets. When this was found to require more sophisticated targeting technologies than the Japanese possessed, they simply replaced computer guidance with human guidance. And so "all jet and rocket-powered craft developed [in Japan] after 1944 had as their goal suicide attacks."[105] The first of these was also Japan's first domestically designed rocket aircraft, the Ōka human-guided rocket torpedo.[106] The vehicle itself had a range of around 25 mi (40 km), and a top speed of up to 685 mph (1,100 kmph). It also failed to live up to expectations.[107] Designed as a short-range device capable only of a single burst of rocket propulsion, Ōka had to be carried to within striking distance of their targets by bombers. These were slow-moving and hard to defend, which made them susceptible to being downed well out of Ōka's effective range. Thus the first air sortie by the weapon on March 21, 1945, was a complete failure—none made it through to its target.[108] Between March 18 and June 22, the US military recorded fifty-seven Ōka attacks, none of which had a major tactical effect, even if they did have a psychological one.[109]

Other high-profile rocket weapons arrived too late in the war to be of any use. One was the IJN-funded Mitsubishi J8M1/Ki-200 Shūsui, largely a retro-engineered version of the German Me-163 Komet.[110] As many as 3,600 were planned to enter service by 1947, joined by a second-generation suicide rocket, the Kawanishi Baika. Much like Shūsui, Baika also had its roots in the introduction of a German technology—the "pulse jet," a design that had very few moving parts and thus was relatively easy to produce.[111] In the end, though, neither went to war. Baika never got beyond the construction of its engine, the Kawanishi Maru Ka10. Shūsui got further—completed models underwent testing in late 1944 in Nagoya, then in Yokosuka, and then in Hakone, moving each time due to B-29 air raids. Meanwhile, engine testing continued in Nagano. Then, on July 7, test pilot Toyohiko Inuzuka took Shūsui on its maiden flight. During the flight, a fuel tank on the underbelly slid out of place, causing it to skew and plummet to the ground. Neither plane nor pilot survived. In the end, neither did the project.[112]

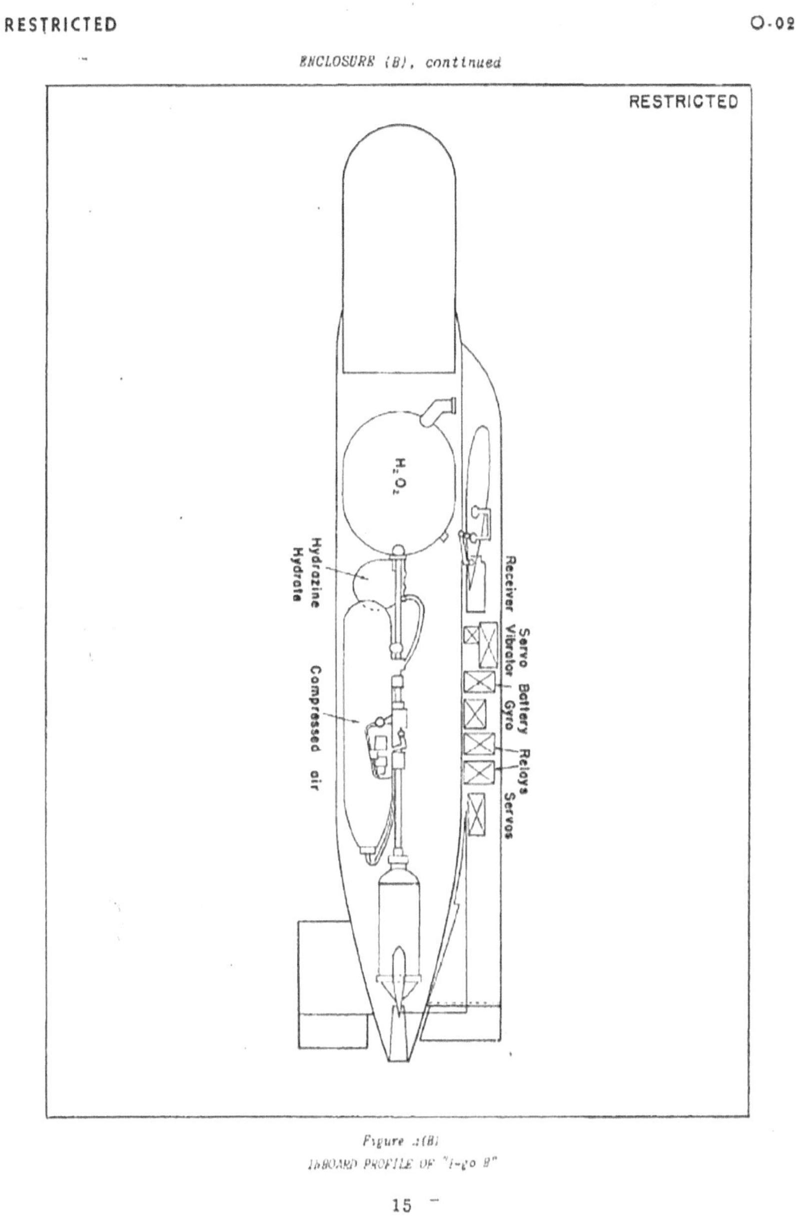

FIGURE 1.2. I-go B schematic showing the basic structure of early Japanese rockets.

Source: US Naval Technical Mission to Japan, "Japanese Guided Missiles," in *Report of the US Naval Technical Mission to Japan* (San Francisco, CA: US Naval Technical Mission to Japan, December 1945), National Diet Library, Tokyo, 15.

SUCCESS AND INNOVATION IN JAPANESE WARTIME ROCKETRY R&D

So far, so disappointing. However, judging Japan's wartime rocketry R&D only by such failures, particularly from the perspective of its postwar legacy, would be misleading. In terms of fundamental research, rocketry was one of the few areas—along with chemical warfare, meteorology, and atmospheric measurements—where the Japanese made any meaningful progress during the war.[113] This becomes clearer when examining not suicide torpedoes and pulse jets, but the simpler and more rugged rocket weapons that were deployed by the Japanese armed forces throughout World War II—for example, the reliable 4FR-110 rocket system used aboard IJN warships from at least 1943. From July 1944 onward, the navy also deployed a 4.7 in (12 cm) incendiary shrapnel anti-aircraft rocket aboard the aircraft carrier *Zuikaku,* as well as the battleships *Ise* and *Hyūga,* systems that were "highly valued for their increasing of fire power of the ship and for their threatening effect."[114] On land, both the IJN and the IJA deployed *funshin,* or rocket artillery. Examples included the Type 4, 7.8 in (20 cm) rocket launcher, a bazooka-style weapon; and the larger, 15 in (40 cm) version launched from a wooden frame.[115] The fuel developed for these would later play a major role in the success of Japan's first postwar rocket, the Pencil.[116] Other weapons were created by the expedient of attaching rocket motors to preexisting bomb designs, as in a 550 lb (250 kg) projectile designed to create clouds of lethal shrapnel on Iwo Jima.[117] The Japanese also made headway in developing their own rocket engines: the Ichi-go guided missiles were driven by Tokurō-1 LFRs designed at Mitsubishi's Nagasaki ordnance plant. [118]

The single most significant burst of fundamental rocketry development in this period was initiated by the return of Eiichi Iwaya in July 1944. After flying from Singapore back to Japan, Iwaya set to work almost immediately at the No. 1 Naval Technological Research Arsenal, located in Oppama, Yokosuka, Kanagawa. His carefully smuggled schematics contained, among other things, the BMW-1003a and Jumo-004b jet engines, and engineer Hellmuth Walter's work on rockets, including blueprints for the aforementioned HWK 109–509. He had also convinced his Nazi hosts to part with miscellaneous materials on how to manufacture manganese nozzles capable of handling high-temperature rocket exhaust.[119] News of his return spread quickly through the ranks of aeronautical fuel researchers and engineers.[120] Within a few weeks, a rocketry association (*roketto iinkai*) was formed within the navy, which brought together specialists keen to explore the new technologies. Alas for them, they realized in short order that there were huge gaps in Iwaya's data.[121] One reason was German caginess—Iwaya had been shown both the Me-163 Komet ("beautiful," in Iwaya's opinion) and the Me-262 jet plane ("grotesque") on April 6, 1944, but had not been allowed to examine either closely.[122] In some places, German nomenclature and phrasing were unclear, meaning the Japanese were often baffled by things such as the ingredients in fuel recipes. And last, but perhaps most worrying, the Germans themselves had been unable to resolve certain fundamental issues associated with their technologies.[123]

The engineers working on the technologies therefore had to innovate. When developing Shūsui, for example, it proved impossible to simply re-create the Me-163 from schematics, and practical issues necessitated work on the aircraft's climb rate, range, and armament, the production of propellants, and manufacturing systems.[124] Developing fuel received particular attention. Japan was already starved for energy—securing access to Indonesian oil fields had been among the country's foremost imperial objectives. Iwaya and his team thus had to develop new fuel recipes, which in turn required redesigned engines.[125] Furthermore, the Walther engine had utilized ammonium sulphate as a stabilizer, which was both too unstable at high temperatures, and too reactive, to be reliable. To compensate, researchers at the naval facility in Ōfuna developed the fuel stabilizer ortho tolyl urethane (OTU), a substance deemed by NavTech "the only point of possible superiority of a Japanese propellant over those of other countries." OTU was "so easy to roll and extrude that [it] . . . had practically none of the danger from fire that had been so troublesome in the American manufacture of rocket propellant."[126] In the meantime, the navy also researched the production of fuels at Ōfuna. By November 1944 it had established a hydrogen peroxide (H_2O_2) fuel production unit at the facility, complemented by the facilities of the Edogawa Technical Company. Academic researchers devised an entirely new apparatus for the purpose of refining H_2O_2, and by the end of spring 1944, they had succeeded in boosting its saturation from 30 percent to 97 percent, producing substantially more powerful fuels.[127] Six months later, an academic researcher from Kyoto working at the IJN's Meguro Technology section produced fuel of 98.7 percent saturation.[128] Japanese engineers also improved infrastructure, for example, by developing a system of observation posts on high ground to measure the trajectory and performance of the Ichi-go 1A, and deploying a variety of sophisticated observation devices.[129]

Subcontractors, in developing the capacity to make the required products, also improved both their production and their own R&D. These teams often innovated by bringing the knowhow associated with traditional domestic industries to modern industrial processes, for example, by using porcelain to develop tanks for storing the notoriously unstable H_2O_2 liquid.[130] The industrial production of H_2O_2 required large quantities of platinum, and thus birthed the first large-scale platinum extraction efforts in Japan.[131] When these proved insufficient, researchers began working on methods that did not require platinum.[132] Developing Ōka required the research and development of a 100 percent reliable fuse for the explosion mechanism, as well as fundamental research on thrusters.[133] The production of OTU, meanwhile, was outsourced to the Hodogaya Chemical Manufacturing Company's Yokohama plant, which produced approximately 54,000 lb (27 tonnes) of the product between 1938 and 1945.[134] The Sanboku Edogawa plant (as of 2022, still working as a gas-refining facility) developed key chemicals, and the relatively small Ōe facility was responsible for assembling Shūsui.[135] Other suppliers included Mitsui Dye-Stuff, Nippon Chemical Synthesis, Sumitomo Chemicals, Yasushigaya Chemicals, and Dainippon Chemicals, which constructed plants in areas such as Nagoya, Kawasaki, and Ōmuta; combined, these companies were producing up to 220,000 lb (100 tonnes) of specialized chemicals for Japan's aviation and rocketry projects per month by the end of the war.[136] Of course, large conglomerates such as Ishikawajima Aircraft Manufacturing, Hitachi, and Nakajima were critical for projects such as Shūsui.[137] However, no single *zaibatsu* was as involved in Japanese wartime rocketry R&D as Mitsubishi. Engineers and managers from the company were among the first to meet with Iwaya on July 20, 1944, and by August they were already working on what would become the Tokurō-2 engine.[138] Mitsubishi staff in Nagoya and Nagasaki also worked extensively on the "difficult" project of producing fuel for the plane, and tested new materials capable of withstanding the 1800°C temperatures produced inside the new engine.[139] Thanks to these intensive efforts, it took the IJN roughly three months from acquiring its patchy data on the Me-162 to producing a flight-ready model of Shūsui.[140] The weapon's failure was not a complete disaster for its producer—in a trend that would continue for the rest of the twentieth century, Mitsubishi had already found other successful applications for the technologies it had developed for a rocketry project.[141]

DEVELOPING SPECIALISTS IN JAPAN'S ROCKET INDUSTRY: KUMAO HINO

In the end, it was not the material products of wartime rocketry R&D that mattered for Japan's spacefaring history—it was the people who had produced them. Two examples serve to illustrate the sort of impact these men (as all were) would have. Kumao Hino was closely involved in the development of rocket fuels in the 1940s, before becoming one of Japan's leading specialists in the field of directed explosions (a technique crucial in mining, among other industries). In this, he is representative of former rocket engineers and aviators, products of Japan's push for heavy industrial independence in the early twentieth century, who were crucial to Japan's economic development "at every point from 1945 through 1964 and beyond."[142] Rocket engineer Tadanao Miki went on to work on the Shinkansen bullet train; the designer of Ōka's wind tunnel test model, Hidemasa Kimura, was one of the few active aircraft designers in Japan in the 1950s; and rocket and jet engine theoretician Fujio Nakanishi continued to produce pioneering work on the tensile qualities of materials like steel in the 1950s and 1960s.[143]

As a naval officer, Kumao Hino was among those assigned to participate in Eiichi Iwaya's work in 1944. He was born on January 30, 1914; his father, Kumazō Hino, had been one of the pioneers of aviation in Japan. As a lieutenant colonel in the IJA, Hino senior had often been at odds with his employer over his lavish expenditure on flight, even being demoted once for excessive spending.[144] Despite this, Hino junior followed his father into the armed forces, albeit in a technical role, and in the IJN. By 1941 he had reached the rank of technical lieutenant commander, and joined the navy's rocket fuel program at Funaoka Naval Explosives Arsenal in Miyagi, 220 mi (350 km) north of Tokyo. He spent most of the rest of the war there working on advanced fuels research.

The full extent of Hino's involvement in Japanese wartime rocket development was unearthed after the war by Fritz Zwicky. Zwicky was particularly interested in Japan's work on rocket fuels, but much of the work survived, he discovered, only in "the minds and notebooks of various Japanese known to Lt Commander Hino."[145] He first reached out to Hino on November 17, 1945, at a time when citizens were starving to death near Tokyo station, and military commanders were being hunted down and killed by disgruntled former grunts. For military engineers who hoped to continue working, there was very little work left that did not involve the occupying forces in one way or the other.[146] Working with the former enemy offered an opportunity to regain some personal pride; as the Allies later observed, people like Hino "took considerable pride" in finishing the work they had undertaken during the war, "to demonstrate the extent of their advancement."[147] In any case, as researchers, many of his status had also chafed against the secrecy and restrictiveness of military stricture, as well as its insistence on applied work. He, at least, had the benefit of being a commissioned officer within the navy. Those working at universities often found themselves treated with such profound suspicion that they were treated "almost as if [they] were foreigners."[148] Few felt much loyalty to their erstwhile employers. Anyone in Hino's position would thus have had every reason to respond to Zwicky's request for help. And respond Hino did, within four days, requesting, in English, a period of six months to complete reports "on the combustion and detonation velocities of solid explosives including the application of the principles". In return, he asked for "Ten thousands [*sic*] Yen, if possible, in advance" and "Three lunch boxes a day," as well as an official title, "such as translator or interpreter in the US FEAF [Far East Air Force]."[149] Five days later, he submitted a proposal for a report "On the combustion and detonation velocities of solid explosives (including the applications of the principles)." The final work, handed in five months later, was exhaustive, beginning with reports on fundamental research on explosives, and moving on to descriptions of applied research into rocket propellants in the third section. Hino revealed that he and his team had worked on low-pressure combustion and stable propellants, and had engaged in theoretical work regarding "the future of the rockets with solid propellants." Unfortunately, he observed with a touch of dryness, his experiments were "interrupted before some definite conclusions were attained."[150]

Still, defeat had not paralyzed Hino; even as he was finishing his report, he received his doctorate from Tokyo Imperial University on October 5, 1945.[151] Nevertheless, he had no future in aviation, for the simple reason that the Allies had proscribed all aeronautical research in Japan. Furthermore, these authorities—often designated "SCAP" ("Supreme Commander of the Allied Forces," the official title of the occupation's chief commander, Douglas MacArthur)—had also begun their stated goal of "demilitarizing" Japan by grounding all military pilots and airplanes.[152] On November 18, 1945, the SCAP's General Headquarters issued directive SCAPIN301, grounding all *civilian* pilots and aircraft as well; meanwhile, nearly all military aircraft in Japan were torched, blown up, or tossed in the sea under

American oversight. SCAPIN301 also included the two following clauses, terminating all aviation research and training in Japan for the foreseeable future:

1. On and after 31 December 1945 you will not permit any governmental agency or individual, or any business concern, association, individual Japanese citizen or group of citizens, to purchase, own, possess, or operate any aircraft, aircraft assembly, engine, or research, experimental, maintenance or production facility related to aircraft or aeronautical science including working models.

2. You will not permit the teaching of, or research or experiments in aeronautical science, aerodynamics, or other subjects related to aircraft or balloons.[153]

In the aftermath, Japan's aircraft manufacturers collapsed. After peaking at more than 24,000 employees in 1944, the mighty Mitsubishi Aircraft Company all but ceased to exist; Nakajima was carved up into a clutch of smaller corporations, none of which made planes; Kawasaki would not resume making planes for a decade. Still, as Richard Samuels has pointed out, the long-term effects of this on Japan's actual capacity to build aircraft was less dramatic than might be expected. By 1948, the Americans were more concerned about the spread of communism than demilitarizing Japan; the resulting "reverse course" in policy necessitated the revival of Kawanishi Aircraft (as Shin Meiwa) in 1948 to supply local technical support for US hardware. Indeed, some 40 percent of aircraft production facilities were maintained throughout the occupation, and 80 percent of aircraft engineers remained in their positions or similar ones.[154]

Still, it was not an industry that had many openings, and Kumao Hino was among the 20 percent who were forced to take their skills elsewhere. He eventually found work at chemical manufacturer Nippon Kayaku, where he used his knowledge of directed explosions (which is what rockets essentially are) in the field of mining. Over the next decade or so, he produced seventeen pieces of research for the company, all focusing on experimental and theoretical explorations of the use of explosives in industrial extraction.[155] They included 1957's "Shock Wave Theory of Blasting with Cylindrical Charge" and 1958's "Theory of Relations between the Detonation Velocity of Solid Explosives and the Thickness of Cases or the Diameter of the Charges," which are clearly connected to wartime research into rocket propellants.[156] Hino also published at least one textbook, in 1959, *Theory and Practice of Blasting*, and spent some time in the United States teaching shock wave theory.[157]

His industrial work produced a clutch of innovations that are still in use; in US patent No. 3111437 of 1963, Hino and his colleagues registered a "cap sensitive ammonium nitrate-fuel oil explosive and a method of manufacturing the same," an innovation utilized as recently as 2012 by Lubrizol, a producer of specialty chemicals for engines.[158] Hino eventually entered management, and became a board member in 1982.[159] Thus, after acquiring his wartime skills in the context of rocketry development, Hino successfully went on to contribute to the development of postwar Japanese industry, and the country's long-term recovery.

DEVELOPING SPECIALISTS IN JAPAN'S ROCKET INDUSTRY: HIDEO ITOKAWA

Then, of course, there is Hideo Itokawa, a man so ubiquitously connected with rocket technology that he has come to be dubbed "Dr. Rocket" in the popular consciousness. Itokawa did some brief work on rocketry during World War II, but primarily worked as an aircraft engineer.[160] His experience is representative of figures who converged on rocketry from other fields of study (even if, in Itokawa's case, it was an adjacent one). These included people such as Ichiro Tani, who, having produced cutting-edge work on fluid dynamics and airflow during World War II, later brought his skills to the Space Aeronautics Research Lab at Tokyo University.[161] This trend would endure into the twenty-first century, as shown by Hideo Shima (head of NASDA, 1969–1977) and Shūichirō Yamanouchi (head of NASDA, 2000–2003, and first chief of JAXA from 2003), both of whom went from being railway engineers to space administrators; Tomifumi Godai, one of Japan's senior rocket designers of the 1990s, who began his career at Fuji Seimitsu; and Chiaki Mukai, the first Japanese woman in space, a trained physician who only came to the space program more than a decade into her career.

Born into an affluent family in Aoyama, Tokyo, on July 20, 1912, Itokawa followed his siblings into the prestigious Tokyo Imperial University.[162] The institution was Japan's leading producer of fundamental aeronautical research; an Institute for Aeronautics had been founded there in 1918, and by 1920 the group had a wind tunnel.[163] However, it is the foundation of the semi-autonomous Tokyo Imperial University Aeronautical Research Laboratory the following year that is regarded by many analysts as the real beginning of independent aeronautical research in Japan.[164] Thus, when Itokawa graduated in 1935 with a degree in aeronautics, he did so from the very best program of its kind, most probably in all Asia—one that, it should be noted, also produced Kumao Hino, Eiichi Iwaya, Hideo Shima, Yasunori Matogawa, and Ryōjirō Akiba. Itokawa's first job after graduation was at Nakajima Aircraft, where he worked on military projects including the Ki-43 Hayabusa tactical fighter, and the Ki-44 Shoki fighter-interceptor.[165] In April 1941, he returned to Tokyo Imperial University to head up an aeronautical research division, where he did some work on missiles.[166] During this time he also established himself as a prolific writer, starting with technical documents published in Tokyo Imperial University's in-house *Tōkyō teikoku daigaku kōkū kenkyūjo ihō* (Report of the Aviation Research Institute).[167] At Nakajima, he also produced a two-part technical piece for its in-house magazine, "Nakajima kenkyū jōhō: Yoko oyobi hōkō no anteisei oyobi sōju sei" (Nakajima Research Report: Sideways and Oientation Stability, and Driveability) in 1939, a densely written work on aerodynamics in moving aircraft.[168]

Despite his pedigree and wartime success—and avoidance of American blacklisting—Itokawa's postwar years were desolate; as he himself put it, "as my equation in those days used to be [Itokawa = Aircraft], I was left with [Itokawa − Aircraft = 0]."[169] Though he was able to retain his position at the reformed Tokyo University's Institute of Industrial Science, the group's primary research relationship had been with the armed forces, which had now been disbanded by the occupation government. Even if funding were available, SCAP's prohibitions on aircraft design effectively denied Itokawa the ability to continue the work he'd spent almost a decade undertaking. For a few years, he busied himself with projects only tangentially related to his specialty. In autumn 1945, a run-down Itokawa visited a doctor, who had asked him if, as an engineer, Itokawa thought it was possible to use "the technology in planes in the field of medicine." Intrigued, Itokawa ended up working for a while with a team to develop a type of "stick" electroencephalograph (EEG) that enabled doctors to measure the effects of anaesthesia.[170] In his spare time he indulged his interest in string instruments.[171] Nevertheless, the future "Dr. Rocket" would later describe 1945–1953 as a period of profound malaise, during which he longed to resume work on flying machines. As his student Yasunori Matogawa was to later observe, his work in the field of music was primarily to do with the vibration, information that was also valuable when designing airframes.[172]

Itokawa's opportunity finally came in 1952, when both the occupation and the ban on aeronautics ended. In January of that year, he found himself in the United States to show off some of his work in the medical field. While there, he was impressed with the theoretical medical groundwork being laid in the country for human spaceflight. The Itokawa who returned to Japan was possessed of the idea that his own country could, and one day should, put a human being in space, too.[173] He subsequently delivered a passionate, two-hour-long speech during a Keidanren-sponsored event in October that year.[174] The former Nakajima engineer even persuaded one of his students, Ryōjirō Akiba, to join a team he was putting together to investigate the technology. Akiba, later head of one of Japan's space bureaus, confessed that he'd had no interest in rocketry until this point.[175] Japan's civilian rocketry project had thus attracted its first outside specialist.

Itokawa commissioned Akiba to design a model rocket to test in a wind tunnel. What happened next is as revealing of Itokawa's character as it is of the history of Japan's space program:

> "'How much can I spend for the model?" the student asked. Itokawa replied, "Nothing." His intention must have been "think of some good ideas yourself." The student's deliberation resulted in a paper rocket. He made a paper cylinder for the body of the rocket. A nosecone and four fins, also made of paper, were attached to complete the model rocket. It is natural that this rocket did not play any role in a wind tunnel experiment, but it did go down in history, much to his surprise. Taking a brief look at the paper rocket, Itokawa murmured "Good," and took it outside to be photographed in the field. The photo was inserted with a caption which read "Domestic Rocket No.1 Manufactured Experimentally at the University of Tokyo," in his essay "Flying Across the Pacific in 20 Minutes—A Dream for a Domestic Rocket." This appeared in the newspaper *Mainichi-Shimbun* [*sic*] on Jan.3, 1955.[176]

It was a gutsy move, since Itokawa's group had neither the funding, nor approval, for a rocketry program. And, on top of that, it was just a paper cutout, not an actual rocket. Nevertheless, the piece attracted the attention of civil servants, who decided to sponsor his work on a *real* rocket as a part of the upcoming 1957–1958 International Geophysical Year (IGY).[177] The project, intended to encourage cooperation across the Iron Curtain, would eventually inspire both Sputnik and the first US satellite, Pioneer.

And so Japan's civilian space programs officially began on February 5, 1954, when Hideo Itokawa initiated the Avionics and Supersonic Aerodynamics (AVSA) group. [178] Less than a decade out from Hiroshima, a team of fewer than twenty (male) graduate students, wartime engineers, and technicians set out to design and build an entirely Japanese rocket with Japanese components on Japanese soil. One participant later recalled that many engineers and academics joined simply because the project was reasonably well funded.[179] This funding totaled ¥3.3 million (around $650,000 in 2023), distributed between AVSA and its supporting companies.[180] It was more money than any of them had seen in a very long time.

In the aftermath of Japan's surrender, physicist Fritz Zwicky of the Allied Air Force Technical Mission opined that "the Japanese had a sound outlook" on the value of rocketry, in contrast to the Allies, whose "awakening" to the technology had been "lamentably and pitifully slow."[181] His observations are a testament to the effort Japan's armed forces, private corporations, and engineers put into rocketry R&D before and during World War II. Emerging from a broader push to develop a domestic aircraft industry, Japan from 1920 to 1945 undertook the creation of an extensive network of R&D and manufacturing infrastructure in locations such as Tokyo Imperial University, Ōfuna Naval Arsenal, and MHI's Nagoya plant, which would be crucial constituents of the country's postwar rocket industry. Once research into rocketry began in earnest in the late 1930s and 1940s, researchers managed to produce a collection of devices that, in most cases, did not have a major impact on the battlefield, but did generate a body of knowhow, infrastructure, and experience in working with the technology. Distributed as it was through Japan's vast, if pulverized, industrial sector, this legacy survived defeat in World War II. This is not to say that there was no major disruption; between the peak of Japan's wartime rocketry industry with the Iwaya documents, and the beginning of Hideo Itokawa's Pencil rocket project in 1954, Japan suffered widespread bombing, a ban on aeronautical research, economic upheaval, social anarchy, and political disruption. The systems described above were bent, stressed, and broken. Yet enough survived that by the 1950s, the engineers and specialists who were developing rockets were able to draw on its remnants—and, of course, new ones that had been developed since the end of the war—so that they could rapidly progress from a cold start to becoming the fourth country to launch a satellite into space.

Yet this is, of course, only part of the story. Although the materials and traditions existed, there would have been no impetus for developing a rocketry program in Japan if there had not been a shared belief that pursuing the technology was worthwhile. After all, Britain, whose own industry was somewhat more intact in 1945 than Japan's, set out on its own rocketry project, before abandoning it for lack of interest. To avoid this outcome, rocketry in Japan had to demilitarize. As we have seen, the proximate rationale behind developing rocketry in Japan in the period from 1930 to 1945 was military deployment. Yet such rationales were worse than nothing in a postwar period darkened by the shadows of mushroom clouds and fleets of B-29s. Instead, Japan's rocketry specialists set out to purge their technology of these wartime associations, and tap into another set of narratives. In this telling, rocketry was an aspirational, peaceful, and futuristic technology—vehicles that would carry not warheads, but people, and their dreams. The survival of rocketry as a technology in Japan would depend to a large extent on whether people bought what they were selling.

TWO

Demilitarizing Rocketry in the Postwar Period

In April of 1945, the battleship *Yamato* set sail to bring a ray of hope into a time of utter despair. We do the same. . . . as long as there is one faint ray of light in the dark—as long as we have even a chance—we must go forward.

That is the destiny of any ship named *Yamato*.

***Space Battleship Yamato*, 2010**[1]

SPACE BATTLESHIP YAMATO, FIRST AIRED in the 1970s, has a premise that would never have made it past US censors just twenty years earlier. In the series, a future Earth, battling the genocidal Gamilas, refloats the World War II Japanese superdreadnought *Yamato*, outfits it as a spaceship, and dispatches it to save the day. The seamen who sank the actual ship in 1945 might well have questioned the description of the ship's final wartime mission as a "ray of hope into a time of utter despair."[2] Indeed to Western eyes, this might, in fact, seem like a bizarre narrative inversion—after all, weren't the Japanese the bad guys? Yet *Space Battleship Yamato* brushes past any such criticisms. Yes, World War II was a terrible time, but the ship itself meant something *good* to the Japanese, and it's that inherent goodness that permits a ship that once fought for the Japanese Empire to now fight for every human being in the world.

Narratives matter when it comes to developing rocketry and space programs. Historian Yasushi Sato points out that it "requires attention to the worldview and self-conception of relevant nations" to understand how and why they choose to invest in developing spacefaring vehicles.[3] On one hand, the public rhetoric associated with rocketry and space programs—"the discourse generated by governmental agencies, journalists, historians, and public commentators"—is critical to ensuring continued public support.[4] On the other hand, such narratives are also important within the industry, as motivating visions vital for attracting engineers, administrators, scientists, and funding agencies. And the discursive overhaul of *Yamato* is pertinent because it mirrors the reconfiguration of narratives around rocketry in postwar Japan. Battleships and rocketry, machines of war both, were whitewashed of their military past, and reconfigured as vehicles for a peaceful future. Both retellings enjoyed national and international acceptance; although the original release of *Space Battleship Yamato* had renamed the ship *Argo* in the West, by the time of the 2010 movie release, *Yamato*'s name was restored, even in the very Allied countries responsible for the original ship's destruction.

In Japan, a similar reconfiguration of narratives about rocketry occurred in two phases between 1920 and 1960. As militarism surged in the 1920s and 1930s, depictions of rockets as superweapons dominated, informed by a preoccupation with how far Japanese technology lagged behind the West, as well as anxiety over the country's own position in relation to Asia. In the aftermath of the country's defeat in World War II, however, these justifications became politically toxic. Japanese industrialists and researchers were faced not just with reconfiguring their wartime infrastructure for civilian use, but with the challenge of marketing their new products.[5] And whereas scooters and sewing machines could easily find a home in Japan's demilitarized imagination and its markets, rocketry was far more expensive to produce, far less obviously useful, and (arguably) far more easily used to kill people. Furthermore, the only sense in which the average Japanese person could "own" a rocket or probe was mediated through the idea of being "Japanese" itself—that is, they had to have a sense of ownership that was *not* generated by proximate daily usage. In short, rocketry was in dire need of a good story.

Fortunately, Japan's rocketeers rose to the challenge, initiating a policy of active public outreach by senior members of Japan's space programs that would last to the present day. Leading the way were wartime engineers and specialists, who laid the reputational groundwork for the technology's resurgence in the 1950s and 1960s. Of particular note was the work of Eiichi Iwaya and Hideo Itokawa, which deployed nostalgia, futurism, nationalism, and technological knowledge, while acknowledging the wartime roots of both the technology and their expertise. Iwaya used wartime memoirs to emphasize the hardships and ingenuity of Japan's first rocket designers; Itokawa argued that the military use of rocketry yesterday should not stand in the way of its peaceful use today. Such arguments tapped straight into the notion that, as Walter McDougall puts it, "through our technology, our ability to manipulate Nature, we are subcreators, a demiurge."[6] For writers like Itokawa and Iwaya, working in the postwar period, it helped that this exalted not only the "technology," but the "we" as well.

Rocketry in the Japanese Consciousness to 1945: Militarism and Disillusion

In March 1936, the aviation magazine *Sora* (Sky) began featuring a column called *Sekkeika no yume* ("Designers' Dreams"—later, *Shinsekkeika no yume*, "New Designers' Dreams"). In it, a variety of aeronautical specialists let their imaginations run riot on designs ranging from airborne carriers capable of extending fighters' ranges, to an assortment of flying wings. Many of the designs were vehicles of war, and military use intruded even into nominally civilian designs. In March 1938, for example, one designer presented a rocket plane that was capable of moving 15 people at 500 mph (800 kmph)—guaranteed, he said, to find a market in Japan, the Soviet Union, and the United States. Still, he could not help himself from pointing out that "when adapted for military use, [the plane's] power should be feared [*osoru beki de aru*]."[7]

The conflict between the military and aspirational sides of rocketry dates back to before the technology had even been physically realized. Two generations before *Shinsekkeika no yume,* the Russian theorist Konstantin Tsiolkovsky's work had inspired an "explosive" interest in space exploration. Science fiction writers such as Aleksander Beliaev, the "Soviet Jules Verne," wrote some of the best-selling novels in the USSR in the 1920s and 1930s; Beliaev's 1928 *Bor'ba v efire* (Battle in the Ether) saw rocket weapons deployed in defense of a communist utopia threatened by the drones of an expanding American empire.[8] The capitalists had their own visions, too. German director Fritz Lang was one of the first filmmakers to depict a multistage spacefaring rocket on screen in his 1929 work *Frau im Mond* (Woman in the Moon). In Britain and the United States, writers such as E. E. "Doc" Smith, H. G. Wells, and Olaf Stapledon wrote stories featuring interstellar rockets. In many of these works, spaceships took the spiritual place of seafaring vessels, bearing plucky Western adventurers to barbarous spacelands. In others, though, they were symbols of hope; in the 1936 film *Things to Come* (directed by William Cameron Menzies and

(11) コメツト 78
ロケツト多座機

躍る赤鷲案

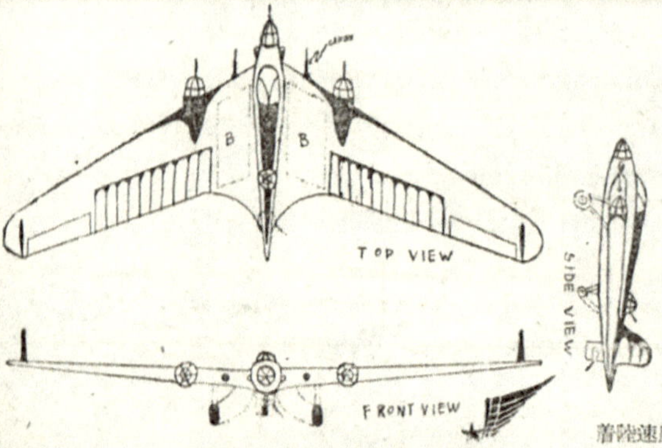

型式	全金屬中翼單葉無尾翼機
用途	戰鬪爆擊機
	完成豫定年(皇紀2650年)
發動機	彗星ロケット 20 氣筒
全長	11 m 20.
全幅	25 m 90.
自重	4000 kg.
全備	7000 kg.
乘員	8 名
搭載量	2000 kg.
武裝	カノン砲 (38mm) 2 門 (固定)
	機關砲 (20mm) 3 門 (可動)
	機關銃 2 基
最高速度	800km/時 (8000 mにて)
	720km/時 (12000 mにて)
	650km/時 (6000 mにて)
	800km/時 (20000 mにて)
巡航速度	500km～320km
最低	1 0km
着陸速度	95km 80km(フラップ使用)
航續距離	10000km

備考
1 翼下面にフラップを付す
2 圖中Bの部分に爆彈を有す
3 攻擊火力甚だ强し
4 無電、着陸燈を有す
5 機關砲座等の視界廣し、下部機關砲を有す
6. 水平速度大なり
7 方向轉換はロケットの出力加減で助ける

設計家の夢 (35) 高橋萬右衛門案

MT.D—4型 旅客輸送機

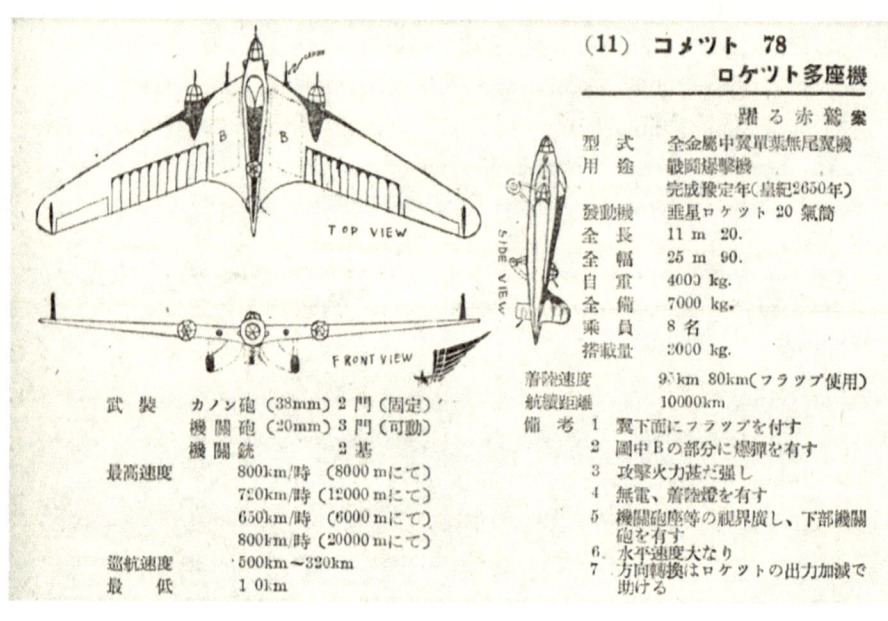

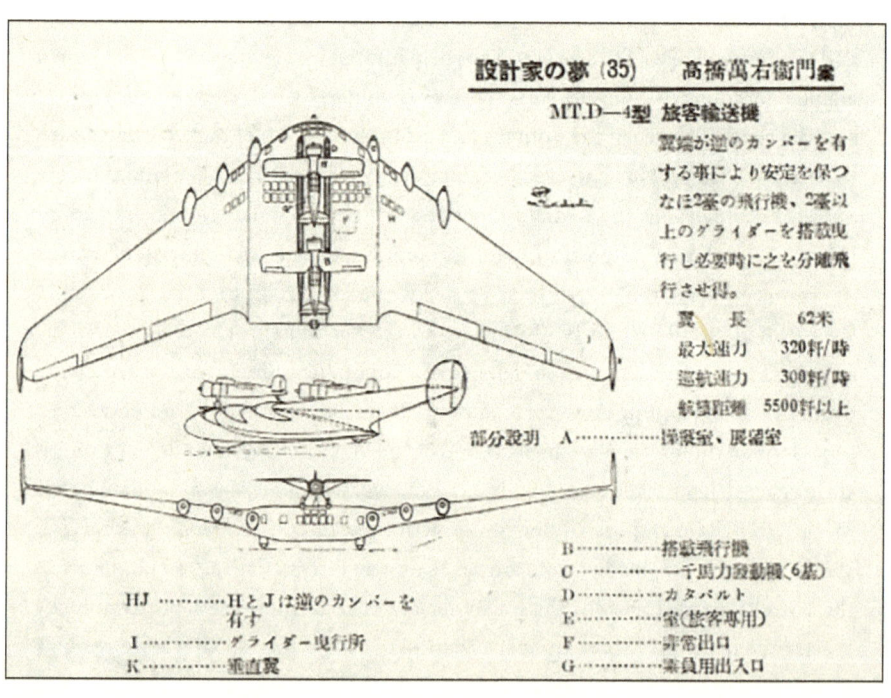

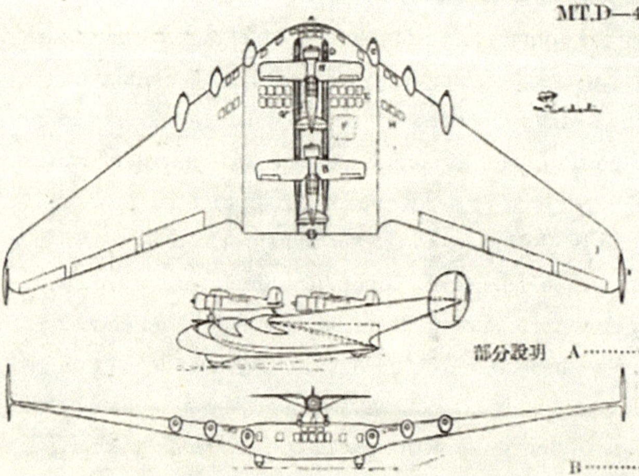

翼端が逆のカンバーを有する事により安定を保つなほ2臺の飛行機、2臺以上のグライダーを搭載曳行し必要時に之を分離飛行させ得。

翼長	62米
最大速力	320粁/時
巡航速力	300粁/時
航續距離	5500粁以上

部分設明 A……………操縦室、展望室

B……………搭載飛行機
C……………一千馬力發動機(6基)
D……………カタパルト
E……………室(旅客専用)
F……………非常出口
G……………乘員用出入口
HJ …………HとJは逆のカンバーを有す
I……………グライダー曳行所
K……………垂直翼

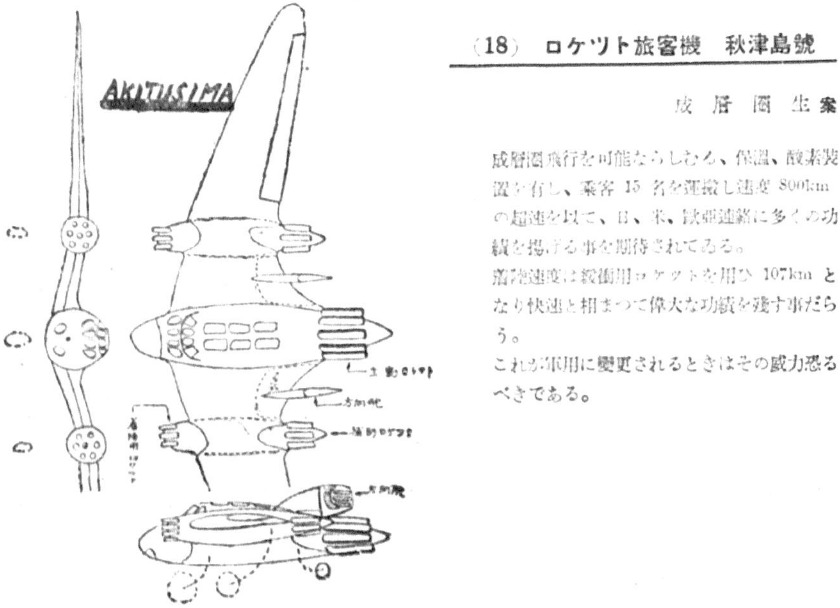

(18) ロケツト旅客機　秋津島號

成層圏生案

成層圏飛行を可能ならしむる、保温、酸素装置を有し、乘客 15 名を運搬し速度 800km の超速を以て、日、米、欧亜連絡に多くの功績を揚げる事を期待されてゐる。

着陸速度は緩衝用ロケットを用ひ 107km となり快速と相まつて偉大な功績を残す事だらう。

これが軍用に變更されるときはその威力恐るべきである。

FIGURES 2.1A–C. Images from *Sekkeika no Yume*.

written by H. G. Wells), for example, a modern-day Adam and Eve escape an Earth ruined by war via a cannon-launched rocket.[9]

The earliest popular discussions of rocketry began to appear in Japan around the same time. Images were often cribbed directly from abroad; Kazuo Harada's 1929 *Dare ni mo wakaru kagaku zenshū dai kyū ken: saikin hatsumei romansu* (Science for Everyone Complete Series No. 9: the Romance of Recent Inventions) included depictions of rockets from a German magazine.[10] Soon afterward, the influential literary magazine *Chūō Kōron* named rocketry among the most important scientific products of the age.[11] In 1934, fuel specialist Shinzaburō Hori thought it important to include an entire section on how the technology worked in his aggressively titled *Sakidzu detarame nise bunka o hōmure* (Bury the Bullshit and False Culture First), a book targeted specifically at the general public. Despite the author's pretensions—at one point, he dubiously claims to the be the only "general scientist" (*sōgō kagakusha*) in the whole of Japan—it's striking that he placed rocketry in his curriculum for essential public knowledge.[12] Equally striking is the fact that military use is only one among many presented in these early discussions of rocketry. In 1930, for example, science writer Keitarō Kaishima commenced his discussion of the subject in *Saikin no hikōki to shōrai* (The Present and Future of Flying Machines) by pointing to the technology's roots in science fiction. Though he does reference military applications, he spends as much time speculating on the technology's efficacy as a means of terrestrial transport—a 1,000 mph (1,600 kmph) rocket, he points out, could "go around the world in 24 hours."[13] His 1933 *Ashita no Hikōki* (Tomorrow's Aircraft) ends on a similarly hopeful note; the final image in the book is that of a rocket heading to "limitless space" (*mugen no tenkū*). In 1931—the eve of Japan's invasion of China—popular science magazine *Hatsumei* (Discovery) speculated that rockets could be used to end droughts by seeding clouds with soot.[14]

Such uses of rocketry intrigued even military officers. In 1936, in an article in *Gunji to Gijutsu* (Military Affairs and Technologies), IJN officer Akio Saito described rockets as the best option to "visit the world of the stars."[15] Meanwhile academic Yūshichi Nishizawa, a prolific author of chemistry and physics textbooks, produced more than one hundred books and articles on topics ranging from fireworks to fragrance between 1910 and 1940.[16] In 1939's *Kagaku sen heiki,* he turned his attention to the various uses of chemicals in war. The book's preoccupation with conflict is apparent from its first image, which features a painting from the *Illustrated London News* of a gas attack during the Battle of Ypres. The second image, however, is one of a rocket plane (*roketto hikōki*) soaring over a peaceful cityscape. The book's chapter

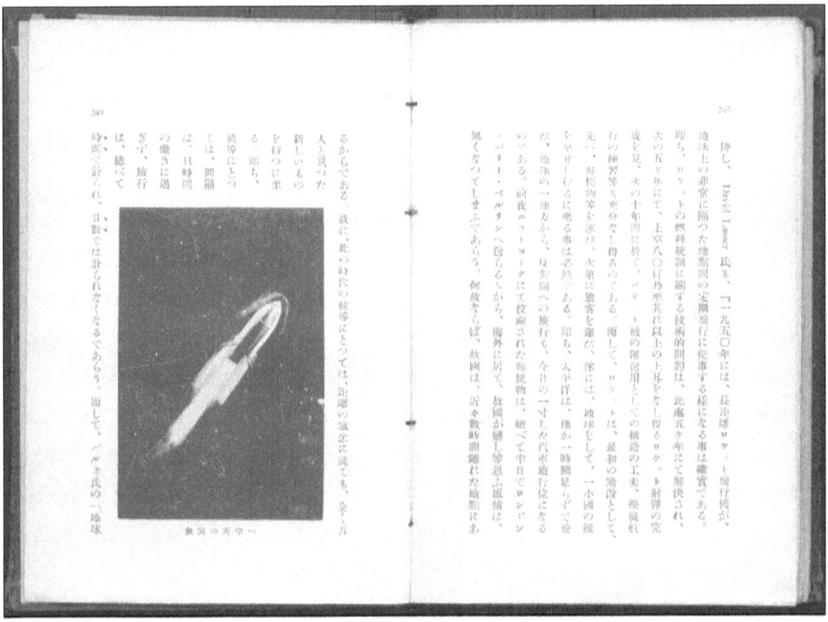

FIGURE 2.2. "To limitless space."

Source: Keitarō Kaishima, *Ashita no Hikōki* (Fukuoka: Kaishima Keitarō, 1933), 132.

breezes past military applications to discussing space exploration. In keeping with the time, scattered throughout are images of rockets lifted from Hermann Oberth's seminal 1929 work *Wege zur Raumschiffahrt* (Ways to Space Travel), a text which itself (as perhaps is evident from the title) was mostly interested in space travel.[17]

Children's texts, such as the 1931 *Sakin kodomo kagaku dokuhon* (A Children's Reader on Recent Science), also provided a crucial venue for the popularization of the idea of rockets as vehicles of space travel.[18] Weapons and military usage often appeared as points of reference for describing basic principles, but not as ultimate usage. A 1936 edition of youth magazine *Shōnen kurabu,* for example, proposed the idea of a "postal rocket," although its description sounded alarmingly like a cannon: "You put mail inside this thing that looks like a torpedo and fire it out of a huge gun!"[19] Such descriptions would have appealed to youths caught up in Japan's prewar mass mobilization training.[20] But the discussion quickly moved on to other, cooler things, like rocket cars. In fact, rocket cars were a particular favorite of science communicators; Yūshichi Nishizawa's above-mentioned *Kagaku sen heiki* ends with a lovingly rendered image of one.[21] German engineer Fritz von Opel's rocket racers take center stage in Mitsuo Harada's 1932 *Kodomo rigaku ebanashi* (Science Picture-Stories for Children). His work begins by elaborating on the first law of thermodynamics by pointing (again) to the recoil of a fired gun, and then describing how the same principle makes rockets work. The next few pages then dive into von Opel's vehicles, depicted in all their multi-nozzled glory.[22]

Such coverage spread beyond the printed page, as Japan's mass media landscape was transformed by the onset of radio in the 1910s and 1920s. Around this time, the national broadcaster, Nihon hōsō kaisha (NHK), launched a program of summertime lecture courses directed at students on break. These featured forty-five-minute broadcasts on various topics related to science; in 1930, they ran from July 27 to September 5. In that year's penultimate lecture, on September 2 and 3, students were treated to a comprehensive update on the design and functionality of rockets from science popularizer and engineer Tarō Senō. Senō covered not only the recent work of Robert Goddard, and a trifecta of European designers—Hermann Oberth, Walter Hohmann, and Max Valier—but also the deep roots of the technology in fireworks and gunpowder. While he did discuss the development of rockets for use in military aircraft, and as bombs, this took up a relatively small part of the lecture; nearly equal time was dedicated to the exciting work of—yet again—Fritz von Opel and his magnificent rocket racers.[23]

FIGURE 2.3. Norakura's space adventures.

Source: Watarō Tsurihashi, *Norakura shōi no roketto tokkantai: Warai no manga bukku* (Tokyo: Taikodō, 1934), frontispiece.

ROCKETRY AS A SYMBOL OF POWER

Despite this early interest in nonmilitary applications, by the 1930s it was rocketry's power as a weapon that became the dominant narrative around the technology. The roots of this shift lay in an increasing sense of anxiety in Japan about what was perceived as encirclement by hostile, and technologically advanced, Western powers. The country's rulers also knew all too well the price technological laggards had to pay on the world stage—they themselves had deployed their more sophisticated technologies to batter and terrorize their neighbors, especially China. Even renowned physicist Nishina Yoshio believed that Japanese technology was crucial to making the Chinese "obedient," and not just by killing them: by building scientific research institutions in China, he suggested, Japan could turn the populace's energies away from resistance, and toward Japan's "great work."[24]

In fact, mastery of technology was seen as one of the key qualities that set the Japanese apart from their Asian neighbors. Mitsuyo Seo's seminal 1943 anime, *Momotarō no Umiwashi* (Momotarō's Sea Eagles), and its sequel, *Momotarō: Umi no Shimpei* (Momotarō: Sacred Sailors), tell stories of a boy-hero fighting the demons of Onigashima (Devil's Island), creatures resembling caricatures of the British and Americans. Momotaro's crew's technology is what sets them apart from both their enemies and the natives they encounter during their campaign. Dazzled by Japanese aircraft, the latter rush forward to touch them as soon as they land. "Something's come," they sing; "it just appeared out of nowhere," containing people who "look like us"—but "just a little."[25] The obverse of this identification with, and faith in, technology was a profound anxiety that Japan's imperial foes were outdoing the nation in R&D. It was an obsession so intense that in this period "hardly a single prewar publication dealing with technology had failed to reflect . . . [the] mind-set [that] Western technology was more highly developed than Japanese technology."[26] The result was that every major Western technological breakthrough or accomplishment was greeted with a mixture of admiration, and anxiety. Hideo Itokawa recalls how, upon hearing of Charles Lindbergh's epochal 1927 transatlantic flight, one of his first reactions was concern that Japan didn't have its own aviators and planes. Four years later, in October 1931, he "received a shock" when Clyde Pangborn and Hugh Hendon Jr. completed the first successful transpacific flight. All this Western success in the skies bred an "irresistible sense of antagonism" in the future Dr. Rocket.[27]

Rocketry found itself in the eye of this storm in the zeitgeist. Authorities noticed the effect on morale that such news was having by the late 1930s, and warned editors and scientists not to show excessive admiration for foreign technology, whether Allied or Axis, in their writing.[28] Despite its roots in Asia, the technology's premier twentieth-century practitioners were European or American.[29] Publications like *Gunji to Gijutsu,* which published several histories of the technology of the 1930s, ignored key developments such as the pioneering metal rockets of Tipu Sultan in the Sultanate of Mysore (now part of India), and focused on the technology's development in Europe.[30] Writing in the magazine in 1934, military officer Akio Saito declared that "there is not a single first-class civilized country that does not have one or two organizations or associations connected to rocketry."[31]

Increasingly, it was this function of rocketry—as a weapon—that came to dominate Japanese discourse on the technology. As mentioned above, it was not a novel narrative; one of the first extensive articles printed on the topic in the country, "Roketto no shōrai" (The Future of Rocketry [1931]), was written by a member of the IJA Technology Department, Kōyaku Matsuda. After summarizing the activities of such luminaries as Robert Goddard, Matsuda argued that "if we are not afraid of the challenges that repose inside rockets, then we will certainly delineate a new element of the conquest of nature."[32] Some of the earliest images of actual rocket launches printed in Japan appeared in the same magazine, nestled between images of heavy artillery and Soviet military parades.[33] As early as April 1933, an edition of *Kagaku Zasshi* (Science Magazine) began by pointing out that "rockets are becoming a fearsome weapon of the future."[34] Yet hitherto it had not been a dominant one.

This changed rapidly in the 1930s. One notable genre that reflected this development was children's literature. By the 1930s, depictions of rockets as weapons for heroes had become the norm. The 1934 manga *Norakura shōi roketto tokkan-tai* (Second Lieutenant Norakura's Rocket Assault Team), for example, commenced with an image of Lieutenant Norakura straddling a rocket, saber in hand, *Dr. Strangelove*–style. Within the text, however, the technology is largely played for laughs, and the emphasis is on the capers of Norakura and his canine followers. Interestingly enough, the rocket they do end up finishing features not a nozzle at its tail, but a propeller, a design commonly depicted in the contemporary popular depictions of rockets globally.[35] By the end of the war, the technology had been shorn of comedy. In 1945 *Shūkan shōkokumin* (The Weekly Young Citizen) ran a short piece on the latest accomplishments of Japan's Axis ally, Germany, in the field of military rocketry.[36] Yet children, perhaps more than adults, still had room to dream; in June that year, articles appeared in *Hatsumei*—which by that time was reduced to printing merely 14 fourteen pages per issue owing to shortages—on the latest developments in rocket turbines.[37] By this point, however, the adults—even nonmilitary folk—were most keen to discuss how the technology could be used to defend Japan. In 1936, a round-table meeting on aviation organized by the venerable Japan Transport Association featured several exchanges on the military use of rocketry, including a discussion of human-guided missiles, a full eight years before Ōka was designed. Interestingly enough, the idea was dismissed out of hand by fuel specialist Tsuneto Ohara, who observed that "we don't need people on a missile [*kūrai*]; one launched from a plane can be piloted wirelessly."[38] At the same time, the country's deep interest in foreign innovations persisted, and even Japan's enemies retained their role as a crucial source of rocket information for longer than one might expect. Ten days after the Japanese attack on Pearl Harbor, American articles on the potential applications of rockets in aircraft appeared in translation in the Imperial Navy Air Service's publications.[39] Hideo Itokawa, in a September 1943 piece in *Asahi Shimbun*, warned that a "fundamental revolution" in turbine technology was underway on US aircraft, initiating the *rokettokei ka* ("rocketification") of conflict.[40] In 1944, he reiterated his warning in the same newspaper.[41] On this matter, clearly, the future Dr. Rocket believed all the attention on rocket weapons was still not quite enough.

DISILLUSION IN POSTWAR PERCEPTIONS OF AVIATION

One of the underlying beliefs in technonationalist thought is that technology represents political power. With so many policymakers, engineers, and researchers invested in the idea, it comes as little surprise that many in Japan in 1945 blamed their defeat on a failure to innovate.[42] Referring to the atom bomb, *Asahi Shimbun* suggested on August 20, 1945, that Japan had "lost to the enemy's science."[43] Nishina Yoshio was appalled not only by the human suffering in Hiroshima, but also by the idea that Japan's scientific establishment had simply been outdone by the Americans. He later suggested to his younger colleague, Tamaki Hidehiko, that "the time has come for us to commit *hara-kiri* . . . [because] Anglo-American researchers won against Japanese researchers."[44] In the midst of this crisis of confidence, their new American overlords were determined to transform the country, for in their eyes there was "no viable past to which Japan might return."[45] Despite the efforts of figures such as Prime Minister Yoshida Shigeru to ensure that the new Japan aligned as closely as possible with the old Japan, pacifism became one of the cornerstones of this new political culture. As a result, as Andrew Robertson has described, technocracy and science had to be shorn of their wartime associations if they were to ever regain an iota of public support.[46]

This was a particular challenge for aviators.[47] The Allies had relied heavily on aircraft in wartime to both destroy and demoralize. Worse, they had taken to occasionally reminding the vanquished nation of this after their surrender. Planes including F4U and F6F fighters, and 462 B-29 bombers, "swooped over Tokyo Bay" during the surrender ceremony. During the occupation, "low-level flights were maintained to intimidate the population"; a steady stream of B-29s "filled the skies" and frightened Empress Kōjun. Rocketry meanwhile suffered from a reputation of its own making. Ōka, the country's highest profile World War II rocket weapon, exemplified the utter commitment wartime Japan had demanded of its people; now, dismissed

FIGURE 2.4. F4Us and F6Fs fly in formation during surrender ceremonies in Tokyo, with USS *Missouri* visible in left foreground.

Source: September 2, 1945, Record Group 80, General Records of the Department of the Navy.

as the "idiot bomb" by the Americans, it became emblematic of the pointlessness of all that sacrifice. Worse, the world leader in rocketry at the time, the Germans, had not only been defeated, but revealed as architects of an industrialized genocide.[48] Things only got worse from there; as the contours of the Cold War became increasingly clear in the late 1940s, the skies of the mid-twentieth century seemed even more dangerous than they had during the preceding decade, filled as they now were with not just bombers and fighters, but inter-continental ballistic missiles (ICBMs), too.[49] The very idea of anyone building rockets worried academics in postwar Japan because there could be no "firm guarantee that [rockets] are not going to be used for the purpose of warfare."[50] They must have realized soon enough that such guarantees would never be forthcoming.

Such anxieties meant that rocketry in postwar Japan was associated with other technologies of war that had dominated a period deemed to be Japan's "dark valley." Central to this notion—now disputed by a variety of scholars—was the idea that this time had somehow been an aberration in Japan's journey towards democracy and prosperity.[51] Yet it was influential at the time, and many concluded Japan's redemption would have to lie in jettisoning the technocratic and authoritarian thinking of the 1920s and 1930s, beginning with excluding from public life those who had perpetuated such thinking. This included engineers and managers, many of whom had witnessed their wartime efforts quite literally go up in flames, before suffering unemployment, loss of status, and concomitant penury—Hidemasa Kimura, who had worked on Ōka during the war, was reduced to selling personal effects like a camera, furniture, and golf clubs just to get by.[52] Worse, technical officers were often blamed by both occupier and civilian alike for the horrors of the preceding seven years. It was a judgment that some, certainly, took to heart: Tadanao Miki, one of the key designers of suicide aircraft during World War II, later stated that he had joined Japan National Railways after the conflict partly because, while the automobile industry could produce tanks, and the shipyards could produce warships, trains could only be used for nonlethal purposes.[53]

Despite this, however, as shown by the work of historians such as Hiromi Mizuno, Aaron Moore, and others, technonationalism persisted in postwar Japan, in that the notion that Japan's technological sophistication was a source and reflection of national dynamism remained—perhaps most famously attested to by Japanese electronics, and large infrastructural projects such as Shinkansen. Moore described this postwar reconfiguration thus:

> As Tessa Morris-Suzuki argues, "the vision of technology as the basis of the Greater East Asia Co-Prosperity Sphere was transformed into a vision of technology as the basis of the new Japan." Janis Mimura notes how "wartime techno-fascism" quickly transformed into "postwar managerialism" as many of the same figures . . . adapted their technocratic "managed economy" approach to Japan's postwar goals of building a middle-class consumer society. This "techno-fascist" system was closely connected to "techno-imperialism," specifically in the form of comprehensive technical projects designed to integrate the empire into Japan's wartime economic system and mobilize colonial peoples. These two components of the technological imaginary—techno-fascism and technoimperialism—adapted to Japan's postwar context of building a prosperous nation at home and exercising soft power abroad through overseas development assistance. Technology continued to serve as a system of power and mobilization as Japan shifted to a peacetime economy and a nonaggressive foreign policy after 1945.[54]

It was on the basis of this narrative change of course that rocketry began its *Space Battleship Yamato*–style pivot away from war. A good place to start was the fact that rockets had never lost their appeal for children; by the 1950s and 1960s, children's magazines such *Ichinen no Gakushū* regularly featured stories such as "Sutekina roketto" (Splendid Rockets) and "Tsuki e iku roketto basu" (A Rocket Bus to the Moon). These not only showed heroes flying aboard rockets, but also took the time to explore and describe the beauty of the vehicles.[55] Similarly, adults banded together to resurrect or preserve traditional rocket festivals. Though their roots probably lay in the usage of *noroshi* rocket smoke signals by warriors during the violent fifteenth and sixteenth centuries, by the Meiji period (1867–1912) temple harvest festivals incorporated handmade bamboo and wood rockets called *ryūsei* (translatable as "dragon's power" or "shooting star").[56] The most famous, the Chichibu Yoshida event in Saitama, had been suspended due to the exigencies of war in 1936, but returned in 1945. There were eventually three such festivals in Japan, each of which circumvented Japan's recent history of war by associating rocketry with a timeless rural arcadia. The Yoshida iteration eventually developed into a major tourist event, big enough that in 1992, organizers were able to invest in their own museum, the "Ryūsei Hall."[57] The event was subsequently deemed an "intangible folk cultural property" by the Japanese Ministry of Education, Culture, Sports, Science and Technology.[58] It continues to flourish: twenty-seven teams participated in the 2023 event, following a brief hiatus due to the COVID-19 pandemic.[59]

Eiichi Iwaya: Humanizing Wartime Rocketry

A key part of the public redemption of rocketry were the writings of two specialists in the technology. The first of these was Eiichi Iwaya. After his adventures during World War II, Iwaya spent half a decade or so out of the public eye; as a former officer in the IJN, his status among both the occupiers and embittered compatriots would have hovered somewhere between bad and terrible. Nevertheless, beginning in 1951, he would become one of Japan's most notable public witnesses to its wartime rocketry program. Iwaya even collaborated with famed aircraft designer Jirō Horikoshi to co-write pieces for the April 1959 special edition of *Kōkū jōhō* (Air News) on the Type 96 fighter—better known as the Mitsubishi A5M, predecessor of the infamous A6M "Zero."[60] His work was republished as recently as 2014.[61] Fundamentally, his work sought to normalize Japanese aviation by placing it in a global context, arguing that aviation in Japan was not only something to be proud of, but also perfectly in keeping with trends around the world. In an October 1958 piece in the aviation magazine *Kōkū jōhō,* for example, he presented a comprehensive overview of nearly all major Japanese military aircraft under encyclopedic headings to celebrate fifty years of aviation in the country.[62] In a 1958 piece on wartime aviation, Iwaya charted a seamless course through Japan's own jet and rocket craft of World War II to the (then) cutting-edge American F11F Super Tiger. Although he made no technological links between the two, the argument is clear: Japan's weapons projects in no way diverged from the global norm. Such arguments rang particularly true from someone with as much knowledge about overseas developments as Iwaya. In a piece from 1952 about German wartime experimental craft, he describes how "German friends in Bonn" had recently dispatched magazines such as *Der Flier* and *Die Flugwelt* to help with his "reminiscence" of his experiences in Germany.[63] Perhaps more notably than having German friends, Iwaya was also the first person to translate jet pioneer Frank Whittle's book, *Jet: The Story of a Pioneer,* into Japanese.[64]

Historian Kenji Ito has observed that postwar accounts written by wartime engineers tend to contain apologia and post hoc justifications of their wartime work.[65] This tendency is also evident in Iwaya's writing. Drawing on his firsthand experience, Iwaya's work sought to reengage the public with prewar dreams of postal rockets and space exploration. One way to do so was to make the point that even though they'd failed, Japan's wartime rocket weapons had been marvels of engineering—an argument he made in a December 1951 piece in *Sekai no kūkō* (World Aviation) about Japan's pulse jet project.[66] The piece marked the beginning of a steady stream of work that, over the next eight years, spread word of Japan's wartime rocketry projects to a public hitherto largely kept in the dark about it. In all of them, Iwaya's writing was clear, accessible, and bereft of jargon; for less technically inclined readers, he included extensive and well-labeled schematics. Throughout, he weaves in anecdotes to illuminate the practicalities of such R&D. Here, for example, is his description of work on advanced fuels:

> [we used] an 80% concentration of hydrogen peroxide (H_2O_2). When this was combined with ammonium sulfate $(NH_4)_2SO_4$ through electrolysis, it distilled into $(NH_4)_2SO_5$. One quality of this fuel was that it became extremely unsafe at high pressures, immediately bursting into flame when touched by organic matter, penetrating skin and causing injuries. Furthermore, because it reacted with copper, lead, zinc, iron, etc., it was extremely important that we were meticulous when storing and transporting it. When handling it, people had to wear protective clothes and gloves. It was the sort of dangerous material that, if it leaked when boarding or flying the plane, we had to immediately dilute it from a hose to below 40% [to prevent fires].[67]

His most prominent action during the war—his journey back from Germany aboard the I-29—was inevitably the focus of much of his writing. His first memoir of the experience was published under the swashbuckling title "Taiheiyō sensō mitsuwa kaitei ichi-man go-senri no misshi" (A Secret Tale of the Pacific War: The 15,000 km Underwater Secret Messenger) in *Gekkan Yomiuri* in January 1951. His full wartime diary followed in a 1952 collection, *Kimitsu heiki no zenbō* (The Full Story of Secret Weapons); a 1976 reprint included a digest of the history of rocketry in postwar Japan. The war itself is a slippery presence in his work, omnipresent but rarely in focus. Instead, the focus is on Iwaya's daily experiences, including his relationship with fellow naval officer Hisao Furukawa, also on assignment in Germany, and Furukawa's impulse to return to Japan to take a more "direct" role in the war. He details the weather (Berlin, Friday, March 12, 1944: cloudy) and life in the Third Reich,[68] including a description of British bombing.[69] His journey home aboard *Matsu* is redolent of a thriller—the cramped quarters, the shortages of fresh water, the Allies sniffing alarmingly close off northern Spain and near the Azores.[70] When discussion of conflict is inevitable, the former officer is careful to emphasize that for Japanese designers, the motivation for working on weapons like Shūsui was to *defend* Japan against attack. In "Jettoki Shūsui ni tsuite" (1952), for example, he emphasizes that the plane was specifically designed to counter American B-29s, which had reduced much of urban Japan to rubble; to clear up any potential doubt on the matter, he even included one of his own doodles, showing the rather squat planes soaring heroically to intercept.[71]

It would be easy to dismiss both doodling and writing as a play for postwar redemption by Iwaya, were it not for the fact that he was one the first writers to seriously suggest Japan might, itself, one day send people into space.[72] Despite his wartime experiences, Iwaya was one of the great optimists of rocket technology in Japan at this crucial time. The motivation for this might be found in the economic miracle Iwaya lived to see the early stages of, or the fact that Japan established its own rocketry program in 1954. Or the answer might be found in the fact that Iwaya himself enjoyed not just rehabilitation, but success, after the war. In the 1950s he served as a highly regarded aviation advisor, eventually acquiring a position as a technical advisor to the Kōkū kōgyō-kai (Aviation Industry Association). He died in 1959, at the relatively young age of fifty-six, with Japan at the threshold of space.

Hideo Itokawa: Popularizing Postwar Rocketry

The rocketry project begun in 1954 was begun by Hideo Itokawa, Japan's rocketeer par excellence. Itokawa was well aware that the work of people like Iwaya was critically important for his own. His AVSA group was, from the very beginning, susceptible to budget cuts; two years into the project he found himself with reduced funding, and no long-term plans for income. Maintaining a high public profile was required to provide something of a buffer from such financial headwinds—public buy-in was required for public funds.[73] Also, like Iwaya, Itokawa would have had personal reasons for wishing to redeem the technology, not least that he was as reputationally soaked in militarism as missiles. Itokawa's training and experience had been almost entirely in designing military aircraft at Nakajima, and it was there that he produced his signature wartime aircraft, a fighter. In fact Itokawa first achieved national fame as an inaugural winner of the IJA's "Invincible Machines" merit award in 1941.[74] His earliest appearances in the press were also intimately tied up with the war. In January 1941, for example, he gave an interview to four middle school students for a children's magazine, *Shūkan shokokumin*. In it, he discussed how to distinguish between fighters and bombers by sight and the sound of their engines; a year later, he returned to the same magazine to advise on the same topic.[75] In 1943, he wrote a series of articles in *Asahi Shimbun* analyzing Allied bombers and published a book on military aviation. He published another the next year, and returned to *Asahi Shimbun* to break down Germany's deployment of the pioneering V-2 rocket.[76] Thus, by the end of the war, Itokawa had established something of a reputation as a specialist on military aviation.

After the quiescence of the early 1950s, Itokawa returned to rocketry in 1954 at a time when things were looking up, reputationally, for the technology. Then, three years into the AVSA's work, came Sputnik. Apart from touching off a panic in the United States, the Soviet Union's pioneering launch also led to concerns in Japan about the People's Republic of China (PRC), which by 1958 had secured Soviet help with rocketry research. This was seen by Tokyo as nothing less than another step along the road to the "mastering and production of armed rockets."[77] Newspapers such as *Mainichi Shimbun* drew attention to the use of Eastern Bloc rocket technologies in Vietnam and on a PRC attack on Kinmen Island in 1958.[78] Politicians, particularly on the right, saw the development of Chinese rocket technology as a threat to Japanese security.[79] Itokawa personally, and the technology he championed, came in for intense criticism.[80] One observer even denounced him for turning from "medicine, which saves people, to rocketry, which kills them" (*hito wo ikasu igaku o yamete hitokoroshi no roketto o yaru*).[81] Still, Itokawa would have had reason for hope. Though rocketry was a source of much anxiety, Japan's innovators and policymakers now saw consumption and infrastructure-building—and the R&D that enabled improvements in these—as essential elements of a successful state.[82] For the bulk of Japan's population, memory of wartime austerity was swept away by consumerism; once forced to ration essentials like rice, families now clamored to purchase TVs, refrigerators, motorbikes, electric fans, and washing machines. The smoking hollows of war-bombed cities were replaced with power stations, ports, airfields, and high-speed rail networks. In the cities this meant jobs; in the countryside, it meant infrastructure and jobs. Certainly, there were downsides; certain kinds of infrastructure, such as nuclear power stations, were resisted as "public bads," which threatened local health and quality of life.[83] Yet for many, technology and industry were understood as conduits that diverted resources from urban centers to localities parched for investment.[84]

FIGURE 2.5. One of Itokawa's early articles covering American bombers: "Sen koto hikkei/bōgyoryoku ni nayami saikin no bakugeki-ki kaisetsu (1)," *Asahi Shimbun*, June 21, 1943.

FIGURE 2.6. Itokawa discusses how to identify bombers with schoolchildren.

Source: "Shin eiki no hanashi—Itokawa sensei kakonde," *Shūkan shokokumin* 3, no. 1 (January 1944): 24.

FIGURE 2.7. Statue of Hideo Itokawa at Uchinoura Launch Center.
Courtesy of JAXA.

Itokawa's personality was such that he almost never saw inaction as a viable option. Thus, beginning in 1957, he picked up where he'd left off in 1945—except that, much like rocketry itself, he too had reconfigured his raison d'être for being a public communicator. Henceforth, Itokawa would be *the* public authority on *peaceful* rocketry. One of his first works was a collection of reflections on foreign research called *Uchū o sampo suru: roketto no zuihitsu*. Though the title is translated into English as *Esseys* [sic] *on Rockets*, a more accurate rendering of the Japanese would be *Strolling in Space: Essays on Rockets*. Over the next five decades he produced articles, books, TV programs, interviews, documentaries, official reports, and academic pieces, all in the service of promoting space exploration and rocketry in Japan. It was at this point in his career that he became Japan's "Dr. Rocket." In the Japanese spacefaring imagination, he thus fulfilled the same role as Konstantin Tsiolkovsky, Robert Goddard, Hermann Oberth, Sergei Korolev, Vikram Sarabhai, Wernher von Braun, and Qian Xuesen did in their respective nations—"founding fathers," whose careers and writings became an essential part of the narratives associated with space technology.[85] And indeed, in Japan, there were few, if any, who could rival Itokawa's status in the space program, and his reach as rocketry's public face. His position as the head of AVSA enabled him to effectively act as its press officer; when the public heard about what was going on at Tokyo University, it was usually in Itokawa's words. Beyond that he had access to data that other writers simply didn't. His 1965 *Roketto* (published by none other than NHK), for example, contained information on the future Mu series of rockets, which at the time of publication had barely finished development.[86] Keeping on his good side was worthwhile for publications keen to have access to the sorts of images Itokawa could provide.[87] Even after he left the space program in 1967, Itokawa clearly had enough inside knowledge to enable him to keep writing about Japan's space projects as if he were involved in them personally.[88]

Itokawa's popularity as a communicator came partly through his ability to present technical information with clarity and humor. His 1941 encounter with the middle schoolers in *Shūkan shokokumin* is a good example of his adaptability: he neither talks down to the students, nor pitches his explanations over their heads. The students who visited him were familiar enough with American raids to recognize that "bombers have a lot of bombs loaded, so their fuselages are fat," whereas fighters were "slim" (*hosoi*). Itokawa then helps the students identify particular models of plane from their silhouette, while also engaging with the children's enthusiasm for aviation generally.[89] Fifteen years later, in *Seisan kenkyū* (1960), he speaks to Japan's legion of *sararīman,* salaried workers in nine-to-five jobs who were the backbone of the country's postwar resurgence. On this occasion he used parallels between the organization of the space project and corporate organization, to elaborate on how things are getting done.[90] His writing is sometimes wry, sometimes ironic, and sometimes jokingly conspiratorial; in *Nihon sōseiron* (1990), he tells the reader that he is about to tell them a "secret" that "even now only a [single] cosmologist in England knows"; in the end, it turns out to be information on the performance of Japan's first satellite in space, Ohsumi.[91] For rocket enthusiasts, there was simply no better source on the technology they loved. For others, there were few writers on the topic who were more accessible.

Itokawa also established himself as a key mediator of overseas technologies to Japanese audiences; one of his earliest published pieces, from 1938, was a co-translated volume on aerodynamics.[92] He resumed this line of work in the 1950s and 1960s, but not just for fellow specialists, such as when he was invited to the National Diet House of Representatives (NDHR) standing committees, in 1957, to explain precisely what the Sputnik launch meant. Dr. Rocket, in his element, happily obliged them.[93] Keeping up with the international Joneses was a running theme in his work.[94] For the public, Itokawa was particularly good at drawing attention to any part of Japan's space R&D that marked the nation out as distinctive. In 1965's *Roketto,* he compared Japan's Kappa rockets not to the giants fielded by the United States and USSR, but to vehicles such as France's Diamant: both were two-stage vehicles using liquid fuel; both were deployed for similar research purposes such as atmospheric observation; and both were launched within three years of each other (Kappa in 1962, Diamant in 1965). Thus, though the French vehicle was twice the size of a Kappa, it was overall a more sensible technical comparison, with the added bonus that the subsequent Mu rocket actually came out a little ahead by comparison.[95] In *Nihon sōseiron* (1990) he emphasizes Ohsumi's uniqueness by pointing out that previous launches (in the Soviet Union, the United States, and France) had utilized modified ICBMs, whereas the Kappa-series solid-fuel rockets were designed specifically for space exploration. This accomplishment was, he proudly declared, "a total surprise to specialists all over the world."[96] He also often emphasized that he had a direct line to these specialists, so he would know what their reactions were.[97] American colleagues, he claimed, generally "use the set phrase 'How are you' when they answer the phone," but his American friends greeted him instead with "How is your Pencil rocket?"[98]

The emphasis Itokawa placed on the nonmilitary roots of the Kappa speaks to another key element of the AVSA chief's outreach—emphasizing the entirely civilian nature of Japan's space program. A great deal of professional calculation was involved in this stance; as a 1962 piece in the *Wall Street Journal* observed, his team "scrupulously decline even discussing military applications for fear of riling Japanese leftists and pacifist groups."[99] Yet many of its most committed adherents genuinely supported the position in good faith. Engineers and specialists were as traumatized as anyone else by their experiences during World War II; as late as 1962, one engineer involved in rocket development described how the noise and fire reminded him of wartime bombing raids.[100] Many had concluded that military technology had failed to secure for Japan the power and prosperity it needed, and that the way forward lay in the embracing of strictly peaceful innovation.[101] Itokawa and his successors were committed enough to encourage the inclusion of a clause limiting space technology to "peaceful purposes" only in the 1969 Basic Space Law.[102] The extent to which Itokawa himself became a genuine pacifist is uncertain (see below), but it's no coincidence that one of his earliest postwar pieces about rocketry, in 1955, discussed not rocket bombs, but rocket buses.[103] When *Asahi Shimbun*'s earliest mentions of AVSA's activities pointed to its German precursors in the V-2 project, he was sure to mention in an interview for the same article that Japan had no such military objectives.[104] In public, Itokawa consistently emphasized the distinction between the Japanese rocketry program and the prominent role of the military in the superpowers' programs.[105]

Despite all his influence, Hideo Itokawa's legacy remains inflected by his foibles and contradictions. Despite his public championing of the concept of "peaceful purposes," Itokawa retained something of his prewar worldview of a Japan beset by enemies; he opined in the 1970s, for example, that much like Israel, Japan "has no allies nearby," and hence technical cooperation between the two would be essential for their futures.[106] Nor did Itokawa object in the 1960s, when rocket component suppliers, such as MHI and Nissan, started producing missiles for the Japan Self Defense Forces.[107] Beyond this, his brusqueness and changeability made enemies, and even his disciples would recall him as a capricious, frenetic, and authoritarian figure.[108] Perhaps the most consequential souring of relations was with *Asahi Shimbun*, which launched an investigation into financial irregularities in the mid-1960s that eventually forced his departure from AVSA's successor, the Institute of Space and Aeronautical Science (ISAS), in 1967.[109] By this point, however, his team was a scant three years away from launching Ohsumi. If Itokawa's objective had been to put Japan's space program on a stable footing, he had succeeded splendidly.

Rocketry in the public consciousness in early twentieth-century Japan was laden with militaristic connotations. Although aspirational ideas about space travel and exploration were popular, narratives emphasizing military rocketry gradually came to the fore in the two decades preceding World War II. Then, in the postwar period, these justifications became anathema, and the technologies associated with them tainted by their association with war. The work of figures such as Eiichi Iwaya served to keep rocketry in the public eye, by emphasizing the human story behind the technical impressiveness of the enterprise. Hideo Itokawa followed suit, becoming one of the prime sources of information about postwar rocketry in Japan. Working tirelessly to promote the technology, "Dr. Rocket" eventually morphed into the iconic figure of twentieth-century Japanese space exploration. Under his aegis, Japan took its first steps toward space; by the time he left the civilian space programs in 1967, the country was almost there.

The writers described in this chapter were successful in that they were able to adapt the stories they told to changing times. They were also the sources of this change themselves, not least in the institutionalization of the Japanese space program in the 1960s and 1970s. Following the AVSA's proof of concept for rocketry in the 1950s, a multitude of ministries and departments each set forth to create their own specialized space programs. The subsequent proliferation of laboratories, space agencies, research hubs, and bureaucratic entities provided room for multiple stakeholding groups to voice their concerns and influence policy. These included aeronautical engineers who set research agendas, created their own work cultures, and generated mythologies justifying and exalting their work; private corporations that pushed hard for the establishment of profitable space enterprises; politicians and bureaucrats who tried (and failed) to centralize the civilian space programs; and everyday people in localities that hosted space infrastructure who wrung as much benefit from their neighborhood facilities as possible. Contestation and institutionalization proceeded hand in hand in Japan's civilian space programs in the 1960s and 1970s, and it is this second phase of development that the next three chapters will focus on.

PART II

The Institutionalization of Japanese Space Research, 1960–1980

THREE

Manager-Specialists in Japan's Space Program

AKIRA MEGURO IS HAPPY TO ADMIT that maybe, yes, he is a little superstitious. In a piece for trade publication *Space Japan Review* in autumn 2005, the veteran satellite specialist was upfront about the psychological strain of running space technological projects. After all, once launched, rockets and probes are impossible to fix; mishaps can render decades of work and giant budgets useless in an instant. "Is it really so mysterious," he asks, "that people at the forefront of research and technology also go to shrines before a launch and hold the amulets tight [during it]?"[1] Certainly, his colleagues around the world would agree, if the cheering and jubilation in any control room after a successful launch is anything to go by.

This sort of intense identification with the technologies they produced led Japan's earliest rocketeers to secure not only an institutional future for the technology, but also positions of power for themselves within it. The work of AVSA and its successor ISAS provided many engineers, and their young protégés, with opportunities for work and camaraderie in the 1950s. Under Itokawa's energetic leadership, AVSA's work had proceeded at a blistering pace at Tokyo University's laboratories in Kokubunji.[2] By April 1954 the Pencil, with a weight of 7 oz (200 g) and a length of 6 in (15 cm) had completed several successful test launches. By August, the team had established a dedicated launch facility where, in 1955, a Pencil rocket reached 600 m in altitude.[3] Experiments then moved to a more spacious beach near Michikawa, Akita Prefecture, where two years later they launched Kappa-8, a 3,000 lb (1,500 kg) device capable of reaching space.[4] By 1966, now reconstituted as ISAS, the group produced the three-stage Lambda 3-H SFR, which was capable of traveling 1,865 mi (3,000 km)—meaning it could put something into orbit.[5] This the Japanese duly did, on February 11, 1970, when a Lambda 3 launched the 53 lb (24 kg) Ohsumi orbiter into space. Japan had become only the fourth country, after the United States, USSR, and France, to put a domestic orbiter in space atop a domestic vehicle.[6] All this was accomplished while AVSA's leaders worked to insulate the group from outside pressures such as budgets, institutionalization, and political overview, a policy that continued with a new generation of manager-specialists such as Akira Meguro and Yasunori Matogawa. This new generation was also particularly mobile, moving between state and private settings, resulting in the creation of a common transinstitutional culture. Engineers, specialists, and managers—who were, very often, the same people—were thus not the "docile, voiceless, and faceless agents of technological transformation" they were sometimes characterized as, but significant players in the formation of Japan's space and rocketry programs in their own right.[7]

Japanese Aeronautical Specialists in the Transwar Period

The pioneers of rocketry in Japan had all had strikingly similar experiences before the 1960s. Many had entered their professions as part of an influx of new blood into Japanese academia, encouraged by reforms after World War I.[8] The new institutions where they had studied and worked, such as RIKEN and military research bureaus, lacked the hierarchical systems of older institutions, and in due course such divisions began to disappear even in conservative institutions such as Tokyo University. Many were influenced by technonationalism and utopianism, and believed that they could create rationalized societies free from the corrosive influence of "old-fashioned" politicians.[9] Technology was key to this vision of social renovation, particularly aeronautics, electronics, railways, and shipbuilding.[10] During World War II, they also found themselves in positions of power far earlier than they could have expected, due to personnel shortages. Until at least 1940, for example, the design of air bases was largely controlled by specialists who had qualified through special examination; by 1942, fresh university graduates were being tasked to help them, and many were commissioned as technical officers immediately upon graduation.[11] Wartime exigencies also required the movement of researchers across state and private institutions—fuel specialist Nobuhei Nakahara, for example, worked on state-sponsored aeronautical fuel research at the navy's Tokuyama and Yokkaichi facilities while he was as general manager of oil refining at Kokura Oil.[12] Hence by the end of World War II, Japan boasted an experienced body of aeronautical specialists who had both specialized technical education and managerial experience, forming the roots of a crucial class of stakeholders in rocketry: the manager-specialists.

Then came surrender. As historian Andrew Robertson has put it, "the Emperor's admission of defeat in a flash reduced years of technical work, invention, and application to insignificance."[13] Worse, the occupation's ban on aeronautics immediately rendered the skills of aviation specialists useless.[14] Some even faced the prospect of being individually sanctioned for their wartime activities; in 1948, the occupation authorities banned 167,035 ex-military personnel, including technical officers, from holding public office.[15] Despite these shocks, however, many specialist-managers came to understand that the American project of liberalizing Japan's body politic actually presented them with new opportunities. For starters, universities were reformed along American lines; the introduction of notions like unions, self-governing academic departments, and independence from military research gradually replaced the "sullen resentment" of wartime with optimism.[16] Many academics now found that, provided they could find funding, they were freer to pursue personal interests.[17] Concurrently, publications such as *Kagaku no Shakaishi* (A Social History of Science), by historian of science Tetu Hirosige,[18] promoted a vision of a "renewed" Japan free from militarism.[19] Hence, as Takashi Nishiyama puts it, "the postwar era provides an illustrative case of ordinary engineers at a grass-roots level actively molding values for modern technological transformations in their society."[20]

Nishiyama also points out that a "radical change in the socio-technical landscape" exerted a "centrifugal force on the aeronautical engineers, releasing them into a volatile employment market."[21] In short, with the banning of aeronautical research by the Americans, many technical specialists who worked on aviation had to find opportunities elsewhere. Some wartime engineers took the opportunity to go work at universities, where they could keep one foot in the aeronautical door by doing theoretical work.[22] Figures such as Shōichi Takayama (who had worked on the Mitsubishi A6M "Zero" fighter) did so as educators; others seized the opportunity to continue or initiate research projects.[23] The Japanese state also helped, for example, through the state-owned Japan National Railways dedicated program of hiring aeronautical engineers postwar.[24] Miki Tadanao, one of the designers of Ōka, joined JNR in December 1945, just a few months after the end of the war; a materials specialist like Naruo Yamana was simply able to continue his work on decay-resistant wood, albeit now in aid of trains, not battleships.[25] Others moved into private enterprise, or set out to create their own businesses, and found success amid Japan's postwar economic boom.[26] Indeed, historian Kenkichirō Koizumi has pointed out that the period after occupation was a creative one in terms of innovating for the market in Japan, as producers concluded that "Japan could turn itself into a manufacturer of highly desirable material things," rather than simply adapting Western products.[27] Nobuhei Nakahara would certainly have agreed; after World War II, he took his wartime experience and established the Tōa Fuel Corporation, which then expanded by borrowing American capital. It survives into the twenty-first century as part of Tōnen General Seikyū, one of Japan's largest fuel suppliers.[28]

Recapturing Status and Entrenching Influence in Japan's Early Rocketry Program

There was, of course, eventually another option for these wartime aeronautical engineers—Japan's space program. Just as for Akira Meguro, for Itokawa and his team the rocketry project was more than just a job. Their work was also an opportunity to regain some of the status and esprit de corps they had enjoyed during World War II. In the early 1940s, for example, Itokawa had found himself living in a shabby boardinghouse with other young engineers. He recalls it as a bonding experience, as the young men worked on their projects, whilst waging their own war against an army of mice who behaved "as if the place was theirs."[29] Their activities featured all the make-do spirit of wartime, and the young and middle-aged men who participated formed a tight-knit unit with strong personal relationships.[30] When in Akita, the AVSA team functioned largely in tents; on experiment days, launches were coordinated on a small blackboard, which occasionally doubled as a display for Hideo Itokawa's haiku poems. Launches themselves, coordinated from another tent, included plenty of physically strenuous preparation. Images from the time of Baby rocket launches in 1956 depict a team of mostly young men, all in sun hats, posing by their rocket in the blazing noon.[31]

The youthfulness of the team, however, and the hothouse environment of their work had the potential for producing ructions. This became a particular concern when, in 1962, AVSA decamped to a launch site in Uchinoura in Kagoshima Prefecture on Kyushu, Japan's southernmost major island. Before they moved, Itokawa told

FIGURE 3.1A. Members of the Uchinoura Ladies' Association greeting Hideo Itokawa and AVSA.

Courtesy of Kimotsuki Town Hall.

FIGURE 3.1B. Members of AVSA enjoying themselves in Uchinoura.

Courtesy of Kimotsuki Town Hall.

his young team to avoid messing with the locals: "The young women in Uchinoura are very beautiful . . . I want you to be careful to avoid issues with women [*josei mondai nado o okosanai yō jūbun ni chūi shite hoshii*]." Yet despite a 10 p.m. curfew, carousing would ensue; future ISAS and JAXA researchers Junjiro Onoda and Koichiro Oyama would later recall a surge of business for local bars, perhaps aided by the fact that "for some reason, shōchu[32] tastes so much better when drunk in Uchinoura."[33]

Beyond the perks of an occasional "drinking party," joining AVSA also proved to be an excellent professional commitment for those involved in terms of their ability to assert control over their research agendas. Japan's rocketeers in the 1960s enjoyed considerable autonomy, aided by the fact that their work had very little military connection. In the USSR, Cold War anxieties about national security and reputation ensured that a great deal of Soviet space organization, down to the very names used to designate projects, was secretive, and intensely scrutinized by the state.[34] In the United States, the army, navy, and air force space programs, and eventually NASA, employed academics in a model based on the wartime Office of Scientific Research and Development. In practice, this meant contracting certain development work out to academics and specialists, but keeping program planning and management within the respective service branches or agencies.[35] In contrast, Japan's earliest space programs were either built and run (AVSA/ISAS) or managed primarily (NASDA) by manager-specialists—figures who dealt not only with the organizational and management systems, but also the design of their space machines. Furthermore, at the beginning, AVSA was not regarded as a major state project that required careful management and oversight; rather, it was at best a "mini-project under the Ministry of Education, Science, Sports and Culture's jurisdiction." It took the best part of a decade, until the early 1960s, for rocketry to attract consistent attention from other bureaucratic and political entities.[36] Even afterward, it helped that AVSA had a fairly successful decade before Ohsumi's launch, vindicating its autonomy.[37] The result of all this was that researchers and engineers were able to assert substantial control over the direction of their research. As Yasushi Sato puts it:

> ISAS professors always decided their plans for themselves, set the pace of research and development on their own, and established their own style for research. They had deadlines and specifications, but they could set those constraints themselves. The circumstances allowed it. The purpose of their research was scientific rather than practical, and they were not subject to demands from external patrons.[38]

Japan's manager-specialists placed a premium on their ability to determine their own research agendas, as shown by the turf wars they fought to protect their independence. For example, relations with the Science and Technology Agency, the state bureaucratic entity in charge of overseeing both nuclear and space technology in Japan, deteriorated in the 1950s. The STA had been responsible for providing AVSA with some funding, and as such felt entitled to keep tabs on, and influence the direction of, AVSA's work. Itokawa and his team disagreed—strongly.[39] They eventually recruited the Ministry of Education and Tokyo University—their direct institutional overlords—to help insulate them from such meddling. Another ongoing battle was resisting American involvement. The only country from which Japan could reasonably expect to import LFRs was the United States, which had kept an eye on AVSA's activities from the beginning. *Stars and Stripes*, the US Army's in-house newspaper, had sent reporters to its earliest rocket tests; later, a contingent of military observers arrived under the aegis of the US embassy to watch launches of Kappa rockets.[40] Dr. Rocket and his team also visited facilities across the United States to learn about SFRs in more detail.[41] Yet a decade later, when they met representatives of TRW, the company responsible for the first US ICBM, the Atlas, things were different. The Japanese later recalled that the Americans were helpful, and their interactions—which featured some gentle ribbing of the setbacks AVSA had faced—were warm.[42] However, when TRW leadership later suggested that the Japanese might, like Australia, avail themselves of American launch services, senior ISAS figures "politely and plainly" rejected the offer.[43] Even after Itokawa's departure from ISAS in 1967, staff continued to argue that the technological importation deals struck between the government and the United States did not apply to them. They would stick with their local suppliers, and would absolutely not countenance having to compromise their research objectives for the sake of overseas suppliers.[44]

Then, in the 1960s, pressure built for Japan to sign a deal to import American Thor-Delta LFR technology. The manager-specialists understood that this meant they'd be forced to rely on American technologies and American personnel, compromising their independence and commitment to developing entirely domestic technologies.[45] Itokawa personally wanted to stay as far away from American technology as possible, and loudly trumpeted AVSA's Japaneseness, at least partly to avoid accusations of theft or copyright infringement; when newer space powers (such as Japan, India, and China) proclaimed the story of domestic development, it was partly to help "inoculate the program[s] against 'accusations of clandestine (or otherwise) appropriation.'"[46] Hence Dr Rocket responded with a rhetorical assault on the idea of US–Japan rocketry cooperation, an attack so intense that NASA administrator James E. Webb blamed his "quite calculated and entrenched opposition" for almost singlehandedly stalling progress on the matter.[47] Nor was it just Americans who were kept at arm's length by AVSA. In fact Itokawa and his team also had little tolerance for cooperating with other institutions—even within his own university. When Kankuro Kaneshige, also of Tokyo University, established the National Space Development Center in 1964 in order to facilitate the importation of American technology, Itokawa attacked the group as mere dilettantes (Kaneshige's specialty was textiles). Itokawa also leveraged his close personal relationship with Prime Minister Eisaku Sato (the two would often socialize together) to ensure that the highest echelons of power remained sympathetic to his institution's independence.

All this activity would eventually have personal consequences for Itokawa. By the early 1960s, it was apparent to most observers that rocketry in Japan needed to be put on a more permanent footing. To this end, it was decided that the group would be expanded by incorporating a few other organizations working in connected fields and creating a new institution, the Institute of Space and Aeronautical Science (ISAS). Of course, Hideo Itokawa was floated as the head of this new organization. However, senior members of one of the groups slated for incorporation, Tokyo University's Aeronautical Research Institute (which had roots dating to well before AVSA's formation), regarded Itokawa as brash and overpowering. The matter swiftly turned personal. Eventually, a leak regarding an audit and untidy bookkeeping at AVSA led to investigations into the groups' cozy relations with the private sector. Though not personally implicated, Itokawa resigned from ISAS in 1967.[48] His personality and overt politicking were later blamed in large part for his downfall.[49]

Even after his departure, however, ISAS successfully maintained its independence. After Ohsumi's launch in 1970, its engineers turned to the project of making a spacecraft "that could escape Earth's gravitation into interplanetary space." Such probes, however, required a brand-new SFR capable of lifting a payload out of Earth's gravity well, as well as new deep-space capabilities, facilities on Earth for communications, tests on the separation of strap-on boosters, and an eight-story launcher at the newly built Uchinoura Space Center. Despite resistance from a variety of institutions and stakeholders who would have preferred Japan develop a more capable LFR, ISAS was able to see the project through.[50] The resulting Mu rocket proved to be a great success, launching as it did interplanetary probes including Hiten (1990), which helped ISAS master the fine control of satellite trajectories, and

Yohkoh (1991), a solar observatory. Mu rockets also launched two probes, Suisei and Sakigake, which intercepted Halley's Comet in the early 1980s.[51] These successes represented a more or less complete realization of the internally generated vision ISAS had formulated in the 1970s.

Perhaps ISAS's most notable bureaucratic victory was its avoidance of being subsumed by Japan's second civilian space agency, NASDA. Founded in 1969, NASDA had as its primary focus the importation and domestication of LFRs, and as such its budget and scale quickly outstripped that of ISAS. The smaller institution nevertheless managed to stay out of NASDA's grasp, even if, on occasion, it made things difficult for itself. For example, in the aftermath of Ohsumi's success, the agency began to develop its own LFR, in addition to the costly work on the same technology being undertaken by NASDA; appalled by the waste of funds, it was forced into a cooperative framework with the other space agency, but avoided losing any other autonomy.[52] Manager-specialists continued to argue that any such alteration would undermine Japan's ability to domestically develop rockets, an argument that rang particularly true, as NASDA's main goal was to import and *then* indigenize technology.[53] Nor were managers above a little bureaucratic chicanery—for example, when some in the 1970s questioned why the ISAS launch site at Uchinoura should continue to be funded when a bigger alternative, NASDA's Tanegashima launch center, had been built so close by, ISAS responded by securing monopoly on rockets less than 4.6 ft (1.4 m) in diameter, which were created entirely for scientific, not commercial, purposes.[54] It was no coincidence that this was precisely the size of rocket that its launch facility at Uchinoura could handle.[55]

Still, conflict was not the default in ISAS's relations with other institutions. There was plenty of overlap in personnel: Kiyoshi Higuchi, future president of the International Astronautical Federation and vice president of JAXA, began his career at NASDA, but was subsequently "dispatched" to join Itokawa's laboratory.[56] Ryōjirō Akiba started off at ISAS, but ended up head of NASDA. And certainly, the group collaborated with other institutions on many joint projects. Yet when one lawmaker opined in 1969 that, when it came to ISAS-NASDA collaboration, "it is impossible to believe that those who had opposed each other so far could easily reach a consensus," he turned out to largely be correct.[57] ISAS maintained its independence for the next two decades, keeping to its own schedule of launching experimental payloads several times a year.[58] The group eventually outgrew its headquarters at Tokyo University and, on April 14, 1981, was transformed into an independent Institute of Space and *Astro*nautical Science (as opposed to *Aero*nautical Science), a form in which it persists even today.

NASDA, for its part, had its own sense of mission, embodied in the form of its own manager specialists. Notable here was Hideo Shima, who had established a reputation as the chief engineer of Japan's iconic Shinkansen. In 1969, he was appointed the first head of NASDA. Despite also being a Tokyo University graduate, he could not have been more different, personally, from Itokawa. As Yasushi Sato describes it:

> Itokawa actively approached political power and used it to achieve his engineering goals as well as his own personal fame. He was a brilliant engineering manager, but at the same time, he acted almost like a politician. Hideo Shima . . . focused on the management of the organization and its engineering activities. He always distanced himself from politics, although he sometimes had to interact with politicians and bureaucrats. Without playing political roles, he adroitly managed the interface between politics and technology.[59]

Whereas the influence of Hideo Itokawa and his successors was focused on preserving ISAS's independence, Hideo Shima's experience as the first director general of NASDA focused primarily on developing liquid fuel technology as quickly and cheaply as possible. Where his independence lay was in his ability to determine the methodology and planning for achieving this. Here, Shima early on took a position that appeared diametrically opposed to Itokawa's Japan-centered approach, arguing that "insularity does not work in the space age," and that national projects need not "consist purely of Japanese parts and components with Japanese flags on them."[60]

Shima's first challenge was to derive Japan's first domestically built LFR, the N-I, from the American Thor-Delta launch system.[61] This required substantial changes to the country's preexisting plans. Until 1967, these had focused on the Q rocket—a 94 ft (28.5 m) tall domestic vehicle capable of putting a 220 lb (100 kg) payload into orbit, up to 620 mi (1,000 km) from Earth's surface.[62] One of Shima's first acts upon becoming head of NASDA was to cancel both the Q and the mooted N rockets, and sign an LFR import deal with the United States. This was by no means a minor decision—thirteen families had already been moved to accommodate new infrastructure for the Q and N at Osaki in Tanegashima, all of which was now rendered obsolete.[63] Furthermore, time and money had already been invested in developing the Q.[64] Yet Shima stuck to his decision—one made not for political reasons, but for considerations of time and efficiency—and in the end met little resistance [65] "Officially he was not the ultimate decision-maker for Japan's civilian space programs," observes Yasushi Sato—that role was officially reserved for the space program's political overseers in the government and in the legislature. "But in reality, he was."[66]

Public-Private Networks in Japan's Civilian Space Programs

The influence of specialists in Japan's space programs was also bolstered by the fact that many engineers and technicians moved freely between state and private institutions. Managers and technicians tended to see themselves as part of a community that transcended their home institutions, and often made common cause across the state and private sectors. As institutionalization progressed in the space program in the 1970s, different centers of power within the state often found their interests aligning more with corporations than with their fellow ministries and agencies.[67]

The movement of workers between private and state sectors in Japan encompassed everyone from technicians to project managers. Take the example of Kazuhiko Hashimoto. After working in a variety of satellite communications positions in Ibaraki, north of Tokyo, Hashimoto moved in 1986 to the newly established Japan Communications Satellites Corporation, where he continued to promote the use of a 30 in (76 cm) satellite receiver developed at his old *state* facility in the same prefecture.[68] One consequence of this intimacy is the ease with which state space entities and the private sector in Japan can coordinate in times of emergency. During the Great Tohoku Earthquake in 2011, NTT (once state-owned, but privatized in the 1980s) deployed two of its communications satellites to provide emergency links for coordinating relief.[69] Both of these, the JAXA-NTT ETS-VIII Kiku engineering satellite, and the JAXA-NICT Kikuna/WINDS, had been produced through public-private partnerships.[70]

This movement of personnel also generated common work culture and expectations across the industry. To be sure, state and private enterprise did not always get along—Akira Meguro recalls that just confirming the minutes of a meeting could take "several hours" when working with bureaucrats.[71] Confrontations, when they came, could be unpredictable and dramatic. In 1977, for example, orders came from a Science and Technology Agency official to turn off a transponder during the launch of the Mitsubishi Electric Corporation–built ETS-2 Kiku satellite. Furious MELCO agents refused, arguing that it was dangerous; eventually, after a two-hour meeting, the STA backed down.[72] Still, by and large what prevailed was a sense of collegiality.

Meguro was, for example, deeply impressed by the diligence of the NASDA staff who in 1994 established the cause of the ETS-VI satellite's failure in just a few hours, and then managed to salvage a tranche of data from the device.[73] The boundaries between public and private were further blurred by the fact that private sector staff who were seconded to, or were otherwise involved in, state projects tended to retain loyalties to their institution of origin. Key academics and engineers sometimes wore uniforms bearing the emblems of private sector employers, and not the government body they were working under.[74] In 1999, more than 200 of NASDA's 859 specialist engineers were seconded from private corporations.[75] It is no surprise, then, that a common work culture prevailed, down to the level of in-jokes shared between NASDA employees and workers at Mitsubishi's Nagoya plant.[76]

Engineers and technical specialists were some of the most influential stakeholders in Japan's early space program. First among them was a well-educated generation of engineers who had come of age during and right after World War II. These figures initiated Japan's postwar space projects, and continued to have considerable independence within the institution they created, AVSA/ISAS. At the same time, even in their rival NASDA, engineers were ensconced at the highest managerial levels, making key decisions based on their own technical evaluations. At all levels, specialists developed their own networks, moving between state and private institutional formations with relative ease. Although they nominally worked for different agencies, engineers were afforded plenty of room by these networks to promote their visions for the space programs. Their tendency to move between state and private enterprise also served to create a joint culture that further consolidated a common identity.

This sense of independence endured into the twenty-first century, as younger generations of engineers not only continued the traditions generated by their forebears, but also produced a key cultural representative to serve as an exemplar of who they were: Hideo Itokawa, "Dr. Rocket."

The ability of Japan's engineers and specialists to assert their interests both within and without the civilian space programs is at least partly a product of the institutional formations they were working in. As seen above, by 1980, Japan's space program was a decentralized collection of often competing agencies, and manager-specialists were by no means the only group to wield power within Japan's civilian space programs. Japan's vast economic interests—large state-owned enterprises like NHK, private conglomerates like MHI and Kawasaki, and smaller specialist manufacturers dotting the country—also asserted their interests within this system. In fact, as one of the key proponents of the importation of LFR technology, private enterprise had a major hand in the creation of NASDA in the first place. Beyond this, it was also created initially not in order to develop its own rockets, but to import someone else's. It is to this peculiar origin story that we will turn next.

FOUR

Influence of Commercial Interest

ROCKETS ARE THE ONLY VEHICLES regularly used by humans that shatter into pieces whether or not they work properly. The expense this entails is justifiable to powers such as the United States and Russia because of the strategic and propaganda value of rockets. Japan's space programs, however, have always been more modest: a single Apollo flight, at roughly $2.5 billion a pop, cost twenty-six times more than the entirety of Japan's rocketry development budget from 1955 to 1970.[1] As entirely civilian institutions were strictly separated from military expenditure and R&D in Japan, funding, especially in the late 1950s and early 1960s, was further limited.

AVSA and its successor, ISAS, kept their costs down by developing close links with Japanese companies—as suppliers, sources of knowhow and technical specialists, and even for testing sites. Although the Science and Technology Agency, and Tokyo University, had provided Itokawa's group with startup funds in 1954, they had offered no long-term funding. By the time this funding ran out in 1959, it was evident to many that the situation needed to change; one member of the Diet Budget Committee opined that "it would be irresponsible" not to provide consistent and substantial funding.[2] Five years later, the worst fears of some were confirmed when Hideo Itokawa organized commercial sales of suborbital rockets to Indonesia and Yugoslavia.[3] Though Itokawa argued these sales showed the commercial viability of his rockets, it was a headache for policymakers—Yugoslavia, though not Warsaw

Pact, was officially communist. Furthermore, as there was no guarantee the Yugoslavs wouldn't surreptitiously use elements of the vehicles for missiles, some on the Japanese left argued that such sales violated Article 9 of the Constitution.[4]

Fortunately for all involved, Itokawa had also found a more palatable way of saving money: sourcing all AVSA/ISAS needs locally. When he was told early on that it would be cheaper to import solid rocket fuel, for example, Itokawa refused and managed to secure supplies of it from within the country.[5] Many of the companies he approached leaped at the opportunity to reuse wartime production capacities. These included Fuji Seimitsu, once part of the Nakajima aircraft corporation, which produced bodywork, fuels, and components. Firing tests of Itokawa's engines took place at the disused pistol range of the Shin Chūō Company in Tokyo.[6] Meanwhile, the Prince Motor Company, once part of Tachikawa Aircraft Company (and one day to become Nissan) fabricated body components for the rockets, and provided its Kawagoe facility for the development and testing of engines.[7] When AVSA started cracking orbital flight, Hitachi furnished it with a HITAC 5020 computer.[8] Corporations also participated in the earliest launches as experimenters, as Kyushu Denryoku did in the 1956 launches of the Baby rockets.[9] Japan's private sector became a critical resource as it expanded during Japan's economic boom of the 1960s and 1970s. Interestingly enough, one company that did *not* agree to participate at this stage was Mitsubishi Heavy Industries, precisely because it couldn't see the profit in the project.[10] Itokawa made the case throughout the 1950s and 1960s that domestic production should be at the heart of *all* space research in the country. At a symposium hosted by Japanese state-run communications monopoly Kokusai Denshin Denwa (KDD) in 1965, he argued that Japan could launch its own satellites within three years even without importing technology, and that importing technology wouldn't necessarily speed up the process much. "Buying" technology from the Americans didn't actually mean owning it, as Japan would not have the right to produce copies; in fact, it actually meant spending more money in the future, as engineers and technicians would have to live in the United States to train. Lastly, Japan would not be able to stage launches as and when it pleased.[11]

Itokawa's vision, however, was soon challenged by some of these very corporations, which developed into major stakeholders with objectives, and large purses, of their own. The extent of their influence differed from that of equivalent entities in the superpower space efforts. Early Soviet space research relied on a "vast network of contractors and subcontractors spanning dozens of institutes, design bureaus, and factories," with next to no role for private capital.[12] Conversely, the Soviet military—and its budgets—played a dominant role in space R&D until 1991.[13] The American civilian space program, meanwhile, was orchestrated by government-run laboratories rather than contractors; government engineers crafted contractor proposals (often in secret) themselves, and employed preferred vendors whenever possible. Thus, following the model of World War II contracting, the government decided what it wanted, and contracted with private entities to provide it.[14] The exception to this was telecommunications companies, which early on took an interest in developing their own space technological projects; already in the mid-1950s, AT&T's Bell Telephone Laboratories estimated that a communications "mirror" in space would be worth a billion dollars to the company.[15] KDD wasn't too far behind in its thinking. It, and other service providers, came to the conclusion in the late 1950s and early 1960s that establishing viable satellite infrastructure in Japan required two things, pronto: a reliable medium-lift LFR, and advanced (that is, imported) satellite components. As a contemporary Tokyo University fluid dynamics specialist put it, LFRs were "superior [to SFRs] in that they can be precisely controlled"; whereas SFRs have set burn times and cannot be throttled, LFRs can, which allows their payloads to be placed in a variety of orbits, with greater precision.[16] In 1961, the Space Development Council dispatched a fact-finding mission to the West to see what their LFRs could do, and what they might be willing to sell; by the mid-1960s, officials had begun considering signing deals to allow this.[17] The problem was that Itokawa's arguments, and the policies they'd helped generate, prevented this from happening—KDD, for example, was forbidden from importing LFR *or* foreign satellite technologies. Luckily for KDD, however, its private sector partners wholeheartedly agreed with its objectives, and had no qualms in pressing the issue. Organizing through the powerful Japan Business Federation, Keidanren, companies including KDD's suppliers and contractors began to agitate for changes to space policy in the mid-1960s. Their efforts would, in short order, culminate in the foundation of NASDA, and Japan's entry into the elite club of nations that could produce LFRs.

Satellite Communications and Commercial Use as Motivators for Space Research

THE RISING IMPORTANCE OF SATELLITES IN JAPAN IN THE 1950S AND 1960S

Ohsumi would last in orbit until August 2, 2003, but the vision of an all-Japanese space program it embodied was, by that point, long dead.[18] The roots of Japan's transition from Hideo Itokawa's vision, to one more favored by certain large corporations and ministries, begins—like so many stories of the Space Age—with Sputnik.

The 1957 launch of the Soviet Union's 185 lb (84 kg) space grapefruit opened an exciting new set of technological possibilities. Itokawa's reaction was initially a little sour — he argued in November 1957 that it would be "wrong to set up a research institute for artificial satellites . . . just because the Soviet Union has launched artificial satellites." Later, though, Dr. Rocket changed his mind, acknowledging that "commercial satellites" would be "a very profitable private enterprise."[19] In 1959, future prime minister Yasuhiro Nakasone described to fellow parliamentarians how easily Japan could create a globe-spanning satellite communications network; all it would need were "three satellites 22,400 mi (36,000 km) above Japan, Athens, and Buenos Aires," to reach the whole world.[20] For now, though, the technology was just too expensive for the Japanese: a single domestic satellite launch, including rocket development, payload, and facilities, was estimated in 1957 to cost around ¥700 million ($21 million in 2024).[21] For comparison, the entire budget for the country's rocket program in 1956 was ¥3,300,000 ($6.5 million in 2024).[22]

Then came the "economic miracle." The average GDP growth rate in Japan from 1945 to 1958 was 7.1 percent; from 1959 to 1970, it was 9.5 percent. Average income in 1945 had been around $2,000; by 1965, it was $9,000.[23] By 1970, Japan had the third largest economy in the world.[24] A notable consequence was an explosive growth in TV usage. The first TV stations, the state-owned Nippon Hōsō Kaisha (NHK) and the private Nippon Television, launched in 1953. When the Committee on Communications of the NDHR began discussing the promotion of color TV in Japan in 1960, the potential use of satellites was a key part of the debates.[25] By 1967, roughly 8 million black-and-white TV sets, and 4 million color TVs, were being produced annually in Japan, compared to practically zero fifteen years earlier.[26] Within three years, state entities had begun investing in the construction of the parabolic dishes and antennas required for satellite relays across Japan.[27] This enthusiasm spread to the public as well. Preparations for the pioneering 1964 satellite broadcast of the Olympic games from Tokyo captured the imagination of the popular press, whose coverage foregrounded high-tech infrastructure like large-scale satellite dishes.[28] Publications also covered the importance of broadcasts *to* Japan via satellite, such as the Mexico Olympics in 1968, as well as pretty much every major development in American communications satellite capacity.[29] Even stamp collectors contracted satellite fever: in 1967, the Post Office issued a stamp celebrating the inauguration of Japan–US satellite communications, featuring the Intelsat II Lani Bird satellite.[30] It wouldn't be the last time a spacecraft graced a Japanese stamp.

The main institutions charged with realizing Japan's communications ambitions were massive state corporations, such as NHK and KDD. Both had been founded in the 1950s as "statutory corporations," semi-independent entities answerable to the Japanese government. They were charged with responsibilities from conducting primary research to identifying where facilities might be located.[31] In the 1950s, in response to the rapid expansion of broadcast media, they initiated plans to establish some thirty broadcast stations, with 80 percent coverage of Japan, by 1962.[32] Their later communications projects were, on occasion, as large as those of AVSA/ISAS; in 1969, the Ministry of Telecommunications alone budgeted ¥300 million ($7 million in 2024) for satellite-related technologies, equal to the annual budget for developing SFRs.[33] By 1973, research expenses for satellite development had risen to ¥800 million ($15 million in 2024), in large part entrusted to KDD.[34] Further expenditure also went toward creating ground facilities; a single tracking station built in Yamaguchi in 1967 clocked in at ¥2.7 billion ($70 million in 2024).[35] Money was also invested in developing domestic commercial satellites, culminating in Japan's first home-built communications satellite, Sakura, in 1977.[36]

In promoting these activities, state-run corporations now found that their interests aligned more closely with those of the private sector than with lawmakers' priorities. For KDD, increasing the speed of communications was one of the key justifications for government funding, and the basis of its public reputation.[37] Cost reductions and profitability were also a matter of public discussion.[38] Crucially, however, state corporations were forbidden from importing foreign components, so they relied almost solely on local producers; the Ministry of Posts and Telecommunications Radio Research Laboratory, for example, relied on up to fifty-two private sector companies to help develop experimental satellites in the 1970s and 1980s.[39]

State corporations could also, on occasion, compete with each other, as when the KDD earth station in Kashima, Ibaraki, competed with the similar state-owned Radio Research Laboratory in the early 1960s.[40] Still, the overall trend was a harmonization of objectives across these institutions and their private sector partners, and together they deployed their influence in ways that undermined or altered preexisting state policy. In the 1970s, for example, though the Toshiba-built BS-2A broadcast satellites suffered a series of malfunctions, NASDA decided to stick with the company—and its US supplier, General Electric—for the foreseeable future. NHK, however, fiercely opposed this move, and campaigned for its cancellation. In due course, the contract was instead awarded to NEC.[41]

PRIVATE CORPORATIONS: WELCOMING AMERICA, WELCOMING PROFIT

Initially through links with AVSA, and then through the growing importance of satellite communications, large private corporations such as Mitsubishi Heavy Industries (MHI), Ishikawajima Heavy Industries (IHI), Hitachi, Toshiba, and NEC now found themselves invested in a burgeoning space industry. One area in which their contributions were critical was R&D. Indeed, Japan's private sector shouldered a major part of the country's research activities in the postwar period. Between 1951 and 1954 private enterprise established some fourteen centralized research laboratories, working on topics from cosmetics to communications satellites; in 1961 alone, a further twenty-five came into existence.[42] In 1952, of some 26,633 researchers active in the country, around 5,000 were in private enterprise, while about 14,000 were at universities. By 1962, around 55,000 worked in private enterprise, while 42,000 worked at universities.[43] Many of these individuals had been involved in Japanese aviation prior to and during the war; alongside the Shinkansen bullet train (1964), the Mitsubishi Colt (1960), and the Kawasaki Jet Ski (1972), the institutions they worked in were also responsible for Japan's space vehicles.

For companies involved in communications satellites, it became evident that waiting for Japan to domestically develop the ability to precision-launch satellites would leave them far behind their overseas competitors. It was far simpler, and cheaper, to import the technology from the Americans. After all, they argued, wasn't Japan already reliant on American technology for communications, given that links with nearby countries like Thailand and the Philippines had to go through

American satellites?[44] Furthermore, companies such as MELCO already had their own deals with American companies; by 1966, MELCO had established joint production agreements with TRW, and had also an agreement enabling it to import technology from American aircraft manufacturer Hughes.[45] Smaller, more specialized companies also looked across the Pacific for a leg up. Skilled professionals were often hired from the United States to help establish satellite development efforts. Indeed some companies owed their start in telecommunications entirely to knowledge acquired in the United States; the entire team of ten engineers and technicians at Japan Satellite Communications (JCSAT) had visited the Hughes offices in Los Angeles for training, before returning and setting up shop in Yokohama to produce orbiters run entirely on Hughes systems.[46] Companies also developed ground-based infrastructure that worked with American satellite technology—Hitachi pioneered a domestic laser-based tracking system in 1968 primarily to track the movements of US satellites.[47] Even companies that worked with AVSA/ISAS found that they could benefit from overseas technology. A good example here is NEC, which had been involved in AVSA's work from 1956, while at the same time developing extensive links with US corporations for technology importation.[48] In 1969, the corporation established a Yokohama plant specifically for its space business, and in 1970, the same year Ohsumi launched, it finalized a technology importation deal with its long-term partner, Honeywell.[49]

These companies channeled part of their efforts for an LFR import deal through Keidanren, the powerful Japanese business federation. Keidanren had long had an influence over Japanese space policy, providing six of the twenty-nine members of one of Japan's earliest space policy bodies, the National Space Activities Commission, in 1960.[50] It established its own Special Committee for the Peaceful Usage of Space in 1961, and Space Development Promotion Council in 1968, both of which were active in promoting the industry by campaigning for concessions such as tax breaks, and securing protections for Japan's fledgling satellite industry.[51] Strengthening the domestic satellite industry was another priority of private industry. At the time of the launch of the first commercial broadcast satellite in history, Early Bird (1965), it was estimated that a domestic communications satellite from Japan, even if launched on a foreign rocket, would cost ¥1.8 billion (around $50 million in 2023).[52] Prices did not drop dramatically in the next fifteen years, meaning that Japanese satellites were both less capable and more expensive than those of overseas competitors. Nevertheless, laws restricting entities like KDD to buying Japanese satellites meant that companies such as Toshiba, NEC, and Mitsubishi were able to expand their satellite divisions, despite their products being uncompetitive in the global satellite market.[53]

The rewards for engagement with the market for space technologies were considerable profits—and not just in Japan. The first satellite communications facility in Zambia, for example, was commissioned from NEC in 1972 for $370,000 ($5 million in 2024).[54] The company was also heavily involved in constructing telecommunications infrastructure in Arab countries, India, and New Zealand.[55] The same year, KDD and Mitsubishi agreed to construct a land station in Yugoslavia, in tandem with two Yugoslav partners.[56] Overseas orders for Japanese hardware also increased, driven by innovation. In 1968, Matsushita Denki developed and marketed pioneering technology that could determine attitude (that is, the direction in which a satellite is pointing) while in orbit.[57] A decade later, an IBM part-subsidiary ordered 200 digital modems from Fujitsu, in the first deal of its kind in Japan's history.[58] Thus by the 1980s, with state help and US technological links, Japan's satellite communications industry was well on its way to becoming a major global player. There was no reason, in the private sector's eyes, that a similar process should not have unfolded in the field of rocketry—apart from the resistance of those who still believed an all-Japanese SFR-based space program should take priority.

Liquid-Fuel Rocketry and the Rise of NASDA

THE ANXIETIES AND ADVANTAGES OF LIQUID FUEL ROCKETS

Despite all this pressure, Japan in the mid-1960s simply did not have a rocket that was powerful or accurate enough to put satellites in orbit. The only ongoing project for developing a vehicle of the required size, ISAS's Lambda series, was an SFR of limited utility and insufficient payload capacity.[59] At the same time, official policy essentially forbidding Japanese commercial satellite launches on foreign rockets persisted, and pressure from entities like NEC and NHK for LFR technology intensified.[60]

The proximate challenge was overcoming the legal, policy, and organizational entrenchment of SFR favoritism in Japan's space R&D. Hideo Itokawa's arguments in favor of an all-Japanese space program had initially resonated with figures like Eisaku Sato (minister for the Science and Technology Agency in the early days of

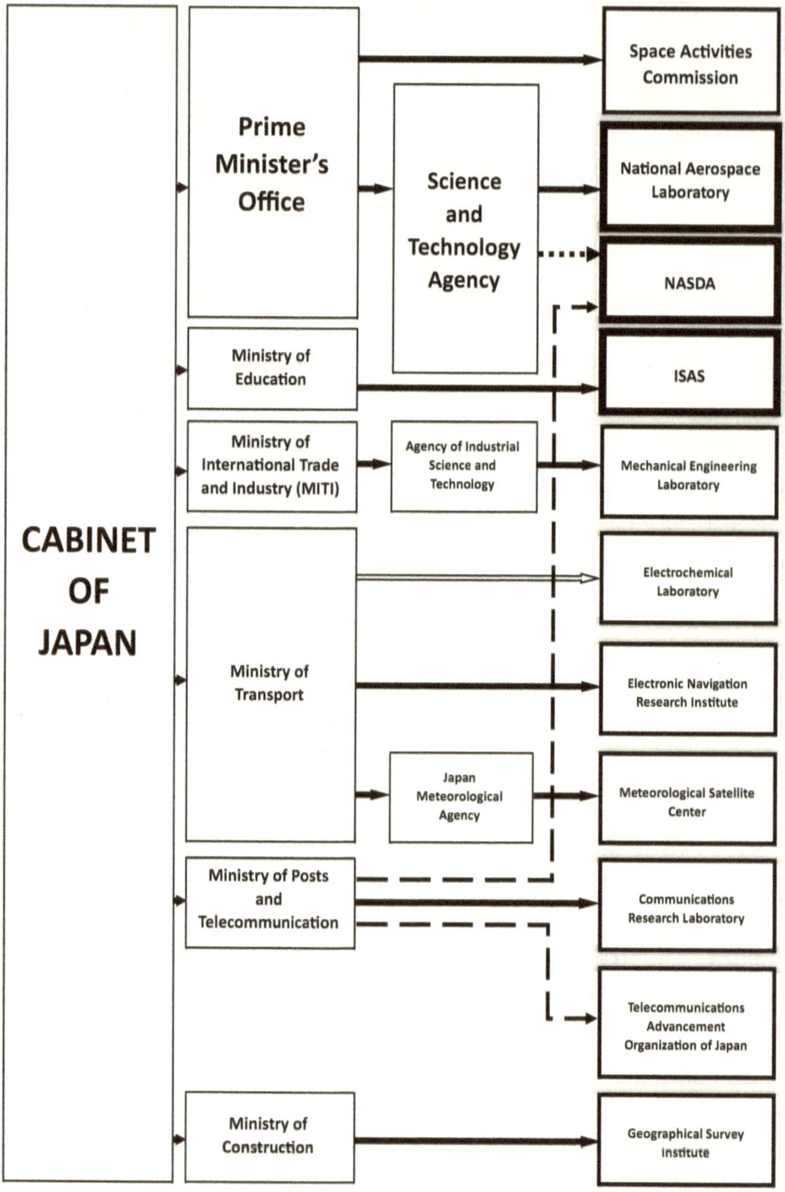

FIGURE 4.1. The administrative structure of Japan's space programs in 1980.

Adapted by author from Steven Berner, "Japan's Civilian Space Programs: A Fork in the Road" (Santa Monica, CA: RAND Corp., 2005), 13.

AVSA and, from 1964 to 1972, prime minister of Japan) and Yasuhiro Nakasone (minister of science in 1959 and also, from 1982 to 1987, prime minister). For these men, and the influential Science and Technology Agency, a successful space program as an international "necktie" for Japan.[61] The more Japanese it could be, the better. Furthermore, even when their attitudes changed, they found themselves stymied by the decentralization of space R&D. So even though lobbying persuaded Sato to order the new director of the STA, Nikaido Susumu, to expedite the development of LFRs, political scientist Hirotaka Watanabe has pointed out that Japanese policymaking remained stubbornly "bottom-up."[62] Lawmakers bemoaned the "wall of administration" that resisted the unification of the enterprise into "one ministry and one bureau," which could make decisions swiftly.[63] "No matter what kind of structure you may tweak," complained another, ". . . you must reconsider all aspects of the civil service."[64]

Beyond this, one particular concern of Itokawa's—avoiding accusations of clandestine military research—was also a source of anxiety for lawmakers and bureaucrats alike. LFRs are infamously "dual use"—that is, they can be used to launch satellites or warheads—and the liquid-fueled R-7 Semyorka rocket that launched Sputnik had been the world's first ICBM.[65] A week after Sputnik, Foreign Minister Aiichiro Fujiyama expressed the concerns of many in Japan that in this manner, "scientific advances are competing to develop murderous weapons," and warned that such technology could cause "the destruction of humanity."[66] A decade later, LFRs still suffered from precisely the same connotations of war and death that figures such as Eiichi Iwaya and Hideo Itokawa tried hard to dispel.[67] This is not to say that Japan hadn't invested in military rocketry at all. By the 1960s, the notion of self-defense had somewhat amended the blanket postwar rejection of certain kinds of weapons, and Japan had begun purchasing missile systems from the United States. The investment in such systems increased—in 1970 alone, Japan budgeted ¥66 million ($2.7 million in 2024) for importing Nike missiles (as compared to a budget of ¥800 million [$32 million in 2024] for developing scientific rockets).[68] These, however, were largely short-range, could not be used to carry nuclear payloads, and in any case had nothing to do with the civilian space programs in terms of budget or administration. Furthermore, the overwhelming majority were SFRs.

As a result, prior to 1969, LFR research in Japan was carried out in piecemeal fashion by various entities, such as the Science and Technology Office, which began testing liquid rockets in Niijima off the Japanese coast in 1967.[69] The same year, the fifth prospectus of the National Space Activities Commission initiated the development of the domestic N class of LFRs, aimed at lifting a payload of 220 lb (100 kg) into orbit. In 1969, another one of its LFRs, the Q, lifted 132 lb (60 kg) into space (but not orbit)—impressive, but still paltry when compared with the Americans.[70]

INTERNAL AND EXTERNAL PRESSURE TO IMPORT LFR TECHNOLOGY

This slow and grudging progress was nowhere near enough for Japan's private sector. In 1966, Atsushi Oya, chairman of Keidanren's Committee on the Peaceful Usage of Space, stated that while "Japan's space development is something we should show off to the world . . . American rockets are 100 times larger." The United States had already launched the first direct-relay communications satellite (Telstar 1, 1962), the first geosynchronous communications satellite (Syncom 2, 1963), and the first commercial communications satellite in geosynchronous orbit (Intelsat 1, 1965). Japan didn't even have a single one. Keidanren also emphasized that a domestic LFR industry could only mean one thing: "jobs."[71] All this required was government permission, and perhaps a little financial lubricant.[72] It would also be nice if the process of securing patents and government contracts were centralized—having seen how US producers, faced with a "chaotic and bewildering" system, had resorted to skulduggery (such as masquerading as NASA engineers to secure deals), they were keen to avoid the same problems.[73] Backing them were KDD and NHK.[74]

This alliance began its charm offensive by targeting sympathetic politicians.[75] At the same time, the gathering threat of Soviet and Chinese rocketry capacity also served to persuade previously reluctant legislators that Japan needed to step up its LFR game.[76] Yet another group saw that signing an LFR import deal would help reduce growing transpacific trade frictions, and also sweeten ongoing negotiations over the return of Okinawa to Japanese sovereignty.[77]

It was at this point that they acquired another key ally—the Americans. The United States had its own, considerable reasons for signing an LFR export deal with Japan. It would also help alleviate American concerns about Japan's space program. Controlling the spigot of technology flow could help the United States limit Japan's ability to develop a competing commercial launch industry, and its own ICBMs.[78] By the late 1960s, the US Arms Control and Disarmament Agency had somewhat hastily concluded that Japan had the capacity to develop an ICBM within three years if it wished.[79] By 1965, there were even reports of potential shenanigans: questions were raised when American companies such as Hughes tried to purchase Japanese rockets from Mitsubishi and Prince, under the dubious pretext that America had insufficient numbers of short-range rockets with which to carry out scientific tests.[80] Some concluded that the Americans were worried about the sophistication of Japanese rocket technology, and were thus trying to secure samples to retro-engineer. Limited technology transfer would obviate the need for subterfuge, boost a key Cold War ally, bind Japan more closely to US technological systems, and perhaps even generate a little money through licensing agreements. Lastly, US–Japan LFR cooperation would prevent Japanese cooperation with the USSR, which had already occurred in a few minor projects—in 1967, for example, the Soviet Kosmos-144 satellite relayed a TV program in Japan, an unusual step that displayed not only the potential of satellite communications, but also the interoperability of Japanese and Soviet systems.[81] Policymakers in Washington were keen to nip such interactions in the bud.[82]

For these reasons, it was US officials who in 1963 made the first official approach to Japan suggesting rocketry technology transfer. They received no response. Matters escalated in 1965, when NASA's head of foreign relations issued a point-by-point refutation of Hideo Itokawa's reasoning for rejecting US assistance in launching satellites. Keidanren then requested that Diet members visiting the United States at the time explore the possibility of signing an LFR deal.[83] In May 1965, Secretary of State Dean Rusk met with Japanese foreign minister (and future prime minister) Miki Takeo, and brought up the issue again; when Eisaku Sato visited the United States in November of that year, President Lyndon B. Johnson brought it up again.[84] Despite all this, however, in 1967, the National Space Activities Commission stated that official policy for Japan was still to develop domestic rockets, and to launch its satellites on its own.[85]

Issues finally came to a head in January 1968 when, at a meeting with Keidanren, US ambassador to Tokyo U. Alexis Johnson took the unusual step of publicly revealing details of the so-called "Johnson Memo" delivered to Japanese authorities.[86] In it, the US government had offered "virtually complete cooperation between both countries on rockets and satellites under three conditions: peaceful purposes, no export to third countries, and compatibility with International Telecommunications Satellite Consortium (INTELSAT)." Also included was an offer to provide Japan with the American-built Thor-Delta LFR launch system. Reports of his speech appeared in newspapers the following day; furious lobbying from Keidanren ensued, supported, now, by Eisaku Sato. On November 14, 1968, the CEO of Lockheed visited MHI's headquarters in central Tokyo to "promote cooperation," while at the same time approaching Kawasaki Heavy Industries.[87] Eventually, an "exchange of notes" in 1969 ended the debate once and for all—Japan would soon be in receipt of American LFR technology.[88] The only regret from those who'd arranged the deal was that it had come after far too much "wasted" time.[89]

NASDA AND THE POWER OF THE PRIVATE SECTOR IN THE DEVELOPMENT OF LFR TECHNOLOGY

The signing of the LFR import deal was accompanied by a bureaucratic reconfiguration within Japan's civilian space programs. Late in 1969, the Space Development Council finally took official control over LFR production in Japan.[90] One of its first steps was shepherding the 1969 National Space Development Agency Law, which established NASDA, replacing the older National Space Development Center. NASDA was placed under the STA, which reported directly to the prime minister, thus placing it outside of Japan's squabbling ministries. These same ministries, when called on to give up their space research agendas to NASDA, then demurred. The Ministry of Posts and Telecommunications, for example, ceded the prestigious Radio Research Laboratory, but held on to other areas of research, and did not consent to share its budget—even though it had been a proponent of NASDA's formation.[91] Conversely, one of NASDA's founding managers was a senior figure from MHI, included specifically for the purpose of representing private industry. Finally backed by the legal and administrative framework they'd been agitating for, private industry now struck a series of deals with US suppliers. In February 1970, MHI came to an agreement with North American Rockwell and McDonnell Douglas; Nissan, which had acquired the Prince Motor Company, signed deals with Aerojet; Kawasaki did so with Lockheed.[92] As the sophistication of Japan's rockets increased, so too did the diversity of companies involved in the space industry. By the late 1970s, these included hundreds, in fields as diverse as plastics, ceramics, and pharmaceuticals. All of these purchased licenses, staff education, and key equipment from the US.[93] They also secured a place at the space policymaking table; by 1980, five members of the Council for Science and Technology were chairmen of major firms.[94]

Keidanren also assumed a role in coordinating the distribution of orders and imported technology between state institutions like the Space Development Council, and its members.[95] The trade federation also publicly promoted its own vision for space development in the media, as in a comprehensive document demarcating four key areas of development (including LFRs) in 1977.[96] It also took NASDA to task when it felt necessary. In the 1970s, for example, there was a growing sense that Hideo Shima was more interested in developing advanced LFR technology than in making practical launch vehicles—two years after the import deal with the United States, the mooted Japanese-made N-1 LFR was four years away, and capped at a 770 lb (350 kg) payload. The rocket was simply not powerful enough; two communications satellites slated for launch that year, including an NHK broadcast satellite, weighed more than it could lift.[97] In the end, NHK went directly to American launch suppliers, an action for which it had to get special dispensation.[98] Dissatisfied with NASDA's responses to its concerns, Keidanren launched a campaign in the 1980s to push NASDA and the Space Activities Council toward more commercially oriented activities.[99]

The federation was also critical of officialdom's timid approach to international cooperation. When the United States made overtures to Japan in the 1970s to join its post-Apollo program of constructing a space base and the Space Shuttle, it was Keidanren that took the lead in investigating the offer, organizing a 1971 delegation (which included NASDA personnel) that concluded that Japan should have some limited participation in future US space projects.[100] Its activities encouraged bureaucrats to examine the issue further.[101] Ultimately, however, NASDA decided to sit the Space Shuttle out, and Keidanren's disappointment at this remained a major gripe well into the 1990s.[102]

Despite these tensions, under Shima's steady hand, NASDA swiftly outgrew ISAS. By 1970, its budget was more than twice that of the older program; by the time Shima had retired in the early 1980s, NASDA was being afforded budgets in the region of ¥8.37 billion ($138 million in 2024), eight times greater than ISAS's ¥1.03 billion ($15 million).[103] The institution developed two LFR families, the N series in the 1970s and the H series from the mid-1980s. The N-1, which lifted the 772 lb (350 kg) Himawari satellite into orbit in 1981, was mostly made of American parts; five years later, the H-1 rocket was unveiled, boasting a payload capacity of 2,500 lb (1,100 kg) to low Earth orbit (LEO), and consisting mostly of locally manufactured components.[104] By 1984, NASDA was confident enough to embark on the creation of the H-2, an entirely domestic, medium-lift LFR. Along with these projects came expansion of space infrastructure. After creating makeshift testing ranges in the 1950s, by the 1960s both ISAS and NASDA began construction on two launch centers in Uchinoura in Kagoshima Prefecture and on the island of Tanegashima, both in southern Japan. These were serviced by an array of secondary and tertiary facilities ranging from fuel depots to tracking stations, both state-owned and privately built, located along the length and breadth of the country.

The second decade of Japan's civilian space programs thus witnessed remarkable change in their scale, organization, and objectives. The first phase of the enterprise was largely conducted under the aegis of the University of Tokyo's AVSA, whose activities were defined by three qualities: it focused on developing fundamental technologies and acquiring scientific data; it only used solid fuel rockets; and it emphasized developing the required productive capacity for components within Japan, without recourse to foreign importation.

The launch of Sputnik in 1957 ushered in a series of changes that would end this status quo. As Japanese corporations—both state-owned and private—began to comprehend the capacity of satellite technologies to change the landscape of telecommunications, they began to agitate for increased investment in the field. They also pushed for the rapid development of the necessary infrastructure. Government investment began to flow to other institutions to fund space technological research outside of Tokyo University's purview, while private corporations swiftly developed links with overseas producers to secure the required technologies to make the creation of an advanced satellite telecommunications network possible. Of crucial importance to this process was the importation and use of liquid fuel rockets. Although Japan did not have any prospect of rapidly building these vehicles, the long-term future of the country's presence in space was increasingly seen as requiring them. In particular, it is impossible to launch communications satellites without LFRs. Various groups within Japan, ranging from politicians to business leaders, banded together to push for deals with the United States to authorize technology transfer across the Pacific. When Japanese officials dragged their heels, organizations led by Keidanren worked to overturn their reluctance, resulting in the establishment of US–Japan rocket technology agreements in 1968 and 1969.

While this process was unfolding, however, another crucial influence on Japan's postwar space program was developing in locations around Japan. As the sophistication of the country's launch vehicles increased, leading corporations and other engineers to flock to the technology, so too did the need for specialized facilities to feed into the project. These refineries, laboratories, testing ranges, and launch sites would have a significant impact in the areas they developed—contributing as they did to postwar Japanese anxieties about runaway industrialization, local government, and pollution. The winding tale of how exactly Japan built its extensive terrestrial infrastructure is also a story about how the people in these localities found a way to assert their rights and priorities within the civilian space programs—and in turn affected their shape, timing, and development. We turn now to this story of angry fishermen, rock-carrying housewives, and children blowing their fingers off.

FIVE

Welcome and Resistance to Japanese Space Facilities

EARLY IN THE AFTERNOON ON September 12, 1968, a group of fishing boats from the Miyazaki Fishermen's Union were plying waters off the coast of the village of Uchinoura, Kyushu.[1] In itself, this was nothing out of the ordinary—the area teemed with a variety of delicious sea creatures (lobsters, amberjack, tuna), and boats from all over southern Japan congregated there regularly to seize their share. This day, though, they weren't supposed to be there—at 2:00 p.m. that afternoon, ISAS would be launching a rocket from Uchinoura Space Center nearby, and the boats were in the exclusion zone.

Yet rather than getting to work, the boats simply sailed in circles for hours on end. Eventually, the coast guard was summoned, but the fishermen refused to move. They were there, they said, to send a message to the workers of Uchinoura Space Center, many of whom were gathered on the hilly, forested coastline nearby: until their demands for compensation for fishing time lost to rocket launches was met, there would be no testing. As the scheduled launch time came and went, the excitement the ISAS staff had felt that morning dissipated, and they were forced to scrub the day's work. "It's a shame we don't have the cooperation of all of the local fishermen," lamented launch leader Yoshio Moridai, "I hope there are no further interruptions to our schedule."[2]

Rockets may head for a vacuum, but aren't built in one. A sophisticated space project requires hundreds of thousands of pounds of infrastructure, spread across many hundreds of acres of land.[3] Given how noisy, dirty, and dangerous many of these facilities are, the Japanese agencies had to situate them in far-flung places, well outside of the economic heartland. Nevertheless, in such a crowded nation, even such remote locations supported small communities; in fact, given the country's geography—a mountainous spine with narrow, flat fringes—both facilities and people tended to be drawn to the same areas.[4] Japan's postwar localities were familiar with the arrival of large industrial facilities, but responses from local populations could be conflicted. Some felt excitement for the economic and investment opportunities afforded by the arrival of a big-spending institution, while others were more concerned over the impact on their health and livelihoods. The story of the arrival of Japan's space facilities in the south is thus part of the story both of Japanese spacefaring, and of the reactions of folk in the economic peripheries to the might of the postwar developmental state. As in the cases of industrial disasters, such as Minamata, or the construction of nuclear power stations and Narita Airport, it provides a fascinating glimpse of how low-level organization and local concerns could alter the trajectory of a huge national project.

Local populations exerted a powerful influence in the development of Japan's space infrastructure between 1960 and 1980. When ISAS and NASDA decided to create facilities in southern Japan, they engaged in extensive outreach to, and consultation with, local citizens. Many of these in turn welcomed the arrival of the rocketeers and the investment and tourists they would bring, but one major group refused to play ball. Slighted by what they saw as a lack of consultation, and an existential threat to their livelihood, southern Japan's fishermen organized between 1965 and 1970 to contest space agencies' plans. Their agitations put on hold ISAS's plans to place a satellite in orbit, as no rocket testing occurred in southern Japan, for years at a time. The compromise that defused the situation would continue to play a huge part in the pace of Japanese space R&D well into the twenty-first century. That all this could be accomplished by a handful of men and women in fishing boats and small town halls is nothing short of remarkable.

Early Encounters: Akita, Niijima, and the Move South

The opposing reactions of welcome and resistance were on full display from the very beginning of Japan's space project. Testing ranges must be big, and thus in Japan they more often than not get in the way of someone else. For example, in the 1950s and

1960s, the STA utilized a Defense Agency site in Niijima, a sparsely populated island off Japan's east coast, for early rocket tests.[5] Remote, and difficult to upgrade, the site was unsatisfactory from the start.[6] The fact that it was testing explicitly military-use devices also rubbed locals the wrong way, while others objected to the enormous tracts it occupied on an island where space was at a premium. Ultimately, five locals opposed to missile testing sued the government for access to the land.[7] Even in such a small and sparsely populated region, locals could assert their power in such ways.

Meanwhile, on May 26, 1955, AVSA began looking for locations for full-scale launch testing.[8] It eventually settled on a strip of land provided by the government-owned Japan National Railways, on Michikawa Beach in Iwaki, Akita Prefecture, in Japan's northwest.[9] Locals initially greeted the group with enthusiasm. In November 1956, seventeen students and teachers at the No. 2 Middle School in Yokote, about two hours from Michikawa, created a star map "about eight tatamis" (roughly 133 sq ft [13 sq m]) in size, featuring illuminated constellations, and costing a hefty ¥6,000 ($180 in 2024).[10] The next year, a cake shop in Akita City produced an elaborate Christmas cake on the theme of artificial satellites, featuring a depiction of the various Sputniks launched up to that point.[11] In July 1957, local agencies and businesses founded the Rocketry Observation Cooperation Association (Roketto kansoku kyōryoku kai) in order to coordinate the community's provision of resources to the experimenters.[12]

At the same time, many would have had some misgivings. Generally speaking. the 1950s and 1960s had seen unprecedented levels of investment in physical infrastructure across Japan generally. Plants, factories, and power stations promised jobs, customers, and greater levels of investment for the areas where they were constructed. This was particularly attractive in villages, because the postwar economic "miracle" had resulted in populations and capital bleeding off into the big cities. Yet these facilities could also be dangerous, as demonstrated by the mercury poisoning of as many as 15,000 people by industrial waste in Minamata, Kyushu. Worse, the people who suffered in such incidents often found justice elusive.[13] Unsurprisingly, therefore, resistance to big projects could be intense—protests of the Narita Airport project actually involved the death of ten people, including three policemen.[14] So effective did local resistance become that institutions often deployed campaigns of "soft power" in advance of commencing construction, to convince local political bosses, businesses, and citizens' groups of the benefits of their presence.[15]

Thus, it did not escape the attention of locals that the Michikawa testing range had, from the outset, several downsides. For starters, the facility was a scant 660

FIGURE 5.1. Early rocket test at Michikawa Beach.
Courtesy of JAXA.

ft (200 m) north of the local railway station, leading at least one engineer to express concern that trains would be arriving and departing while rockets were being launched.[16] Some homes on Michikawa Beach were so close that AVSA had to vacate them even for horizontal engine tests.[17] Local curiosity also became a liability. Although it was certainly encouraging to see large groups gathering to watch, the rocketeers were less pleased when some decided to observe from boats at sea, putting them at risk of being hit by debris.[18] The perils of this intimacy became painfully apparent on the night of May 26, 1962, when a two-stage Kappa 8–10 arced over the experimenters and crashed up range. Debris scattered over a large area, and started a fire nearby; fortunately, there were no casualties.[19] The same cannot be said of the phenomenon of "rocket play" (*roketto asobi*), in which local children tried to make their own rockets using bottles, gunpowder, and other bits and pieces. Between September 1967 and January 1969, four youths were injured doing this, including one person who blew their fingers off. Another child died.[20]

Yet of all the problems AVSA faced in Michikawa, the most prophetic was the group's relationship with fishermen. Their power and influence were so well known that Hideo Itokawa and his team sometimes met with them first thing upon arriving for the next round of launches.[21] A major area of concern were exclusion zones. Large areas of sea were forbidden to fishermen during launch tests, for fear they might be hit by debris. Even launches of the tiny Pencil had required exclusion zones of around 2,000 ft (610 m); by the testing of the third-generation Kappa rockets in late 1957, this had grown into a "fan-shaped" area 38 mi (61 km) wide, extending 40 mi (65 km).[22] Given that many fishermen in Akita refused to put to sea in a much larger area during tests, just to be sure, testing by AVSA was a huge inconvenience. Historians Daniel Aldrich and Martin Dusinberre have shown how reactions to the siting of "public bads" can have a tremendously divisive influence on local social structures.[23] Though the term "public bad" was usually applied to sites such as nuclear power stations or waste disposal sites, space facilities, with their noise, explosions, and disruption, could be expected to arouse some suspicion and resistance wherever they were built. By the 1960s, it was evident to Japan's rocketeers that even if they were not necessarily unwelcome, locals were concerned about negative impacts on their welfare and livelihoods. They were running the risk of becoming a "public bad."

It was in these circumstances that both ISAS and STA decided they needed to build permanent launch facilities somewhere else. Southerly locations are closer to the equator, and hence far better suited for launching satellites into orbit—the surface of the globe moves fastest at the equator, and the added speed dramatically reduces the need for fuel and engine power.[24] Luckily, Japan is extremely long. Okinawa (just 26 degrees north of the equator) was still under US jurisdiction, and thus out of the question. Both agencies eventually settled on the area at the southern tip of Kyushu—far enough away from major centers of population, without being too hard to get to. From here, rockets could be launched southward, over smaller islands with lesser populations, or out into the Pacific, away from Japan's touchy neighbors on the Korean peninsula. In February 1962, both *Yomiuri Shimbun* and *Asahi Shimbun* reported that Hideo Itokawa and his team had settled on a small tract of land near Uchinourachō in Kyushu for their new facility. Meanwhile, by 1966, the STA began laying plans for building and testing an LFR in Japan.[25] That April, it made the snap decision to construct its own new launch site on the island of Tanegashima, less than 63 mi (100 km) from Uchinoura.[26] Construction was to proceed on the understanding that the government would "avoid competition with the current facility in Uchinoura as much as possible."[27] For the next few years, however, it wasn't costs that would dominate the story of these facilities—it was fishermen.

FIGURE 5.2A. Launch complex at Tanegashima Space Center. Note the proximity of the facility to the sea.

Courtesy of JAXA.

FIGURE 5.2B. Rocket launch at Tanegashima Space Center.

Courtesy of JAXA.

Tanegashima, Uchinoura, and the "Fishing Problem"

ARRIVAL AND WELCOME

Once the decisions regarding location were made, both Uchinoura and Tanegashima underwent rapid development. Uchinoura's Kagoshima Space Center hosted its first AVSA rocket launch in August 1962, even before the official opening of the site in December 1963; seventy-two further rockets would be launched there in the next four years.[28] The facility added some 108,000 sq ft (33,000 m^2) between 1964 and 1966, and other expansions cost more than ¥400 million ($8.9 million in 2024) by the end of the decade.[29] As in Akita, large sections of the local population reacted positively to the facility's arrival; as one observer put it, "if it weren't for the cooperation of local people, space research wouldn't advance at all."[30] Locals in Uchinourachō formed a society to coordinate the facility's activities and development. One particularly strong source of welcome was local women. AVSA had several female assistants, mostly helping with menial work. In Uchinoura, the local Ladies' Association (*fujinkai*), headed by the formidable Kimi Tanaka, provided "heartfelt support in the form of paper cranes and cooperation in securing safety during the launch"—in addition to food, refreshments, and guidance.[31] The group even made up for a labor shortage by helping clear debris in order for roads to be constructed. Indeed, the relationships established between the space program and the women of Uchinoura were so significant that the *fujinkai* were among the first groups thanked during ministerial visits.[32] Kimi Tanaka became a fixture at ISAS events, becoming a sort of maternal figure to the rocketeers; some later recalled, for example, how she yelled out, "Thank you so much!" at the meeting where it was announced that Japan's first satellite would be named Ohsumi in honour of the area around Uchinoura.[33] Yasunori Matogawa, later head of the Kagoshima Space Center, also recalls having spent an "enjoyable time" with her during his tenure. Meanwhile, many members of AVSA were regularly hosted by another member of the Ladies' Association, Masako Hashimoto, at her own home (perhaps to prevent excessive carousing). Hashimoto also became something of a local celebrity, regularly interacting with students in the area to explain the origins of Japan's space program, and its arrival in their village. Her red-tiled house near the local primary school has become something of a shrine to ISAS's time in Kimotsuki, and still stands today.[34]

Part of the locals' enthusiasm came from the influx of tourists—4,000 people

FIGURE 5.3. Celebrations in Uchinouracho following the successful launch of Ohsumi, 1970.

Courtesy of JAXA.

gathered to witness the launch of the first small-scale rocket in 1962, a substantial number for a village of around 10,000.[35] In time, the village's reputation became intimately tied up with the launch facility, as shown by a 1966 petition to open a museum commemorating the activities of the rocketeers working there.[36] Another area that benefited was infrastructure. When Hideo Itokawa had first arrived in the area, a taxi driver had refused to take him up to see the site, because "the roads are so bad." At the time, Itokawa had simply commandeered the car and driven himself. Once construction began, however, the first part of the project completed was a 2,400 ft (735 m) access road winding up and along the hilly coastline.[37] This also happened to provide a much shorter route through the hills; bus journeys that previously took two and a half hours now took half as long. By 1968, a paved "rocket road" wound its way up along the coast, providing not only better access to the facility, but speedier transport for locals.[38] As in Akita, children also indulged in *roketto asobi*, though the arrival of the space age inspired them in other, less dangerous ways, too. Primary school fifth-grader Mikiko Kanadome, for example, wanted to be a stewardess one day—not aboard a plane, but on a rocket.[39]

Meanwhile, the STA's new facility in Takesaki, Tanegashima (which passed to NASDA in 1969), expanded even faster, to even greater size.[40] By 1969, a ¥1.1 billion ($27 million in 2024) "rocket boom" had overhauled much of the area's infrastructure.[41] A new port was built to handle ships of up to 6 million lb (3,000 T); 34 mi (54 km) of trunk roads were added; railways now ran between major locations on the island; and the port at the village of Minamitanechō was expanded with reclaimed land.[42] Tanegashima Space Launch Center would eventually grow to encompass some 28 million sq ft (8.6 million sq m), and become the mainstay of Japan's civilian space programs.[43] As in Michikawa and Uchinoura, many locals welcomed the facility, and the start of construction was greeted with banners and Japan's rising sun flag displayed over doorways.[44] In 1966, a festival in celebration of the new space center attracted so many people that the island's scant transport infrastructure was completely overwhelmed.[45] The outlook seemed bright for both space agency and locals. "If the construction of this space center proceeds well," one attendee observed, "it will be of great benefit to people in the locality."[46]

FISHING UNIONS AND ORGANIZED RESISTANCE

The abovementioned jubilation did not extend to the powerful fishermen's unions of Uchinoura and Tanegashima. The Uchinoura union contained some 240 members, a legacy of the fact that the economy of the area relied largely on the fishing of *buri* (Japanese amberjack).[47] They were supported by the larger, more powerful Miyazaki Fishermen's Union, which in turn made its presence felt even more fiercely in Tanegashima, farther south.[48] Miyazaki lies directly north of Kagoshima; its own fishermen were reliant on fishing grounds that fell in the Tanegashima and Uchinoura exclusion zones.[49] Particularly irksome to the fishermen was the fact that though the space agencies voluntarily limited launches to forty-five days a year, they had reserved the right to extend if they deemed it necessary—meaning that sometimes fishermen were kept out of the exclusion zone for up to three months.[50] This was in addition to the disruption already caused by US military training grounds, and a separate Defense Agency launch site in the area. Thus the arrival of the rocket facilities intersected with two explicit reminders of their association with war. The rise of rocketry in the communist world further heightened the sense of threat from the technology.[51]

These concerns arrived at a time when rural populations in Japan were growing increasingly concerned about the impact of large industrial facilities. In three separate incidents in 1956, 1961, and 1965, corporations, including chemical manufacturer Chisso, and Showa Denko, had caused major outbreaks of disease in Kumamoto, Niigata, and Mie Prefectures, respectively. Each incident was caused by the improper handling of industrial waste; each had required extensive mobilization by locals to secure justice; each had resulted in long-drawn-out legal proceedings, which in some cases weren't resolved for a generation.[52] Meanwhile, state developments could be even more arbitrarily damaging—in one incident in 1952, a government official summarily informed the population of Kawarayu in Gunma that a new dam would be constructed just downriver. In all, 1,200 people were to be displaced, more than 340 homes destroyed, and 50 acres of farmland flooded. Daniel Aldrich describes the aftermath:

> Homeowners who refused to sell faced expropriation of their land at unknown prices. Residents organized against the plan and managed to stall it but had difficulty drawing in regional or national allies and maintaining resistance over the four decades of planning that ensued. As of late 2007, relocation plans were being finalized for many local residents, and construction had begun on the main site. The state offered land to displaced families, but approximately half the village, frustrated with what they saw as higher prices for the "replacement" land, abandoned the area altogether.[53]

As a result of such experiences, many locals, while cheered by the possibility of economic benefits, were also keen to ensure that they extracted maximum benefit from the arrival of the space program, and had robust guardrails in place to preserve their health and livelihoods. Fortunately for them, the Miyazaki union was part of a collection of combative labor organizations in Japan with a long history of successfully slowing or preventing the construction of major facilities.[54] Its chief, Kazuei Fukuda, became one of the major political voices of the movement.[55] Bolstering the fishermen was support from Japan Socialist Party legislators, such as Sueo Kodama and Yoshio Miki, who spoke out against what they saw as the "infringement of fishermen's rights in the name of scientific progress."[56]

The resistance of fishermen can broadly be understood as developing in three stages: resistance, rambunctious negotiations, and finally reparations. Given AVSA's experiences in Michikawa, officials who supported the space facility in Uchinoura had laid some rhetorical groundwork even before construction began, pointing, for example, to greater demand for local seafood, thanks to the team and tourists.[57] After construction at the launch centers had begun, negotiations with the unions continued through a centrally coordinated team, in which Japan's various space entities pooled their resources. [58] Then came the major sticking point: the launch windows, which the space agencies were reluctant to compromise.[59] The demands the fishermen were making in return were deemed unrealistic. In Uchinoura, relations began to actively sour, and the fishermen began their campaign of launch interruptions.[60] In 1966 and 1967, despite at least nineteen visits from officials of the STA, protests from fishermen resulted in a delay to a planned launch of the Lambda-4S-3 rocket, in effect setting the Ohsumi project back by two years.[61] In response, authorities offered the unions compensation to the tune of ¥100 million ($26,000 in 2024). Despite this generating a temporary truce, fishermen soon stepped up their campaign, primarily by sailing into the exclusion zone at times when rocket launches were planned. Joining them, despite some internal divisions, were fishermen from Miyazaki.[62] A committee established to determine what had gone wrong was told that the union had received less than what it had been offered. Further attempts at compromise ended in failure, and the fishermen shut down activities at Uchinoura entirely for at least another eighteen months.[63]

Tanegashima fared little better. Villagers who lived nearby seemed mostly concerned about the potential use of the site to develop weapons, and worried that they might be forced to move as a result of construction and expansion.[64] Officials' belief that they could "obtain the necessary cooperation from local people" was soon dashed, as protests erupted in Tanegashima before the facility had even finished construction.[65] Again, objections came from fishermen, who dubbed themselves "victims of the rockets."[66] In addition to being annoyed that they had not been informed before construction began, they were also worried that the huge ships required to move construction materials would disrupt their fishing grounds.[67] In November 1966, the Miyazaki union's leader visited to lend his support; that same month, at a meeting with the government, local union members demanded a full inquiry into the impact of the facility.[68] Unconvinced by the response they received, they decided to act, and shut down work at the site from April 1967 to 1968.[69] So extensive was the disruption, and so dim the prospects for resolution, that ISAS even permitted NASDA to launch two rockets from Uchinoura—a rare show of solidarity between the two space institutions at the time.[70]

SCHISM, NEGOTIATION, AND SETTLEMENT

The fishermen's activities were, however, far from universally popular. In places like Uchinourachō, in the absence of launch activities, many tourist-reliant businesses went into the red.[71] Some villagers continued to lend their support to personnel working at the space sites, despite the fishermen's objections.[72] In Minamitanechō, protesters sometimes faced hostility from their own neighbors, who questioned why they would interfere with the "policies of the national government" and opined that the protesters needed to "shape up."[73] Others dubbed the protesters "interlopers" and "uninvited guests," possibly hinting at the fact that the protests were largely organized from Miyazaki.[74] In 1968, the mayors of both Uchinourachō and Minamitanechō gave a joint interview emphasizing the importance of the facilities to their local economies and infrastructure.[75] Kōtarō Hamada, mayor of Minamitanechō, was particularly strident, suggesting that the fishermen's "gloomy" prognostications about disruption to the local economy were wrong.[76] One protester fired back that "such an attitude is largely for [Hamada's own] convenience."[77] Nevertheless, Hamada remained one of the facility's biggest local supporters, dismissing the fishermen's protests as a "public nuisance," and gleefully attending the first launch of rockets after work resumed there in 1968.[78] Many in Minamitanechō were similarly delighted when work resumed on the site.[79]

These tensions spread up the political ladder, as both sides pulled supporters into the confrontation. ISAS, the STA, NASDA, and other entities had pulled together to coordinate their response, but lawmakers became increasingly concerned about their inability to resolve the matter. Some worried about delays to the Ohsumi project; others fretted about the impact on rocketry R&D generally. A central bureau was established to examine the issue, involving the cabinet, the prime minister's office, the deputy chief cabinet secretary, the director general of the Radio Regulatory Bureau, and the director general of the Fisheries Agency, and "all relevant ministries." The governor of Miyazaki joined negotiations, then the governor of Kagoshima Prefecture. The combined pressure of this network was such that in early 1967, even as the shutdowns at Uchinoura and Tanegashima were well underway, politicians expressed hope that the Miyazaki Union had "at least begun to understand the importance of rocket experimentation."[80]

A breakthrough finally seemed to arrive in August 1968, when the Miyazaki Union was persuaded by an offer consisting of subsidies for fishing infrastructure, help finding new fishing sites, and a hard limit of no more than ninety days of launches annually.[81] Local groups demanded that they receive extra compensation—more, that is, than fishermen from faroff Miyazaki—due to the fact that the inability to fish affected them disproportionately. Thus, the union in Minamitanechō was also offered direct compensation to fishermen, a ¥14 million ($331,000 in 2024) lodging for them (complete with built-in hot spring) in nearby Yamakawa, and a collection of small fishing ports built prior to the commencement of the planned 1969 rocket launch season.[82] It was enough to persuade the local unions to hold off their protests for a little while.[83] One quibble remained, however—the unions wanted subsidies to help buy boats with displacements of over 154,000 lb (70 tonnes), which would help them reach fishing grounds farther away during launch windows.[84] Central authorities soon acquiesced, and began surveys on how the infrastructure for such ships could be built on the island.[85]

By late 1969, a subtle shift in rhetoric indicated that sentiment among the resistance had begun to soften. After a Lambda launch failed, one fisherman pointed out that such incidents meant that his day's fishing had been abandoned in vain. It was also, he opined, "embarrassing before the rest of the world.[86] A long-term settlement was finally reached in 1970.[87] Over the next two decades the relationship between local fishermen and the space center gradually morphed into one of cooperation. [88] By the 1980s, ISAS functionaries were even recruiting local fishermen to collect the upper stages of rockets that had fallen into the ocean.[89] In return, ISAS and NASDA continued to consider the fishermen's priorities when planning their activities; the timing of launches for two probes designed to reach Halley's Comet in 1985, for example, required extensive consultations to get the timing right.

Resistance had largely petered out by the 1980s, and the facilities seem to have avoided the designation of "public bad." On one hand, this was due to the incentives offered to locals to tolerate the rocketeers' presence. The public baths (*onsen*), subsidies, and facilities had mollified them to a great degree, even if from the perspective of the space programs, it was the launch windows that were the greatest concession—the restrictions these placed on the timing and frequency of launches would go on to be a serious consideration in rocket development. On the other hand, it is also a sad truth that as time went on, there were simply fewer people with the will and inclination to resist in these areas. Uchinourachō's population dropped sharply after the arrival of the facilities, declining from 11,792 in 1955 to just 3,215 by 2015—a 73 percent decline. Minamitanechō's population in just the five years following the space center's arrival (1965–1970) dropped by 20 percent. Those who remain are older: in 1965, only 6.5 percent of Minamitanechō's residents were over sixty-five; by 2015, this figure was 33.1 percent.[90] What these statistics reflect is the departure of young folk for large cities; whatever economic benefits the space centers brought were short-lived.[91] Those who remain, the elderly who had fought for fishing rights in the 1960s and 1970s, have dwindled. This presents another stark contrast with space programs around the world—whereas the arrival of space facilities in the Soviet Union, the United States, and in French Guiana had all resulted in population and economic booms for their localities, Japan's proved insufficient to stem the tide of rural decline.[92] Arguably, therefore, the plucky fishermen who sailed into the exclusion zone off Kimotsuki's coast in 1968 were smart to extract as much as they could from the authorities, while they could. The rocket facilities would certainly outlive them. What they could not have expected is that they might outlive their hometowns, too.

In the 1950s and 1960s, Japan's civilian space programs had to negotiate with local populations to construct essential infrastructure. Although many people in areas like Michikawa, Tanegashima, and Uchinoura welcomed the investment and excitement these facilities brought, others were less sure. Some were motivated by suspicions about the arrival of large-scale industry in their neighborhoods more generally. Others, however, had specific complaints about rocket launches, and no group was more vociferous or effective in their resistance than fishermen. Angry that they were not consulted before construction began and concerned that exclusion zones limited their ability to work, they staged an effectively organized protest in the late 1960s that stopped Japanese rocketry research in its tracks. Their cooperation was only secured through cajoling, subsidies, and voluntary limitations to the timing of launches.

The space programs' experiences on Kyushu and Tanegashima were to be singular. Henceforth, space facilities would receive a much warmer welcome across the country. Locations such as the ISAS Kakuda facility in Sagamihara were fêted with local celebrations, and anniversaries are highlighted with a variety of souvenirs, including, on at least one occasion, a locally issued specialty frame holding space-commemorative stamps.[93] This acceptance is at least partly attributable to the fact that after launch facilities were constructed in the south, space sites elsewhere tended to be research campuses or observation centers—places that were not associated with the noisy and dangerous consequences of actual launches.

PART III

The Challenges of Advanced Spacefaring, 1980–2003

SIX

Disseminating and Debating Japanese Space Policy

IN AN INTERVIEW WITH *Asahi Shimbun* in 1977, NASDA's Hideo Shima had a few words for those questioning why Japan's progress in developing an LFR wasn't going any faster. The breakthroughs the country had achieved, he argued, shouldn't be diminished by comparison to missions like Apollo. Even if they weren't as spectacular, projects like the recent engineering satellite Kiku represented critical milestones: "There is no shortcut to technology." Developments were ongoing, even when it looked like nothing much was happening. Rushing things by simply buying advanced overseas components, rather than learning to manufacture them locally, should be avoided; Shima compared the idea to "saying, if you have trains, you don't need cars." Rather, "if there's technology more advanced than yours, it's important to understand it, and learn how to use it."[1]

The impatience of Shima's audience was in keeping with the times. As mentioned in the previous chapters, Japanese economic growth between 1955 and 1973 tangibly improved the quality of life of its population. It also resulted in a surge of capital, industrial, and infrastructural growth.[2] By the 1980s, at least in the eyes of Sony's founder Akio Morita and future Tokyo mayor Shintarō Ishihara, Japan was ready to move past the postwar order. In their 1989 book, *"No" to Ieru Nihon* (*A* Japan That Can Say "No"), Morita and Ishihara argued that the country's recent success was due to the superiority of Japanese culture and national character. The country would soon join the ranks of world superpowers, they argued, and so needed to be more assertive in promoting its interests.[3] This included space: after all, who had ever heard of a great power without a space program? On the other end of the spectrum was a new generation for whom television had normalized the idea of space travel. Sputnik was a creature of radio, with its iconic stuttering beep broadcast around the world. Neil Armstrong's first steps on the Moon, however, had been broadcast (as close as possible to) live on TV, a full year before Ohsumi's launch in 1970. Just as influential was the popularity of science fiction TV programs such as *Star Trek* and *Space Battleship Yamato*. Not only did coverage of space exploration become more frequent, but it also involved more spectacle. Footage of spacewalks from both the USSR and the United States dazzled viewers, while Space Shuttle launches soon displaced mere rockets on TV. By this point, Japan's space programs could point to major accomplishments in the form of infrastructure including two launch facilities, a booming (if coddled) satellite industry, and SFRs capable of putting small satellites in a variety of orbits, alongside the fact that they were making headway on LFRs; in 1975, Mitsubishi completed the first N-series rocket, capable of putting up to 1,600 lb (730 kg) into geostationary orbit (GEO). Yet these achievements encountered a public weaned on the pioneering successes of Ohsumi. It was far more difficult to inspire interest in the far more technical and piecemeal development of technologies in the 1970s and 1980s; just as NASA found that "first man on the Moon" was a more exciting proposition than "first reusable space plane," so too did Japan's space programs find the "first three-axis stabilized satellite" a far harder sell than "first satellite." The excitement of new space facilities being constructed had faded also; by the time Shima spoke to *Asahi Shimbun*, the civilian space programs and their industrial base had become a well-established part of Japan's technological landscape. Worse, there was a persistent belief in the 1970s and 1980s that Japan was slipping behind communist China in its exploration of space technologies.[4] On one hand, this further heightened fears from the 1960s regarding ICBMs, and situated rocketry ever closer to warfare in the eyes of the public. On the other hand, the Japanese press's conviction (a few decades early) that China would put a man in space soon was seen as symbolizing a more general slippage of Japan's status with respect to its giant neighbor.[5]

Japan's space institutions had long committed to countering these perceptions wherever they could; by the early 2000s, even the Space Communications Corporation's (Uchū tsūshin kabushiki kaisha) Superbird satellite had its own publicity officers.[6] Entities such as the Japan Federation for Commercial Satellites (JFCS) published semispecialist publications such as *Space Japan Review*, featuring interviews, insider perspectives, and accounts of recent successes; ISAS, NASDA, and other space agencies also produced their own newsletters, with content for both specialists and casual observers.[7] As figures like Hideo Shima departed the space program (Shima retired in 1983), the responsibility passed to a new generation of manager-specialists to keep up the work. This new cohort differed in several key ways from their forebears. For starters, few had experienced World War II as adults. Some, such as Ryōjirō Akiba, had been in their teens during the conflict; others, such as Yasunori Matogawa, were born during it; still others, such as future JAXA vice president Kiyoshi Higuchi, were born afterward. Perhaps for this reason, and because of Japan's growing confidence on the world stage, this younger generation were more willing to question (but not reject) elements of Japan's doctrine of "peaceful usage." Another significant change was the space program's gender balance, as more women began to enter the ranks of engineers and specialists. The overwhelming majority of these, too, were born in the optimistic decades of the immediate postwar period. Chiaki Mukai, Japan's first woman in space, was born in 1952; educated as a doctor, in 1985 she flew to space aboard the Space Shuttle Endeavour.[8] Naoko Yamazaki, born in 1970, was a space specialist from the very beginning—much like Itokawa, Matogawa, and Akiba, she graduated from the University of Tokyo in 1993, and joined NASDA in 1996.[9] Increasingly, there was room alongside her for people from other institutions, too; Kiyoshi Higuchi graduated from Nagoya University.[10]

Just like their forebears, this new generation was well aware that "justifications are central to space narratives because they preemptively try to insulate discussions about space travel from critics both internal (i.e., domestic and institutional) and external (i.e., international and public)."[11] As mentioned previously, space funding in Japan (particularly for ISAS) was always limited; on the eve of Ohsumi's launch, Japan's expenditure on space research lagged behind even Italy's.[12] Given that the connection "between space exploration and the nation . . . endured both in reality and in perception," ensuring support among the public and policymakers was thus crucial to maintaining whatever little funding they had.[13] To accomplish this, Japan's new generation of space manager-specialists undertook outreach activities focused on a clutch of key narratives. First, they generated a mythology for the space program, centered primarily on the figure of Hideo Itokawa as the founding father of Japanese space research. Second, they redoubled efforts to connect space technology to the everyday lives of Japanese citizens.[14] Third, influenced by Japan's growing confidence on the world stage, manager-specialists emphasized not only the raw technical accomplishments of their work, but also its global significance, particularly in areas such as disaster management and Earth observation. And lastly, particularly from the 1990s, they were vocal in questioning the country's commitment to "peaceful usage"—a discursive turn that contradicted their forebears, but reflected changing priorities in the Japanese population.

Hideo Itokawa and the Myth of the Founding Father

Russia's space histories laud figures like Sergei Korolev and devices like Sputnik; in the United States, figures like Neil Armstrong and the Apollo lander receive similar treatment. As historian of science Slava Gerovitch puts it, such tales are a species of "collective memory—culturally sanctioned and publicly shared representations of the past," which provide "narratives through which individuals publicly describe themselves, remember the past, and interpret the present."[15] The idea of a "founding father" is an essential element of such space mythologies because one of their purposes is to "reinforce deterministic explanations for space history (for instance, "Korolev did X, therefore the Russian space program is like Y")."[16] For Japan, this figure was, perhaps inevitably, Hideo Itokawa. Key to understanding the myth of Dr. Rocket is the fact that his public reputation was actively generated and tended to by both himself, and his successors.[17] The work they produced generally avoided hagiography, but did contain two notable characteristics: an emphasis on Itokawa's personality, which served to both distance him from his wartime roots and emphasize his extraordinary characteristics as a man; and a tendency to confer on Itokawa a generalized intellectual authority, which took his successes in the field of rocketry and engineering, and extrapolated their insights across a variety of other topics. This, in turn, emphasized that engineers are specialists, whose extraordinary insights extend well beyond the remit of their chosen technology.

For starters, discussions of Itokawa's motivations played an important role in distancing him from his association with war. Biographers acknowledged that "Itokawa is properly criticized for being a participant in the war," but pointed out that his writings do not necessarily give readers the impression of someone actively *enthused* about war.[18] Others completely dissociated his wartime work from his postwar legacy. In 2003, for example, Japan launched the Hayabusa probe to an asteroid named for Dr. Rocket, 25143 Itokawa. "Hayabusa" was also the name of

Itokawa's most notable contribution to wartime aviation—the Nakajima Ki-43, a warplane. Despite the obvious parallels, however, official sources maintained that the two names "have nothing to do with each other."[19] Diverting attention to particular areas of Itokawa's activities (and not others) is also evident in discussions of his personality. Works such as *Ningen Itokawa Hideo hakase to wa* (Dr. Itokawa Hideo, the Person; 2003) featured anecdotes from a variety of people who had worked with or known Itokawa.[20] Their claims are impressive: Itokawa was moved up several grades in school on account of his sheer brilliance; Itokawa could finish an hour-long exam in fifteen minutes; Itokawa chose aeronautical engineering simply because his brother said it was the most difficult course at Tokyo Imperial University. Haruo Niiyama, an early member of AVSA, claimed that he "never once heard Itokawa say 'I'm sorry' "—a truly staggering revelation in the Japanese context. Instead "he would always respond with 'So, do you mean this?' and proceed logically." Biographer Daizō Kusayanagi attributes this to Itokawa's doctrine of *jikō kōchiku* (self-creation), a philosophy that emphasized the ability of the individual to transcend their circumstances and inclinations by committing to programs of self-improvement. It was inspired, according to Itokawa himself, by the realization at age twelve that his bookishness would lead to a sedentary and unhealthy lifestyle. He immediately committed to physical activities, like bodybuilding, and sunbathing; the improvements to his health convinced him of the power of the will to overcome the vagaries of circumstance.[21]

For many of these writers, these works served a dual purpose. By moving Itokawa away from the business of death, and emphasizing his personal genius, they could use Itokawa to exalt rocketry; after all, if he was essentially a peace-loving genius, wouldn't it be believable that the rockets he designed were peaceable, too? Personal association also helped create a clear line of succession, thereby presenting Japan's space institutions and their managers as heirs to Itokawa's greatness. This approach is evident in the work of one of his most important and articulate students, Yasunori Matogawa, one of the major figures of late twentieth-century Japanese astronautics. Matogawa recalls when he and his family first heard of Itokawa's activities: while his mother and father were "deeply moved," for him it was a revelation. "Wow," he recalls thinking, "Japan's started making rockets too!"[22] He subsequently encountered Itokawa in person while studying for his master's degree at Tokyo University; "it was the beginning," he wrote in 2006, "of an education that lasted, roughly, over thirty years."[23] This potent combination of institutional seniority and social intimacy meant that Matogawa's writing could turn the history of Japan's space program into a profoundly personal one. For example, here is his description of Itokawa's experience during the occupation, from a 2007 piece in *Acta Astronautica,* the official journal of the International Academy of Astronautics:

> After Japan was defeated in the Pacific War, research on airplane *[sic]* was prohibited, and Itokawa was at a loss because he lost all motivations *[sic]* to live on, but he at last recovered from the shock . . . [and] made up his mind to take the initiative of rocketry in Japan. This determination cultivated the way towards space efforts in Japan thereafter."[24]

Here, we see in its most essential form, the fundamental component of the Itokawan foundational myth: that without Dr. Rocket, there would be no Japanese civilian space programs. Similarly, in many ways, without his successors, there would be no Hideo Itokawa.

The exaltation of Itokawa by his successors, however, was not the only way in which Dr. Rocket reached a position of public intellectual in Japan. Itokawa had a roving mind, and produced work on topics ranging from the cello to ballet, sports, Japan's education system, and the role of women in society. A significant portion of this was Itokawa applying a systematizing approach that took engineering insights and applied them to nonengineering contexts.[25] In doing so, he raised the technocratic credibility of his profession: if aeronautical engineers could design both a better engine and a better education system (for example), then they truly were some of the most valuable intellectuals in Japan. By 1969, observers were suggesting that Itokawa's view of management and systems "furnishes modes of thought and methodologies that must be incorporated into Japan's industry, technology, education, communication, and everyday life." At AVSA, for example, he had devised a structure that allowed the institute to draw on a variety of specialists from both private and state entities to create ad hoc problem-solving groups. This approach, writers argued, could be readily applied to instances where industrial issues required quick resolutions—for example, when pollutants from a factory were being expelled into a local water supply.[26]

After his departure from ISAS, Itokawa ranged even further in his writing into topics beyond the ken of space engineering. One long-held area of interest was comparative sociology. Prior to Japan's surrender, Itokawa had written a pair of articles for *Yomiuri Shimbun* on January 1 and 2, 1944, analyzing the main points of difference between the Allies and the Axis powers. The rising powers of Japan and Germany shared "single-mindedness." However, "the West has its own traditions," based on the influence of the Bible—which is why, for example, a student who was paralyzed while attempting to rob his own school in America successfully sued the institution for his suffering. If such a judgment seemed strange to the Japanese, Itokawa opined, they should keep in mind that it was a similar inability to understand Western culture that had caused Japan's reverses during the ongoing war. Victory would thus require not just a close analysis of the West's military capacity, but a proper understanding of their society as well.[27] Such themes were still apparent nearly half a century later, in *Nihon sōseiron* (On Japanese Creativity, 1990), which ran the gamut from philosophy to international relations to a discussion of a special edition of a German magazine on "Japanese women" that had caught his eye in Switzerland. In between, Itokawa speculates that if Cambridge University's organization and educational policies were replicated in Japan, the country could also consistently "create figures like [Isaac] Newton, [Paul] Dirac, and [Stephen] Hawking." In keeping with his already well-established role as a mediator of foreign knowledge and knowhow, he also takes time to explain Cambridge's inner workings, such as its college system and "town" versus "gown" divide ("they don't really get along").[28] Nor was his application of "rational" principles limited to human behavior: his postwar musical studies were inspired by a determination to design the "best violin" ever made.[29] Itokawa's interests were so broad that one writer eventually dubbed him "Japan's Da Vinci."[30]

Yasunori Matogawa and the Promotion of Space Research

As mentioned above, one of the key figures in promoting space research in Japan in the latter half of the twentieth century was Yasunori Matogawa. Born in Hiroshima Prefecture three years before the end of World War II, Matogawa initially dreamed of playing for the Hiroshima Carp baseball team.[31] He completed his graduate studies at Tokyo University in 1970, and joined ISAS in 1981, where he became a senior member of the team building two probes that would inspect Halley's Comet, Sakigake and Suisei. He was subsequently appointed head of Kagoshima Space Center in Uchinoura, vice president of the International Astronautical Federation, and eventually, head of ISAS. After retirement, he served as associate executive director of the Japan Aerospace Exploration Agency (JAXA), and as director of JAXA's Space

Education Center. With a reputation for efficiency and skill in balancing the interests of Japan's various space stakeholders, Matogawa was arguably more important to Japan's civilian space programs than Itokawa, in that he not only stayed in the industry for much longer, but spent a good deal of that time in positions of power. Small wonder, then, that he had an asteroid, 6526 Matogawa, named for him a full eleven years before Dr. Rocket got his own.

Matogawa's numerous contributions to popular journals and newspapers continued his mentor's tradition of relaying detailed technical information to the public in an accessible and engaging style. Between 1984 and 1999, he produced at least nineteen books on topics related to space exploration. Works such as *Uchū kaihatsu no ohanashi* (The Story of Space Exploration, 1991), *Otona no tame no hontō no uchū no hanashi* (Real Space Tales for Grownups, 1992), and *Uchū de kurasu tame ni 69 no chishiki* (Sixty-nine Pieces of Fundamental Knowledge for Living in Space, 1999), were explicitly targeted at the general public.[32] Complementing these were hundreds of articles in mainstream newspapers, trade publications, and other professional forums, as well as regular columns like "Matogawa kyōju no uchū yomoyama hanashi" (Professor Matogawa's Miscellaneous Space Tales) in publications such as *Sekai Jōhō*.[33]

Much like Itokawa, Matogawa wrote in accessible, jargon-free prose. In some pieces, such as *Nihon uchū kagaku no saikin no kekka* (Japanese Space Science and Its Recent Results, 1994), he provides an overview of current Japanese activities alongside analyses of high-tech objects like the Asuka X-ray observatory satellite.[34] He also folds in humanizing insights into his colleagues, particularly Itokawa.[35] As an insider, he is able to provide accounts of behind-doors discussions within the space program. One example is an account he provides of AVSA's debate over the naming of vehicles that succeeded the Pencil. Suggestions apparently included "Momijigari," the poetic term for watching leaves fall in the autumn, the sort of name Matogawa himself might have favored—why shouldn't Japan name its own rockets after local deities, "names like Susanoo-2 and Izanagi-1"? he wondered.[36] His technical knowledge, and his ability to explain it to a lay audience, is evident in his writing about the failure of an M-V rocket launch in 2000.[37] His position also afforded Matogawa the opportunity to put a positive spin on such mishaps; as head of ISAS, for example, he made appearances in the media explaining how the M-V rocket, despite being only twice the size of preceding devices, needed to be three times more powerful—and this in turn meant it was far more difficult to get the technology right.[38] Lastly, his insider knowledge enabled him to provide details on canceled missions like the Lunar-A probe, and failures such as the 1998 Nozomi Mars probe.[39] Similarly, as Japan's massive H-2 rocket project faced difficulties in the 1990s, Matogawa was often called upon to explain in a comprehensible way what exactly was going wrong.[40]

Matogawa also took on Itokawa's role as an important source of information on overseas space activities.[41] Here, he could be a voice of comfort, assuring people, for example, that there was no real likelihood that the Cassini probe's 72 lb (32.7 kg) plutonium power source would spread radiation over the surface of the Earth in the event of a failure.[42] Russia was one of his particular interests, a fascination dating back to Yuri Gagarin's epochal journey to space; when post-Soviet Russia was selling off its space equipment, Matogawa correctly predicted a "revival" of Russia's space program in the coming decades as a launch provider.[43] In 1994, evaluating experimental reusable craft such as the McDonnell Douglas DC-X Delta Clipper, he correctly predicted that their practical use was still some way off.[44] Conversely, Matogawa was also one of the Japanese space programs' chief representatives abroad. In a series of talks at the International Astronautical Congress between 1996 and 1999, for example, he covered topics from wartime rocketry development in Japan to rocket festivals in the countryside.[45]

Matogawa also had an eye for garnering attention. His ideas could be as off-the-wall as his mentor's—such as his suggestion that a commemorative whiskey be issued in honor of Japan's participation in the 1984 Halley Fleet.[46] His outreach included events like *Asahi Shimbun*'s "Easy Science" (*yasashii kagaku*) workshops aimed at middle- and high-schoolers, which he joined in 1990.[47] He regularly gave public lectures and participated in roundtables on topics ranging from astronomy to the history of space exploration, in places ranging from planetariums to the halls of power; the transcripts of some events were published, in a Japanese tradition, in the *Science and Technology Journal*.[48]

The Everyday Benefits of Space Research

Matogawa was just one of many of a new generation of manager-specialists who were writing about the space program for the public. Nearly all of them concurred on communicating just how much space technology had improved the lives of the average Japanese person; from the 1990s onward, they also emphasized that a growing space industry meant more jobs.[49] Others emphasized that space exploration and infrastructure accrued "great profit to humankind" as a whole.[50] Two major narratives appeared in their work: first, that space research was a justifiable source of national pride; and second, that space exploration would contribute to Japanese safety and prosperity, both now and in the future.

NATIONAL PRIDE

In the 1980s and 1990s, defenders of Japan's space programs, modeling overseas outreach, foregrounded the notion that creating satellites, rockets, and other space paraphernalia was representative of Japan's national dynamism.[51] Some simply went for the emotional jugular, emphasizing the aspirational elements of space travel—particularly in the 1990s, as Japan's long postwar boom ended, and the mood turned gloomy.[52] Many of the new generation who had entered Japan's space programs, such as ETS-VI/Kiku designer Hiroshi Sakamoto, had been inspired by the visceral experience and wonder of spaceflight.[53] For folks like him and satellite specialist Susumu Kitazume, the "flash of wonderful orange light" at ignition, and rockets that then "rose with a roaring noise" were all part of the allure.[54]

Others instead argued that all of this was, in a way, Japan's birthright. Hideki Mizuno, editor of *Space Japan Review*, pointed out that Japan's rocketry technology in 1945 had been at "one of the highest levels in the world," implicitly positing that successes such as Ohsumi were only a continuation of an earlier—interrupted—trajectory of accomplishment.[55] World-class innovation had been inherent to Japan's nascent space program, despite its small size; AVSA, for example, had pioneered the first mostly plastic sounding rockets.[56] It did not matter that subsequently, Japan was largely following in the footsteps of the Soviets and the Americans; all this served to do was make up for the country's late start.[57] In publications such as *Space Japan Review*, "Space Japan Milestones" articles vaunted the successes of numerous experimental Japanese technologies, from successfully capturing images of Earth to testing the H-2A rocket in Tanegashima.[58] The writers were also keen to point out that, in some ways, Japan had already equaled the superpowers: in 1985 Sakigake, for example, became the first interplanetary craft launched by any country other than the United States or USSR.[59] In the 1990s, Japan's efforts in "following stars from their birth to their dramatic death" through satellites such as Hinotori (1981), Tenma (1983), and Asuka (1993) made the country (some claimed) the "leader" in the observation of high-power waves in space.[60]

Nor were Japan's unique contributions limited to rockets and satellites. Nobuyoshi Fugono, former head of CRL, recalls how once, on his way to a meeting with colleagues from the United States, he complained to a colleague that he didn't have anything interesting to talk about. His colleague suggested talking about his work on atmospheric dynamics, but to use the phrase "space weather"—Americans tended to like snappy phrasing. In the end, both the subject matter, and the term, were a "home run" with his audience.[61] One observer even proudly argued in 2003 that Japan was now a world leader in radar, a technology in which it had notably lagged during World War II:

> for radars, the National Institute of Information and Communications Technology (NICT) has a wealth of research experience and has developed various radar observation instruments. JAXA has the knowledge to develop the most advanced functional equipment in space and the ability to put together the right team to produce needed technologies and observation equipment. JAXA will take the lead in creating satellite technologies to make all Japanese people proud.[62]

Other writers emphasized the connection between Japan's international status and its responsibility to "serve mankind" with technological "progress." Shigefumi Saito, high commissioner of the Space Activities Commission, opined in 1989 that "space is recognized as the common frontier of all humanity," and that Japan's exploration of it was a responsibility to the world.[63] Yasunori Matogawa presented a version of this argument in 1994, in "Uchū kaihatsu no rekishi to Nihon no yakuwari" (Space Development and Japan's Responsibility), declaring that "at a time when Japan's international responsibilities are growing," its space R&D required commensurately increased funding. In particular, human space travel had to be normalized; not to do so would, by implication, mean that Japan was both failing to maintain its status as a leading space power, and shirking the global responsibilities that come from being a member of this elite club.[64] Hideki Mizuno ended his memoirs in *Space Japan Review* in 2003 by praying that "our country . . . [does] not stop leading the international community in terms of technological development."[65]

Just as the civilian space programs could be used to present nationalistic narratives, nationalism could also be deployed in defense of the civilian space programs. Japan's rocket program came of age in the 1980s with the design of the H-1 and H-2 LFRs, but writers about space technology faced a new dilemma: these rockets had primarily been made with satellite launches in mind, and there were no large-scale research projects like Apollo or Mir that would utilize them. The public could only get so excited about the launch of yet another communications satellite. Writers in this period thus developed a tendency to emphasize the *Japaneseness* of the technology. Yasunori Matogawa pointed out, for example, that AVSA's efforts at Tokyo University were "unparalleled in the world" in having "satellite and rocket groups working and cooperating intimately" throughout the process, whereas in nearly all other countries, these two groups were separate. Similarly, he pointed out that at ISAS's Sagamihara Space Research Center, students from the science department and engineering department worked and studied together, and that "there are no other examples of this [institutional structure] elsewhere in the world." All this was because—again, uniquely—Japan's space rockets were designed for science and science alone. This promoted an "intimate" (*gacchiri*) relationship between scientists who developed payloads and engineers, one completely unlike the crypto-military work of the superpowers.[66] On a less serious note, the 1994 "space food" competition that Yasunori Matogawa judged was initiated to celebrate the first trip to space by a Japanese woman, Dr. Chiaki Mukai; the ultimate goal was that Dr. Mukai would, in fact, eat these dishes in orbit. "There is no example," one newspaper proudly reported, "of publicly inviting a space food to be placed on the Space Shuttle in Japan, or in the history of space development in the United States." The winners included recipes from a junior college student in Toyama (steamed soybean *gomoku* rice), a fourteen-year-old in Yokohama (*takikomi* rice with vegetables), a forty-three-year-old housewife in Saitama (*niku-jaga* stew), and a thirty-year-old office worker from Hyōgo ("space Takoyaki" octopus balls), a spread indicative of the broad demographic that such initiatives reached.[67] Yasunori Matogawa was one of the judges.[68]

SPACE TECHNOLOGY AND HUMANITY'S "BRIGHT" FUTURE

In his seminal 1961 work, *Japan's New Middle Class*, historian Ezra Vogel describes the rise of the concept of "bright life" (*akarui seikatsu*) in postwar Japan, characterized by "leisure time, travel and recreation, and few binding obligations and formalities." Central to this new way of living was convenience, as epitomized by consumer goods like refrigerators, toasters, washing machines, and electric fans.[69] In the decades that followed, rice cookers, vacuum cleaners, televisions, cars, air conditioners, and houses had joined this pantheon as well.[70] Supporters of Japan's space program were keen to emphasize the contribution of the space program to the "bright life" as well. Events such as the 1980 reunion of a man who had returned from Manchuria without his family in 1946, with his long-lost son in China—orchestrated via a space relay by TV Asashi—effectively communicated to the public the potential offered by satellite communications.[71] Figures in Japan's satellite industry tried to capture the public's imagination through projects such as placing the first pay phone on Iwo Jima in 1993, so that the JSDF soldiers stationed there could "make a call for the same price as a public phone in Tokyo."[72] A year later, Yasunori Matogawa pointed out that transport uplinks, ship guidance systems, and communications satellites had become integral to modern logistics, and that none of this would have been possible without rocketry. These developments were presented as having a huge impact not only on "earth observation and space observation," but on "human life."[73] The use of satellite technologies like GPS, which meant that supply chains and drivers alike could function more safely and efficiently, was often presented as one of the great tangible benefits of space research.[74]

The impact of space infrastructure on disaster management was also a significant part of this charm offensive. The Kashima Space Research Center, Ibaraki, which focuses on communications R&D, regularly dedicated publicity materials to detailing its contributions to relief efforts during earthquakes and other natural disasters.[75] Indeed, satellites such as Midori (1996) were designed and launched partly to complement Japanese disaster management. As more devices, including privately manufactured ones, entered orbit, Japan's space institutions pointed out that it wasn't just scientific satellites that could help in times of national crisis. The ETS-VI Kiku satellite in 1994, for example, despite not quite functioning properly, pioneered portable satellite phones; the uplink devices developed for use with it were retooled for other satellites and deployed after the Kobe earthquake in 1995, delivered by people riding bicycles some 19 mi (33 km) from Osaka.[76] The fact that all of this was accomplished at a lower cost than overseas competitors' technology was also a source of pride. *Space Japan Review* often drew attention to Japan's money-saving space manufacturing processes, such as installing off-the-shelf components on microsatellites.[77] Well into the twenty-first century, publicity for an annual amateur rocket festival held at the former Noshiro launch range mentioned that Japan produced "cutting edge" technology by efficiently using "fairly limited funds and human resources."[78]

The "bright life" of the 1980s was characterized by growing confidence in the population, and many communicators appealed to the same human instincts in the public that had brought *them* to the space program in the first place. Yasunori Matogawa pointed out that just having such aspirations was critical in encouraging new generations of engineers to enter Japan's space programs.[79] The corollary to this outlook was that Japanese space communicators were also able to entice the public with visions of accomplishments yet to come, and of places they hadn't been to yet. His mentor Dr. Rocket was keenly involved in this strain of futurism, writing about hypothetical light-driven rockets and other future technologies.[80] "Space dreams" such as a space elevator, he argued, weren't meaningless speculations, but preliminary work on an "ideal transport technology."[81] Meanwhile, Matogawa appealed to the sort of dream that had motivated Tsiolkovsky: "Don't most people," he wondered, ". . . want to go see what's *over there* . . . at one time or another?" [82] Producers in the commercial satellite industry similarly foregrounded the cutting-edge nature of their own work.[83] These narratives were, as William Wray points out, "idealistic," and blurred the line between contemporary technology, and future wonder.[84] In fact, the notion of humanity ultimately expanding beyond Earth was a regular theme in Matogawa's writing. In 1994, he argued that with the decline of Russia, and crises in Europe and America, it was up to Japan to "connect" humankind to a future in space.[85] This vision achieved an eschatological significance in his 1996 piece, "Issho ni uchū e—jinrui shinka no akutei ni okeru yūjin hikō" (Together to Space: Manned Flight in the Course of Human Aviation), which takes the impressive step of linking humans in space with ancient fish in the shallows. Matogawa points out that our aquatic ancestors faced the same challenge as our more proximate ancestors in East Africa: either fade along with their habitats, or "find new ways of living appropriate to the changing world." "For the fish in drying rivers, there was the land [to flee to]," observed Matogawa. "For monkeys in the shrinking forests, there was the savannah. But now, for humans, who have spread to every nook and cranny of the lands . . . there is no place to 'flee' to apart from space."[86]

Challenges to "Peaceful Usage"

Of the major narratives and principles associated with Japan's civilian space programs, one of the most enduring had been the 1960s commitment to "peaceful usage." This was enshrined in the 1969 law creating NASDA; a detailed official analysis of its creation emphasizes twice that the activities of the new institution will be peaceful—and in fact dedicates the penultimate paragraph of the piece (consisting of one long sentence) to reiterating this.[87] The first clause of the law, as it was eventually passed in 1969, reads as follows:

> The National Space Development Agency of Japan (NASDA) shall be established for the purpose of peace, and will comprehensively, systematically and efficiently develop, launch, and track satellites and launch vehicles for launching artificial satellites, contributing to the promotion of space development.[88]

It was this purpose of peace—"peaceful purposes" or "peaceful usage"—that leading institutional figures in the space program continued to emphasize when speaking of the activities of NASDA and ISAS to the public throughout the 1960s, 1970s, and 1980s.[89]

The generation of people who moved into the space program who had grown up after World War II, however, proved somewhat more willing to discuss the military aspects of space technology than their forebears. Describing his trip to Iwo Jima to help install the aforementioned payphone, for example, satellite specialist Hideki Mizuno recalls viewing Japanese entrenchments on the island. His thoughts turned not to geopolitics, but to the human drama. What had the men in the trenches been thinking of—their loved ones? what was for dinner? the Americans facing them? "After all," he ventures, "did they not individually have something to defend?"[90] Meanwhile, Yasunori Matogawa frequently had pieces appear in publications such as *The Japanese Scientist* (*Nihon kagakusha*) on the military history of rockets.[91] "It would not be going too far to say," he observed in 1982, "that the history of . . . rocketry, is the history of a weapon." Similarly, to Matogawa, Japan's wartime work on the Shūsui was a "marvelous and outstanding achievement in a technical field that the Japanese engineers accomplished outside of their area of expertise," made all the more impressive in that "they did it with only one year."[92] Later, in a series of pieces at the symposium of the International Academy of Astronautics in the 1990s and 2000s, Matogawa presented an authoritative English-language discussion of Japan's wartime rocketry development. The future associate executive director of JAXA also supplied images of the wartime rockets' components, manufacturing processes, and the transporting of components.[93] Nevertheless, until the 1990s, Japan's official policy remained clear on the matter. A strict interpretation of the policy of "peaceful usage" remained in force, even when it conflicted with Japan's long-held desire to become a player in the rocketry game. In the 1980s, for example, negotiations between McDonnell Douglas and Mitsubishi about using a newly developed Japanese rocket motor for an improved upper stage of the Delta launch vehicle foundered, because a primary mission of the Delta was to launch the Department of Defense Global Positioning Satellite—a military satellite, in the eyes of policymakers.[94]

However, fresh geopolitical and economic challenges in the 1990s resulted in further strains on the policy of "peaceful purposes." At the beginning of the 1990s, Japan's economy entered a long period of stagnation known as the "Lost Decades"; concurrently, both China and North Korea surged in their rocketry and space capabilities. In this fraught context, some began to openly attack the "taboo" of military usage; the idea that hostile neighbors were forging ahead with military strategies predicated on rocket technology caused a great deal of anxiety. North Korea not only began testing nuclear weapons in the 1990s, but also developed a series of medium-range missiles, which it tested by firing over Japan. Then in 2003 and 2004 China successfully put a man and a woman in space. With these developments, as Paul Kallendar-Umezu observes, space had suddenly "joined the air and maritime domains as global commons and global infrastructure but also as a military domain."[95] By the 1990s, Japan had, in fact, been arming for the better part of fifty years, something the United States actually encouraged, so as to have an ally in countering communist influence in Asia. Hence, even as it was trying to put a satellite in orbit in the late 1960s, the Japanese state initiated an ambitious five-year military plan, at the cost of some $8 billion ($1.29 trillion in 2024), expanding its inventory with tanks, armoured cars, submarines, and patrol craft, all built domestically.[96] Then, beginning with the JSDF's participation in the United Nations Transitional Authority in Cambodia (UNTAC) in 1992, public attitudes toward the armed forces also began to shift; whereas previously the public had regarded the JSDF with "almost complete detachment and disinterest," there was now a "reemergence of a political will to have service members kill and die in the name of the Japanese nation-state."[97] In this context, the military taboo around the use of rocketry began to fade.[98] For satellite specialist Takeo Ueda, it was high time for such a change; as he pointed out in 2004, "the majority of national space policies form part of the security strategy" not only of countries such as India and China, which had their own launch capabilities, but also of nations like Australia, which did not.[99] Japan's ongoing refusal to take steps to protect its space infrastructure—steps construed as "militarization"—was in fact a result of a misinterpretation of Japan's pacifist policies. "Peaceful usage" surely only meant no *offensive* capacity; after all, if pacifism meant the rejection of all use of force, even in self-defense, why did the Japan Self Defense Forces (JSDF) exist?[100]

Takashi Iida, the head of the JFCS, similarly argued that leaving defense considerations entirely outside the purview, coordination, and resources of the space programs was undermining Japan's security.[101] And, as satellite specialist Susumu Kitazume pointed out, wasn't "peaceful usage" really just a matter of perspective? After all, "even a normal car could be used for military purposes."[102]

The use of rocketry as a weapon was one element of this debate; the use of satellites as spy tools was another. The 1969 commitment to peaceful usage was generally seen as having forbidden space-based surveillance. The August 1998 "Taepodong shock," when North Korea tested an intermediate-range ballistic missile of the same name, worried many lawmakers, who were not only caught off guard, but were annoyed that the American systems they relied on had failed to raise any flags. In response, the Diet had, within four months, established the legal framework to launch spy satellites. By the early 2000s, Japan had begun to create its own supplement to GPS, the Quasi Zenith Satellite System, which not only enabled closer surveillance of North Korea, but also could serve as a backup if there was any disruption to GPS.[103] There was also an increasing acceptance that nonmilitary satellites could have military uses: ALOS-1 (2006), for example, nominally a land observation satellite, could also be used for surveillance.[104]

For all these changes, however, "peaceful purposes" continued to have powerful defenders within the new generation of managers and specialists, and these figures were able to use policy and influence to fend off military involvement in their work—particularly in ISAS. Ryōjirō Akiba, for example, was determined "never to commit military research."[105] As a result, even when the limits of "peaceful usage" were tested, the testers were keen to emphasize it was for defensive purposes.[106] What were deemed by some to be spy satellites were presented as devices "to watch for natural disasters, and to fight smuggling and illegal immigration."[107] This enduring policy commitment was further bolstered by significant nonstate stakeholders, which remained deeply committed to pacifism. The Sasakawa Peace Fund, for example, was one of the chief sources of funding for both local students and students from Africa and Asia seeking to study space engineering in Japan. For administrators of the program like Rieko Hayakawa, Japanese activities were *defined* by their peacefulness: "Japan takes the original position that pursued only the peaceful use of a satellite in a developed nation."[108] In the end, Japan did not amend its laws to allow for the full militarization of space until 2008.[109]

As the scale and confidence of the Japanese rocketry and space programs increased from 1960 to 2003, so too did public narratives expounded by those who worked within them. The issues of the immediate postwar period—disassociating the technology from its martial origins—became less relevant. Instead, senior managers and specially employed outreach officers focused on extolling the benefits to the average Japanese of the technologies their institutions were providing. Space technology, they argued, provided the Japanese citizen with communications, safety, and national pride. At the same time, martial narratives came full circle. Peaceful usage, some argued, had compromised Japan's national interest, and caused Japan's technological standing to slip. It was also out of step with the increasingly fraught regional security situation, which, some argued, required an amendment of the policy. Still, official inertia and resistance to such changes from senior managers, leftists, and nonstate institutions came to the policy's defense, and ensured that no major amendments to it occurred until well into the twenty-first century.

The influence of geostrategy on Japan's space policy in the 1990s is indicative of the profound impact that relations with other countries had on Japan's civilian space programs. Of these, perhaps no country had a greater impact than the United States. As mentioned previously, in the 1970s, the Japanese had begun working on developing LFRs from imported American Thor-Delta vehicles. However, as Japan's space ambitions grew in size and ambition, the country's development goals moved out of alignment with what the United States was willing to provide. A consensus rapidly emerged among key stakeholders in the mid- to late 1970s that the country's policy of importing LFR technology from the United States was not viable in the long term. The result was a decision in 1984 to embark on the expensive and ambitious task of developing a "purely domestic" medium-lift vehicle: the H-2. Japan, thirty years after building its first rocket, was making its big play to join the major space powers of the USSR, United States, and Europe in fielding such a large rocket. The rocket resulted in an utter transformation of Japan's space program—but not in the way intended.

SEVEN

Growing Ambitions and Difficulties with the United States

IN DECEMBER 1990, VIEWERS ACROSS Japan tuned into the Tokyo Broadcasting System (TBS) TV channel to enjoy an odd spectacle being beamed to them directly from the Mir space station. On screen, mumbling somewhat, was a glum-looking middle-aged man—Toyohiro Akiyama, TBS's former Washington bureau chief. Having reached the space station on December 2, he had spent a day or so addressing the nation upside-down, and complaining about the effect of the journey on his stomach. Thereafter, he began a series of nightly broadcasts, featuring updates and various activities carried out with the grim patience of a harassed parent. Japan's oceans look "muddy" from space, he observed; the frogs sent to him by some school children loved being weightless; getting rid of bodily fluids was a chore, but he still wished he'd brought legendarily stinky Japanese bean paste, *nattō*, with him. That, and some cigarettes; as soon as he got back to Earth, he told viewers, he was going to have a smoke.[1]

The 1980s that had preceded Akiyama's trip saw Japan's confidence in space reach a pinnacle. The civilian space programs expanded in terms of infrastructure, fundamental research, satellite sophistication, and launch capacity. It was also a period in which specialists and managers within the programs had, in keeping with the times, generated ambitious plans for the future. Of particular importance for them was the creation of a domestic medium lift vehicle (MLV) that did not rely on American parts. Though Japan's N series LFRs had been based on imported Thor-

FIGURE 7.1. Toyohiro Akiyama.
Source: Wikimedia Commons.

Delta components, the Americans, wary about Japanese technological competition, had restricted access to key rocket technologies through blackboxing (the practice of providing components without technical knowledge about how they work, or how to alter or fix them). Still, Japan would have to develop its own MLV if it was to crack the international launch market, a key goal in the 1980s, as rockets developed with American components were prohibited from being used as commercial launch vehicles. Lastly, doubts about the Space Shuttle and America's long-term technological leadership also encouraged some to argue that sooner or later, Japan would have to break free of its reliance on US technologies. Having secured its actual independence in 1952, Japan set out to claim spacefaring independence in the 1980s. It was an ambition that would result in the total transformation of the space program.

Success and Growing Ambitions, 1970–1993

EXPANSION AND ACHIEVEMENTS

Thrilled with the reputational and technological boost of Ohsumi in 1970, the space program's political overseers in the Diet and managers in the bureaucracy funded a major expansion of Japan's space industry through robust governmental support when negotiating technology imports and bankrolling their activities. After the foundation of Uchinoura and Tanegashima space centers in the 1960s, for example, NASDA and ISAS continued a construction spree that reflected their growing importance and institutionalization. Building continued at the launch centers, with new facilities being put in place to cater to larger and more sophisticated rockets.[2] NASDA established its flagship Tsukuba Space Center in June 1972, while ISAS built the expansive Sagamihara Campus in 1989; both remain significant parts of Japan's space program today.[3] New tracking control networks allowed ISAS and NASDA not only to follow their satellites, but also to provide telemetry and tracking for other entities like private broadcasters, which were using satellites in growing numbers. Emblematic was the Usuda tracking center, completed in 1984, which featured a 1,980 ton (2,011 tonne) communications dish with a diameter of 210 ft (64 m).[4]

Japan's space institutions also established a strong track record of producing cutting-edge fundamental and applied research in universities and industrial labs, where Japanese scientists worked on technologies such as the interface between humans and robotics in space.[5] Even Japan's smaller space agencies had a claim to producing world-class research. For example, celebrating its thirtieth anniversary in 1985, the National Aerospace Laboratory could proudly point to a slew of recent breakthroughs by its researchers. This work included systems for improving the performance of attitude control units on satellites, a significant global first, in that it enabled orbiting objects to maintain a particular alignment (for instance, always having the same side facing toward Earth) by using the internal magnetic field of the planet itself. Other work was conducted on an ion engine for use on geostationary satellites, which promised to reduce the weight of certain satellites by up to 660 lb (300 kg).[6] This is not to say the big boys were slacking off; perusing ISAS's *Science Report* for December 1989 alone reveals research on a wide array of aerospace issues. One article describes work based on observations of the Cygnus X-1 black hole using a B50 balloon, introduced in 1970, and capable of reaching heights of up to 25 mi (40 km). By 1990, ISAS was producing the hefty B500, which was capable of heaving experiments of up to 2,800 lb (1,270 kg) to the same altitude.[7] Another piece described novel observational work on the behaviors of microbes in zero gravitational milieus during an experiment carried out in 1987 on a domestically built S-520 sounding rocket.[8]

The development of satellites also proceeded apace. Between 1975 and 1987, Japan successfully launched nineteen satellites without a single failure; by 1994, NASDA's tally alone amounted to thirty-three.[9] Of particular note was a series of engineering test satellites, imaginatively dubbed the Engineering Test Satellite series.[10] Through these, Japan's space programs had mastered geosynchronous orbits by the 1990s; satellite engineers would identify it as the most important satellite-related technology the country had thus far developed.[11] The satellite series also helped hone crucial technologies such as three-axis stabilization (that is, holding the satellite still in space); transfer orbit insertion (getting it to the right place); and launching devices of increasing size—where ETS I had weighed 180 lb (82 kg), comparable to Sputnik, the 1994 ETS VI came in at a hefty 4,400 lb (2,000 kg).[12] In 1977, the Science and Technology Agency laid out long-term plans to develop Japan's satellite technology by creating a network of Earth-observation satellites that would be tailored to Japanese needs, enabling the country to break free of the cost and inconvenience of relying on access to US devices. The result was the Himawari series of geostationary meteorological satellites (GMS).[13] By 1984, NASDA's telemetry and satellite tracking systems were among the busiest in the world.[14] Meanwhile, companies like Mitsubishi benefited from robust trade protections that enabled them to develop a domestic satellite industry; by 1985 the company's CS3 type satellites had cornered a staggering 80 percent of the domestic market.[15] The same protections turned Japan into a major player, for a while, in the global satellite market; in 1989, Japanese exports of satellite parts totaled ¥83.4 billion ($642 million in 2024), and ground equipment sales clocked in at ¥129.9 billion ($999 million in 2024).[16]

As these satellites grew larger and more sophisticated, so too did Japan's demands of its rockets, and their producers rose to the challenge. MHI had established an engine factory in Komaki on the outskirts of Nagoya in 1972, and a rocket engine testing ground at Tashiro in Akita in 1976.[17] Its main Nagoya Aircraft Works (located near old wartime aeronautical facilities) was, and remains, probably the most important hub for advanced space engineering in Japan.[18] The company found itself doing so much space-related work that in 1989 it spun off a new subsidiary, the Nagoya Guidance & Propulsion Systems Works.[19] It was in Nagoya that Japan made its first run at creating a domestic LFR, the H-1. The design of this vehicle went back to the 1970s when NASDA had begun work on the N series of rockets, assembled largely from components purchased from the United States. Still, its design and construction had enabled MHI to master key technologies, and boost Japan's payload-to-orbit capacity from 285 lb (130 kg) to 770 lb (350 kg).[20] Specialists initially recommended that the next generation of Japanese rockets be able to put at least 800 kg into GEO, but Hideo Shima decided that such a leap in capacity was overambitious, and NASDA committed to developing a more modest payload capacity of 1,200 lb (550 kg) in the H-1.[21] At the same time, the new vehicle would stick to the other requirement of being as Japanese as possible.[22] Preliminary work on the project began in March 1974, with both NASDA and ISAS agreeing to contribute. At that time, producing an entirely domestic first stage (the largest, most powerful, and hence most difficult part of a rocket to master) was out of the question. Nevertheless, the rocket proved to be a fertile test bed for developing local technologies, such as solid fuel boosters and fuel priming materials, which had heretofore been imported from the United States. Beyond that, H-1's upper stages were propelled by the all-Japanese LE-5 engine. The result was the largest rocket Japan had ever made, around 16 ft (5 m) longer, and 11,000 lb (5 tonnes) heavier than its predecessor, the N-2.[23] Initial testing was completed by 1986, and the rocket launched nine times without failure between 1986 and 1992.[24] Meanwhile, ISAS also continued to plow its own eccentric furrow, launching a series of remarkably capable solid fuel Mu rockets. These began with M-4S in 1971 (payload to LEO, 400 lb [180 kg]), and progressed through the M-3C in 1974 (430 lb [195 kg] to LEO), the M-3H in 1980 (660 lb [300 kg] to LEO), and M-3S in 1980 (also 660 lb [300 kg]).[25] Although ISAS abandoned rocket design in 1984, it still worked intimately with companies such as Nissan to ensure it had a steady supply of vehicles that met its needs. Projects such as the H and Mu series were also crucial in helping engineers and managers such as Eiji Sogame, future head of engine development at NASDA, come to grips with the all-important liquid oxygen and liquid hydrogen fuel systems that larger rockets require.[26] One of the engineers to work on the Mu rockets, Tomifumi Godai, would go on to head a major Japanese medium-lift rocket project in 1984.

The fact that Japan had scored these achievements with a fraction of the amount spent by the United States and the Soviet Union was a source of great pride to the space program's overseers. Japan's annual space R&D bill came to ¥3.9 billion ($1 million in 2024) in 1966, ¥88 billion ($134 million in 2024) in 1976, and ¥117.3 billion ($932 million in 2024) in 1986. Historical exchange rates make it hard to pin an exact dollar figure on these, but certainly the budgets of both NASA and the Russian space program were easily many orders of magnitude greater. For example, ¥3.9 billion at a 1971 exchange rate of ¥360 to the dollar comes, inflation-adjusted, to roughly $100 million in 2023. In contrast, NASA's adjusted budget for the same year topped off at around $53 billion, whereas the Soviets spent an estimated $5 billion per year on their space program between 1966 and 1970.[27] Even better, from the point of view of lawmakers, was that despite growing capacity and complexity, Japan's space budget actually leveled out in the early 1980s. This run was only broken in 1984, when, despite working on projects such as a module for the mooted International Space Station, and a new medium-lift rocket, total expenditure on space technologies in Japan ever so slightly *de*creased.[28] In 1991, NASDA's budget stood at 10 percent of NASA's, and whereas NASA boasted 27,000 nonmilitary workers, the sum total of all space specialists in Japan rarely exceeded 10,000.[29]

At least part of this relative cheapness came from factors such as labor costs (which were lower in Japan than in the United States until the mid-1980s) and the adoption of lean manufacturing practices by producers such as Nissan.[30] There were, however, several distinctively Japanese characteristics that helped too. In the case of ISAS, the unusual nature of the institution itself—a largely self-contained, low-turnover, somewhat scholarly collection of researchers, who more often than not solved problems on the fly, without what they saw as the restrictive procedures of industry and bureaucracy—helped. A good example here is the MUSES program. In 1987, ISAS proposed an ambitious series of Earth- and Moon-observation satellites, but was awarded far less of a budget for it than it had hoped. Its compromise was the Mu Space Engineering Spacecraft (MUSES) program, a series of three probes that, sequentially, explored the Moon (Hiten/MUSES-A), deep space (Haruka/MUSES-B), and, in 2005, an asteroid (Hayabusa/MUSES-C). Despite the tight funding, Kuninori Uesugi, organizer of the program, was able to get it up and running within three years, a process that involved the construction of a new 3m dish in Uchinoura, and accounting for the delayed receipt of critical components.[31]

Additionally, Japan was seemingly able to accomplish so much for so little due to two careful strategies deployed by the space program's managers. The first was the committing of resources to areas and projects with tangible economic benefits. The second was obviating the need for expensive space technological capabilities by forging overseas links, particularly with the United States, to maximize the impact of Japanese contributions. These decisions were necessitated by the simple fact that no one, not even the managers of the civilian space programs, set out to go toe to toe with the Soviet Union and the United States in terms of the biggest, most complex, and most prestigious projects. As Yasunori Matogawa explains:

> I don't think any figures expected Japan would stand shoulder to shoulder with the U.S. or ESA regarding the level of space activities. It is because Japan started much later in this field and could only have [a] much smaller budget as compared with them. Personally speaking, however, only Hideo Itokawa used to have such an ambition, but none of us could succeed him, [and attempt] breaking through . . . the limit[s] of our national power. The difference was too big. What we have been pursuing is not to be *number one* but to be [the] *only one*.[32]

In this sense, the different scale of Japan's civilian space programs and their Soviet and American counterparts absolutely reflected the gulf between their budgets. For starters, despite the impressive progress that the Japanese had made in indigenising LFR technology, their biggest launcher as of 1993—the Thor-Delta–derived H-1—could lift a maximum of 7,100 lb (1,100 kg) into LEO.[33] Compare to this the somewhat older Soviet Energia system, which could lift up to 220, 000 lb [100,000 kg] into LEO. Japan's probes were also smaller and less capable: Hiten/MUSES-A clocked in at 434 lb (197 kg) at launch in 1990, whereas the 1989 US probe Galileo and its orbiter weighed 4,900 lb (2,223 kg), and the Soviet Union's chunky Phobos Mars landers each clocked in at an impressive 14,000 lb (6,200 kg).[34] And while both the United States and the USSR had produced space stations in the 1970s— Skylab and Salyut, respectively—projects of that scale remained well beyond both Japan's ability and budget even twenty years later. Rather than try to compete with these behemoths, however, Japan's civilian space programs instead shepherded their resources into specific areas of development, and committed to excelling at those. The goal, again in the words of Yasunori Matogawa, was to find a "self-reliant and rather unique way of life in the respective field of space activities."[35] The result was a commitment to creating telecommunications (beginning with the 1981 ECS satellites), weather forecasting (such as the Himawari series), and Earth observation (such as 1987's Marine Observation Satellite) systems, providing the country with a suite of useful infrastructure in space. Not only could the costs and functioning of such systems be carefully regulated by Japanese institutions, but they also produced—and could be sold to the public as producing—tangible economic benefits. NASDA, in particular, also displayed a willingness to participate, often as a junior partner, in major international projects. The result was that the Japanese could often contribute highly complex and novel technologies, such as the International Space Station's Kibō module, with its Exposed Facility (essentially an external laboratory,) without having to invest in the means to haul such a huge device into space themselves.

GROWING AMBITIONS FOR THE FUTURE

The aforementioned successes bred a growing confidence in Japan's space program, and a concurrent expansion of ambitions. This was reflected in an August 1990 US National Science Foundation report on the state and future of Japanese space technology. The report provides one of the best summaries of just how "aggressive and forward-looking" the country's objectives in space had become by this point:

> Japan's goals for the present decade include plans for continuing its already strong thrust in scientific space research, bringing its satellite and launch technologies up to international standards, creating the infrastructure for space station activities, and developing the basic technologies required for its own manned space activities. These near-term goals include the promotion of advanced satellite technologies such as its Engineering Test Satellite (ETS) series, and communication, broadcasting, and navigation satellites. Importance is also given to the development of scientific satellites with supporting efforts in space sciences, facilities, and tracking and control systems. These scientific areas are seen as being particularly appropriate for international cooperation. Japan's near-term plans also reaffirm its significant participation in the U.S. Space Station through the Japan Experiment Module (OEM) module . . . Japan's long-term vision for the first decade of the new century and beyond include the implementation of its own manned space capabilities, the launch and operation of a geostationary platform, the development of an orbital servicing vehicle and an orbital transfer vehicle, and the ultimate development of its own space station.[36]

Such observations were derived at least partly from the stated ambitions of Japan's space programs. In 1978, the Space Activities Commission declared that "Japan has completed its first phase of space development activities, during which the emphasis has been on the establishment of a firm foundation."[37] A few years later, in 1984, it presented its vision for the next phase of Japan's space development, based on the following principles:

1. Space activities shall be carried out solely for peaceful purposes, striving to meet various social needs in harmony with national resources.

2. Efforts shall be directed to developing Japan's own advanced technological capabilities to allow the country to freely conduct various space activities.

3. Japan's space activities shall be conducted in harmony with space development worldwide. There shall be appropriate sharing of responsibilities as well as international cooperation for space development based upon the technological potential developed through such activities.[38]

By 1987, official objectives had expanded to include an ambitious program of life science experimentation, astronaut launches, and participation in international projects such as NASA's International Microgravity Laboratory Program, and the International Space Station.[39] The same year, NASDA also pitched plans to develop "space infrastructure" consisting of "various space facilities in orbit, easily accessible space transportation systems and their support systems," as well as a "space factory."[40] Specialists within the field regarded Japan's eventual participation in a manned Moon base, planned to be up and running by the early 2000s, "only natural" (*tōzen*), and a wise economic investment in Japan's perennial search for cheap sources of raw materials to boot.[41]

They also foresaw the need for human-rated spaceflight. A symposium on manned space exploration held in Tokyo in 1985 attracted some 300 attendees, and senior figures from NASDA and Tokyo University not only discussed the future of Japanese technology, but the status of contemporary technologies in other countries as well.[42] Officially, NASDA began training three astronauts in 1985; by 1994, it had trained four more, with plans underway to considerably expand Japan's human space program.[43] Any hopes of making one of these carefully selected astronauts the first Japanese person in space, however, were crushed by private sector broadcaster TBS, which paid somewhere in the region of $30 million (exact figures remain hard to come by) to send Toyohiro Akiyama to the Mir space station aboard a Soyuz. Akiyama's journey was later deemed one of the most significant events in Japan's twentieth-century history.[44] In the end, Japan's own brief flirtation with human-rated space travel, described below, was eventually ditched for being too expensive, and as of 2024, the country has no plans to develop its own capability in this area.

Instead, reverting to the strategy of relying on foreign suppliers when necessary, NASDA was happy to purchase tickets on overseas launches; using this method, as of 2024, Japan has sent fourteen people into space, behind only the United States, the Soviet Union/Russia, and China.

Many of these projects were given a serious hearing by lawmakers because of a firm belief that they, too, could be achieved more cost-efficiently than equivalent efforts overseas. The key was to be cost-saving through innovation. For example, to ensure that multiple annual launches wouldn't cost too much, NASDA initiated the H-2 Orbiting Plane (HOPE) project in the early 1990s. This space plane was to include autonomous cargo transport, rendezvous and docking capabilities, and an automatic microwave landing system, capacities shared with the USSR's remarkable, but aborted, Buran space shuttle. On top of this, it would be able to lift up to 33,000 lb (15,000 kg) to the International Space Station (ISS), land on a runway less than three-quarters the length of the Space Shuttle's, and conduct atmospheric flights and missions "utterly like the airplane." HOPE was expected to be up and running by the end of the 1990s, and to provide a base from which Japan could develop independent human-rated space vehicles, and a space station, by 2010. [45] Theoretical work was even undertaken on the holy grail of many a launch nerd: a single-stage-to-orbit spacecraft.[46] "If Japan is unable to transport personnel on its own until 2010, and is dependent on Europe and the US," observers warned, "it will be impossible to compete with [them economically, and] we will be forced to change our space program due to European or US circumstances." If Japan could crack human-rated space travel, however, space hotels, Moon bases, and Mars exploration—"weighty responsibilities" for Japan—would all be realities by the first half of the twenty-first century.[47] Yet for some lawmakers, even these soaring ambitions were too timid; in the early 1990s, Councilor Atsuo Ōta of the Kōmeito Party wanted to know why, exactly, Japan wasn't contributing more to international projects on Mars exploration.[48]

Underlying all these expansive plans was not just a desire to make money, but also a good dollop of national pride. When announcing Japan's commitment to launching a two-ton engineering test satellite aboard a Japanese rocket in 1985, for example, the Ministry of Science and Technology dubbed it a "Hinomaru satellite launched atop a Hinomaru rocket"—with *hinomaru* (circle of the sun) doubling as a reference to Japan itself.[49] Workers in the space industry were confident in their abilities; a survey published in 1985 found that more managers than otherwise felt that in many cases, the technological level of Japanese space corporations superseded

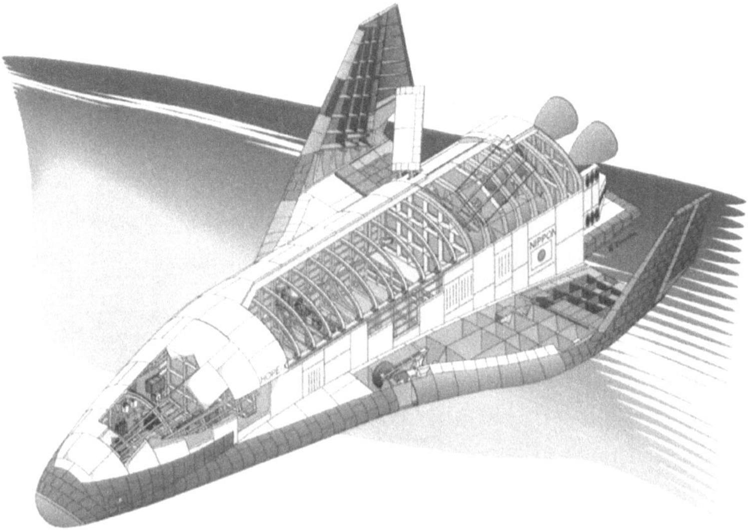

FIGURE 7.2. Schematic of proposed HOPE-X space plane.
Courtesy of JAXA.

that of the United States. Virtually none felt that the United States would continue to hold its lead in the field into the 1990s.[50] In fact, some argued that Japan was already ahead: in the 1980s, Hajime Karatsu, director of the Matsushita Telecommunications Company, blamed inferior US–made chips for delaying the launch of a Japanese rocket. Even in such a fundamental area, he argued, the Japanese producers were already simply better.[51]

Deciding to Go Independent

GROWING DEMAND AND COMMERCIAL OPPORTUNITY FOR SPACE LAUNCHES IN THE 1980S

The upshot of these plans was that in the 1980s, the Japanese space agencies had to make provisions for putting a lot of stuff into orbit on a consistent basis. Official estimates projected demand for the "construction and operation of space infrastructure" at ¥300 billion ($5.5 billion in 2024) annually in the 1990s, and up to ¥500 billion ($8.8 billion in 2024) in the 2000s.[52] When, in the mid-1980s, the United States approached Japan about potentially participating in its mooted International Space Station, the Japanese understood that total launch requirements for such project were estimated at 220,000 lb (100 tonnes) per year by 2000, with a further 11,000–17,000 lb (5–8 tonnes) a year required for resupplying. Even ongoing projects would require higher launch capacities—in 1985, for example, NASDA projected that its ETS program would, within a decade, be producing satellites of up to 4,400 lb (2 tonnes).[53]

Beyond these visions of space infrastructure, the influence of private corporations and their ministerial allies ensured that commercially viable launches remained a consistent objective of NASDA. By 1987, the powerful Ministry of International Trade and Industry had identified three areas in which the civilian space programs could be of commercial benefit: preventing reliance on importing communications hardware and knowhow; establishing quid pro quo relationships with other nations (for example, by using satellite sensing to detect oil fields, and selling that knowledge for exploratory concessions); and developing exportable technologies. Selling access to sensors and equipment aboard already launched satellites could also earn income. The profitability of launches thus became one of the mainstays of articles published by NASDA and other space agencies, such as National Aerospace Laboratory (NAL).[54] In fact, the potential for profit at this time was tantalizing enough that Arianespace's head Frederick d'Allest traveled to Tokyo in 1985 to convince local enterprises to "entrust their launches" to his company.[55] His arrival was seen as a sure sign of growth to come, as the Europeans were fiercely committed to their position as one of the world's leading suppliers of commercial launches. Around 70 percent of ESA's budget in the 1980s was dedicated to developing commercial launch vehicles, meaning that even though its overall budget was a fraction of NASA's, it was able to successfully compete with the larger entity in this field.[56] Furthermore, demand for global space launches showed every sign of growing dramatically in the next few decades, as the development of wide-range, high-speed satellite networks and disaster warning systems necessitated even more hardware in space.[57]

Thus it became increasingly apparent that the key to realizing the ambitions of the 1980s boiled down to a simple pair of priorities: increasing lifting power, and decreasing cost.[58] Yet the H-1, Japan's most advanced LFR at this point, was too small and too expensive to be commercially viable. Its continued reliance on US technology through licensing also made launches prohibitively expensive; putting 2 lb (1 kg) into geostationary orbit on an H-1 cost around $33,000, around $10,000 above a competitive commercial rate.[59] The same licensing agreements also forbade the H-1 being used as a commercial launch vehicle, which meant there was a very real chance Japan's domestic launch market would indeed come to be dominated by ESA and the United States.[60] What Japan needed was a domestic vehicle that could compete with Russia, ESA, and whatever the post–Space Shuttle United States came up with.[61] Hence in practice, the issue boiled down to just one critical issue—redefining Japan's space technological relationship with the United States.

TENSIONS IN US–JAPAN SPACE RELATIONS IN THE 1970S AND 1980S

The 1970s and 1980s were watershed decades in which Japan's relationship with the United States saw increasing tension. President Richard Nixon's unexpected 1972 entente with China furthered a sense across the Pacific that perhaps the United States didn't need to cater to Japan's demands quite so much. In the 1980s, while the likes of Morita and Ishihara preened over Japan's success, the United States fretted over the fact that by 1987, Japan owned $39 billion of America's cumulative $152 billion trade deficit. Economic concerns overflowed into popular culture; Michael Crichton's bestselling novel *Rising Sun* (1992) depicted nefarious Japanese businessmen murdering prostitutes in the United States, and ended with a screed against the dangers of Japan as a global power.[62] Anti-Japanese sentiment also became commonplace in American political discourse; at least one presidential hopeful in 1991 made explicitly anti-Japanese pronouncements during his campaign, and a contemporary CIA document, dismissed by historian Bruce Cumings as "drivel," posited that Japan was a threat to nothing less than the "American way of life."[63]

The sort of ultranationalist self-congratulation of the likes of Shintarō Ishihara in the 1980s also didn't quite account for just how much Japan relied on US technological imports in certain areas. At least in the field of space technology, the United States remained far ahead of its East Asian ally, and the ultimate cultural seal of approval for a Japanese innovation remained the idea that it outperformed the American equivalent.[64] Yet from the very beginning, despite having several reasons for doing so, the United States still had profound reservations about precisely how much technology it should share with the Japanese. Not only did the Americans not help with Hideo Itokawa's initial rocketry project, but the armed forces had refused to let Hideo Itokawa test his rockets on US ranges.[65] The 1969 agreement providing LFR technology to Japan had come partly because the Arms Control and Disarmament Agency believed that the carrot of assistance with LFR technology would help rein in Japan's SFR program.[66] It would also help prevent Japan's rocketry capabilities reaching the level of domestically manufactured ICBMs.[67] It was on the basis of such considerations that the United States had signed the 1969 deal providing LFR technology to the Japanese.

By the end of the 1970s, however, dwindling US–Soviet tensions seemed to make the arrangement both less wise, and less necessary, from the US perspective. To quote historian Yasushi Sato: "when the cold war abated around 1970 and Americans became aware of their declining position in the world they quickly became reluctant to share their technology."[68] American policymakers were also concerned that sophisticated rocket and satellite technology would result in Japan leaving the framework of a US military alliance or, at the very least, becoming more assertive within it. Japan's expenditure on military application devices such as spy satellites would also divert funding from areas where the Americans would have preferred to see Japan's resources deployed, such as "host nation" payments to maintain US forces on Japanese soil.[69]

Meanwhile, others in Washington were more concerned about what they saw as the "seedbed of a long-term technological challenge" to US dominance in the field of commercial launches. In 1972, NASA had warned Nixon of potential competition from Japan's rocketry program, particularly after his administration's decision to sell US rocket launches for cash to all (non-Soviet) comers. This warning hung over attempts by Japan to secure more technology from the United States in the 1970s, and indeed a new, limited agreement in 1976 banned the use of US rocket technology in commercial space launches by the Japanese.[70] Another clause forbade Japan from exporting any vehicles based on US technology, at least partly to prevent important US innovations from falling into unfriendly hands during Japan's pursuit for profits; Hideo Itokawa had already proven willing to sell AVSA's proprietary SFRs to countries that, while not in the Warsaw Pact, were nevertheless communist (Tito's Yugoslavia, 1960–1964) or left-leaning (Sukarno's Indonesia, 1964–1965). The United States ensured compliance with these measures through the liberal use of blackboxing. The practice had been present from the very beginning—the N-1, for example, had certain issues that could only be resolved by US contractors, who had to be flown in at great expense.[71] In addition to obstructing maintenance, blackboxing also meant the Japanese were in no position to "harden" (that is, safeguard) their devices and networks against hostile actors. The only way for the Japanese to get around blackboxing was either to break the terms of the import agreement or to develop a domestic replacement. Such limits were welcomed by American labor union leaders and politicians, who decried the impact of these exports and overseas licensing on US jobs. In 1973, Andrew J. Biemiller, legislative director of the powerful AFL-CIO trade union federation, testified to Congress that the export of Thor-Delta and other aerospace technology was having a "devastating impact" on the economy. Such sales, he argued, "would profit stockholders of the corporations involved at the expense of US aerospace and automobile workers." Some wanted even further restrictions; Abraham Ribicoff, a Democrat senator from Connecticut, decried the fact that rocket technology "created at great expense to the US taxpayer, is being sold to foreign companies for production abroad."[72]

All these conditions posed a problem for the Japanese, for whom developing a profitable local industry had been the whole point.[73] As historian John Logsdon puts it, "the U.S. government has seen space cooperation as a means of demonstrating in a highly visible way its claims to global political and technological leadership; Japan has used cooperation (and not only in space) *as a way of learning from a more advanced partner as an interim step to independent, often competitive, Japanese capabilities.*"[74] In other words, while space cooperation was primarily a geostrategic investment for the United States, Japan viewed it as economic and industrial in its efficacy. Space specialists had thus long accepted that some misalignment between the United States and Japan could well be an enduring part of the country's spacefaring future.[75] Their determination to indigenize the technology they received is evident in the construction of the N and H series of LFRs. The N-1 featured a Japanese-produced LE-3 second stage engine, and between 53 percent and 67 percent of its parts were made domestically.[76] The follow-up N-2 used between 54 percent and 61 percent, but in the subsequent H-1, this figure leaped up to as much as 98 percent, through the domestic production of licensed components.[77]

Yet in practice, despite this success, blackboxing put Japanese engineers in a peculiarly similar position to their forebears working on the Iwaya documents during World War II. "As was the practice in technology transfer to underdeveloped countries . . . there was no knowledge at all about what was not written [in the manuals]," one engineer complained, "and it was honestly impossible to infer anything." Worse, "there were no explanations written of the basis and essentials" of the technologies being handed over.[78] Blackboxing would remain a significant force in Japan's satellite industry for years to come; as late as 2004, senior figures bemoaned the fact that it wasn't just a reliance on blackboxed overseas technology that was holding Japan back, but the country's own inability to produce quality components.[79] The same year, in an article directed at nonspecialists, industry insiders identified the real culprits of this trend as Japanese endline users (that is, the companies that purchased these satellites), which cared more about profit than about the strategic danger of blackboxed technology.[80]

American reluctance to share technologies also had a major impact on finances and research, as shown during the development of N-1 and N-2.[81] These were derived from the American Thor-Delta launch vehicle: the N-1 was essentially a Japanese-built version of it, while the N-2 was the same, but had a larger first stage.[82] However, when the Japanese tried to purchase Castor II solid fuel boosters for the N-1 in the early 1970s, the Americans argued that the items should be bought as a component in toto, without any replication or maintenance being carried out in Japan. The Japanese responded that this violated the 1969 agreement between the two countries; in the dispute that followed, the US State Department intervened in Japan's favor, but only at the cost of Japan footing some of the booster's development bill. Subsequently, when the N-2 required updated components and designs from newer versions of the Thor-Delta, such as the Thor-Delta 2914 (1974), the Americans again responded that this was not part of the deal. Despite several visits to the United States by Japanese space technology administrators, including Hideo Shima, a new agreement was drawn up in 1976 that did give Japan the components they wanted—but blackboxed crucial functions such as inertial guidance, stabilization, and cryogenic propulsion.[83] In fact, so tightly did the United States control the technology it ceded for the N-2 that some in NASDA regarded the rocket as a "step back" for Japan's civilian space programs.[84] Similar issues bedeviled the subsequent H-1 rocket's development: although preliminary testing began 1977, development didn't really get going until a renegotiated import agreement for crucial US technologies was finalized in December of that year. On this occasion, the United States required Japan to commit to its INTELSAT satellite communications system, ensuring that Japan did not create its own.[85]

Such experiences led some managers to drift toward the institutional outlook of ISAS, which was that US invitations to cooperate in critical technologies were more for American interests than Japanese ones. A good example was the thinking behind

Japan's declining to contribute to the development of the Space Shuttle. NASA's third administrator, Thomas O. Paine, had discussed US plans for a vehicle of this type with Japanese lawmakers in 1970. That same year, a committee of aerospace experts was convened to evaluate the information being sent about the project to Japan.[86] Their activities encouraged bureaucrats to examine the issue further, while Keidanren also took a keen interest in promoting Japanese participation.[87] Specialists were particularly excited by the prospect of regular access to manned spaceflight, and the potential to use the Space Shuttle as a platform for experiments, particularly in life sciences.[88] At the same time, however, some lawmakers pointed to the expense of the Shuttle's development and maintenance as unacceptably high.[89] Concerns were also expressed by planners regarding the exact structuring of control and stakes in the project (read: concerns about American dominance), and whether or not Japan had the technological capacity to meaningfully contribute to its development. This was particularly important to observers who expressed concerns that Japan, as only one of many participants, would not gain access to manned Shuttle launches as and when it desired, due to competition for slots from other countries.[90] Furthermore, the country couldn't possibly hope to invest equally in domestic rocket development *and* work on the Shuttle—which, some suspected, might have been the whole point of the invitation. As Sonoyama Shigemichi, director general of the Research Coordination Bureau of the Science and Technology Agency, observed in 1978, relying on Shuttle launches for elements of infrastructure such as weather satellites would make Japan "completely dependent on other foreign countries," proposing (a little dramatically) that as a result "the very stability of Japanese social life becomes uncertain" (*shakai seikatsu no anteisei sonomono ga fuan ni naru*).[91] Specialists also pointed out that the Shuttle was less than optimal for their experimental needs, preferring larger scale space probes, or planetary observation platforms, as objects of investment.[92] Thus, despite support from figures like STA director general, Tsaburō Kumagai, it was decided that rather than contributing to America's overpriced space plane, Japan would be better off developing its own space vehicles.[93] The decision seemed vindicated following the Challenger disaster in 1986, and the fact that as early as 1987, even US observers were calling for the need for a "reborn" space program, free of the Space Shuttle.[94]

ECONOMIC TENSIONS AND THE DEATH OF JAPAN'S SATELLITE INDUSTRY

The consequences of these cumulative tensions, and the capacity of the United States to torpedo economic developments it didn't like, is starkly illustrated by the fate of Japan's satellite industry in the 1980s. From the launch of Ohsumi in 1970 until 1984, only state agencies could operate satellites in Japan, and these were restricted to purchasing from NEC, Toshiba, and Hitachi. These companies had US partners they relied on for certain components and advice, but were protected from overseas competition as the only purveyors of complete devices to local customers. In 1984—under US pressure—government-approved companies were also permitted to join the sector, but their purchases were still regulated. Bolstered by this protectionism, Japanese companies dominated the Japanese satellite market.[95] The result was a lucrative, but inefficient, satellite industry, wherein producers sold less capable devices at prices up to twice the price of US equivalents. For this reason, they made very few exports of whole satellites—in 1988, of the 76.3 billion yen spent on satellites in Japan, only 3.6 percent came from overseas customers.[96] Worse, perhaps, was the fact that despite investment in local innovation, their technologies remained stubbornly behind those of the United States. This was a reality that worried even Japan's political class; in 1988, for example, members of the Diet bemoaned the fact that Hughes and local Japanese agents were working together to buy and launch a Ford-built satellite, while the STA's "toothless" (*ha no uku yōna*) policies had thus far failed to create viable domestic alternatives.[97]

Equally concerned, and more determined to do something about it, was the United States. For it, the situation not only shut out American satellite producers, but fit into the growing narrative that Japan possessed unfair economic advantages. Already, the United States had exerted pressure to resolve issues such as the artificial weakness of the yen, which gave Japanese exports a leg up; this system ended in the Plaza Accord in 1985. Then in 1989, it targeted Japan's coddled satellite industry, evoking the "super" 301 clause of the Trade Act of 1974 in response to what it deemed to be "exclusionary government procurement practices in the satellite and supercomputer sectors in Japan, which bar foreign suppliers."[98] This enabled Congress to punish such "exclusionary practices" through targeted tariff increases on Japanese exports. After negotiations in the 1990s, the Japanese state capitulated, agreeing to limit its policy of buying locally only to "research and development satellites."[99] Commercial producers and purchasers, cut loose, pointed out that the far more advanced technology of US producers meant Japanese competition was more theoretical than real, and that the US Department of Department of Defense and NASA funneled twenty times as much money into US commercial producers. Why was *that* kind of state support acceptable, but not Japan's model?[100] Politicians were also quick to point out that the agreement left "a very strong possibility that the US, which has an established satellite mass production system, will dominate the Japanese satellite market."[101] And indeed this is precisely what came to pass. As a 1999 report from the International Institute of Space Law puts it:

> As a result [of Super-301], the Japanese satellite industry, thus far, has not won any contract as prime contractor, public or commercial, either abroad or domestically . . . [this] has resulted in the U.S. satellite industry being awarded all satellite procurement contracts by the Japanese Government and its related entities. The Japanese industry not only lost in such open bids but also lost their chance to acquire satellite technologies and become internationally competitive. The Japanese Government is also forced to suppress its strategy to get autonomous satellite technologies due to domineering and unilateral U.S. strategies. In short, such strategies have worked very well just as the U.S. expected.[102]

Such moves by the United States were seen in Japan not necessarily as expressions of strength. As mentioned, many specialists had already concluded that the US technological advantage was slipping; in this they concurred with a 1980 ESA report that identified a "relative stagnation in the field of conventional launchers" within the United States. For the Japanese, this meant that in the long term, there would be less difference between their launch capacity and that of the United States, and so it would behoove them to look for other, perhaps more pliable partners.[103] Thus in the 1980s, various space agencies put feelers out to ESA. Beginning in 1992, NASDA and ESA signed an agreement to make their hardware mutually compatible, and to train astronauts together for eventual deployment onto the space station. The two space programs would continue to move closer throughout the 2000s.[104]

Meanwhile, the decision was made to move in the opposite direction when it came to the United States. NASDA committed in 1984 to develop, from scratch, an entirely domestic LFR system, the H-2.[105] When asked why Japan would undertake such a colossal project—not just building an MLV, but designing one from scratch—project head (and future JAXA senior vice president) Godai Tomifumi's answer was blunt: the Japanese space programs wanted "freedom."[106] He could not possibly have foreseen that it would in fact bring the exact opposite. H-2 would become one of Japan's signal spacefaring failures. In its wake came swathes of resignations, the complete reorganization of Japan's civilian space programs, and an end to dreams of Moon bases and space planes. In 1984, however, this all remained far in the future. As far as specialists in Japan's space industry, observers, and the public were concerned, a permanent Japanese presence in space was in the offing. All they needed was a big enough rocket to get them there.

KIBŌ AND THE INTERNATIONAL SPACE STATION

A reader might be forgiven for concluding from the preceding section that the US space technological relationship with Japan was primarily characterized by tension. However, as with so many fields of international relations, countries that squabble viciously over one thing can be firm allies in their handling of another. Accordingly, throughout their existence, all of Japan's space programs—even crotchety ISAS—have cooperated in research projects, academic exchanges, experimentation, and fundamental research with American colleagues. Of all these, perhaps no project better illustrated the tremendous potential of US–Japanese cooperation in space than their most ambitious joint project: Kibō, the Japanese module of the International Space Station (ISS).

Spearheaded by the United States in the 1970s and 1980s as part of NASA's post-Apollo program, ISS was conceived as a multipurpose space laboratory and observatory capable of hosting multiple inhabitants at the same time.[107] Japan's Space Activities Commission responded with alacrity when the idea for a permanent human facility in space gathered momentum in the United States in early 1981, perhaps to appease those still unhappy with the country's refusal to join the Space Shuttle project. Political interest in the ISS project also picked up immediately after President Ronald Reagan's 1982 announcement that the United States would be soliciting partners for its construction. By 1983, the Japanese had submitted an interim report to NASA detailing potential areas of cooperation and co-experimentation. Soon after Reagan's formal announcement of the project on January 25, 1984, NASA administrator James Beggs met with Japanese prime minister Yasuhiro Nakasone to discuss details; subsequent negotiations produced a memorandum of understanding—essentially, a preliminary agreement for cooperation—on May 9,

1985.[108] Japan's contribution was to be an impressive structure, featuring not only internal experimental spaces, but an external "exposed facility" for research to be conducted in the vacuum, and a manipulator arm so that astronauts could carry out their work without the difficulty and danger of a spacewalk.[109] The robotics this entailed were particularly complex, but detailed studies and plans had already been completed before the agreement was made.[110] By 1993, the Tsukuba Space Center even had a purpose-built Space Station Integration and Promotion Center dedicated to coordinating experimentation on ISS.[111]

However, before any of this could happen, proponents of the project had to overcome resistance focused on the two abiding concerns about Japanese space technology and its usage—namely, cost, and potential violations of Japan's doctrine of "peaceful usage."[112] Of these, it was the latter that particularly exercised lawmakers. Reagan's announcement in 1983 that the new Strategic Defense Initiative would rely on space-based weaponry to protect the American homeland—hence its popular nickname, "Star Wars"—raised the specter of a militarized cosmos.[113] When questioned regarding the dangers of this, Yasuhiro Nakasone was forced to concede during a 1984 meeting of the Diet Budget Committee that when it came to the overlap between civilian and military space technology usage in the United States, "the reality is that there isn't always enough information" (*genjitsu mondai to shimashite, nakanaka jōhō mo kanarazu shimo jūbun de naku*) for the Japanese to go on.[114] Members of Japan's leftist parties were further concerned by comments made by NASA specialist Frank Culbertson regarding the potential military activities on the ISS, and the fact that some 30 percent of the Space Shuttle's budget had, in fact, come from the Pentagon.[115] They had none of Nakasone's doubts: "military purposes," observed Japan Socialist Party (JSP) representative Katsu Shinmura in March 1984, "are an integral part of American [space] projects." How, then, could Japan participate in the project while still maintaining the integrity of its postwar pacifist constitution?[116]

Proponents of cooperation responded by emphasizing the benefits of cooperating with the United States—and not just on ISS.[117] From the perspective of researchers and space administrators, it was crucial to maintain access to American satellites and facilities for the purposes of research.[118] Major joint projects such as the Tropical Rainfall Measuring Mission, finalized at the height of tensions over Japan's domestic satellite industry, had assured researchers that at least in certain areas, the two countries could engage in fruitful cooperation.[119] Many were also keen to point out that tensions in areas such as the satellite industry had nothing to do with their own research plans.[120] Furthermore, the Japanese and US space programs were bound by joint infrastructure, such as tailored data exchange systems, created to ensure more efficient collaborations.[121] When it came to Kibō, administrators were concerned that *not* accepting the invitation might have negative consequences for future plans. For example, despite its original reservations, NASDA was coming to regret not participating the Space Shuttle project, which ended up being a crucial part of Japanese human spaceflight; half of the fourteen Japanese citizens to visit space, including the first woman to do so, Chiaki Mukai, did so on Shuttle missions. The project was also receiving attention and support from outside the space community. No doubt intrigued by the practical and reputational benefits, even lawmakers concerned about cost and military usage understood that Japanese participation would be "important for the development of science and technology" in the country.[122] This line of thinking was shared by administrators such as the director general of the Science and Technology Planning Agency, Kimio Fukushima, who assured lawmakers that the ISS was to be used for entirely peaceful purposes, and that Japan would only spend money on its own module. If other contributors had military functionality in theirs, it would have nothing to do with Kibō.[123]

Any concerns on the matter were finally put to rest in the March 14, 1989, memorandum of understanding between the United States and Japan, which initiated the seventeen-year process of designing, building, and launching the Space Station. Part of the agreement was that Japan could veto experiments from being carried out on Kibō. By this point, the project had even acquired some public interest; a feature article in *Asahi Shimbun* that year delved into its advanced capabilities, and highlighted the fact that the facility would enable scientists to carry out their space experiments in the comfort of a pressurized space, which would allow them to work in just their shirts.[124] On April 14, 1989, the House of Representatives approved Japan's participation.[125] Ultimately, Japan would invest around $6 billion in the project, a price tag that included not only the module itself, but also the extensive ground facilities required to run it; alongside the H-2 rocket, it was to be the country's most expensive space project.[126] The module itself, completed in 2008, weighs nearly 49,000 lb (22,200 kg), and remains easily the heaviest Japanese-made object in space—though, it should be noted, smaller than Skylab.[127]

Toyohiro Akiyama's mission may not have delivered the commercial success TBS anticipated, but its significance in Japan's space history cannot be understated. In hindsight, his journey marked the culmination of a momentous era of advancement and ambition in Japan's space ambitions, characterized by substantial growth in infrastructure, technological capability, and international standing. The period from the 1970s to the early 1990s saw Japan's space programs demonstrating their prowess in launching sophisticated satellites, developing complex space systems, and expanding their launch capabilities. This development was also evident in the establishment of facilities like the Tsukuba Space Center and the Sagamihara Campus, along with significant private sector contributions from companies such as Mitsubishi Heavy Industries, underscoring Japan's commitment to building a robust space industry.

However, this period was also marked by significant challenges, particularly in the realm of international technology transfer. Japan's reliance on American LFRs was a double-edged sword, providing essential knowledge and components, while simultaneously imposing limitations through blackboxing and restrictive agreements. The geopolitical and economic tensions between Japan and the United States often stymied Japan's efforts to achieve greater autonomy in space. Yet despite these hurdles, Japan's space program remained resolute in its pursuit of independence, as exemplified by the decision to develop the "purely domestic" H-2 rocket MLV. If successful, the vehicle was to put Japan in the upper echelons of space explorers by providing it with a reliable, capable, and potentially lucrative launch vehicle.

The foundations laid during these decades have had a lasting impact. The strides made in satellite technology, launch vehicle development, and international collaboration set the stage for future advancements. Indeed, Kibō was by no means the last major collaboration with the United States in space exploration. As Yasuhiro Kato of the Science and Technology Agency pointed out in 1995, "the United States is the most advanced country in the field of space, so cooperation with the United States will greatly benefit Japan."[128] Japanese space plans continue to involve detailed examinations of precisely what the Americans, for their part, had planned for the future.[129] The two countries' ongoing cooperation is exemplified by the role Japan has secured in the mooted American return to the Moon in the 2030s. According to plans announced by NASA in April 2024, US astronauts will tool around the Moon on a Toyota-built moon buggy, the Lunar Cruiser, named for the company's Land

Cruiser vehicles. In return for its provision, the United States has agreed to carry a Japanese astronaut to the Moon as the first non-American to fly on the new Artemis program. The notion of a Japanese astronaut driving around the Sea of Tranquillity in a Toyota will surely warm the cockles of many a Japanese space engineer's heart.[130]

EIGHT

Success and Failure in the 1990s

ON JUNE 18, 1992, NASDA hosted a press conference to discuss progress on its impressive new LE-7 engine. Just a few days earlier, the device—crucial to Japan's ambitious new MLV, the H-2—had successfully fired for 365 seconds. Yet during a follow-up test in Tanegashima that afternoon, it had burst into flames, crashed through its testing frame, and fallen 75 ft (23 m) to the ground. Such engine failures may be par for the course for space development, but one thing was particularly noticeable about this event: when the red-eyed public relations officer stepped up to the stage to make the announcement, it was evident he'd been crying.[1]

The failure came at the beginning of what would prove to be a tough decade for Japan. In the 1980s, the country's biggest trade partner, the United States, had struck a series of blows against what it saw as unfair competitive advantages held by the Japanese. These included the 1985 Plaza Accord, which resulted in a stronger yen, which squeezed Japan's exports across the Pacific. A concurrent loosening of investment rules in Japan led to an economic bubble, as Japanese capital flowed into investments such as real estate—at one point, an acre in Tokyo's Nihonbashi district was valued at more than the entirety of Central Park in New York. With the economy overheating, in 1989, the Bank of Japan raised interest rates from 2.5 percent to 6 percent.[2] Worried consumers and companies moved to protect their cash; by 1995, spending had plummeted, the stock market had crashed, and unemployment had increased from 2.1 percent to 3.2 percent. In the decades that followed, interest rates dipped into the negative, and unemployment reached 5 percent.[3] Average income by 2002 was 6 percent lower than in 1990, and 10 percent lower than it had been in 1992.[4] And there were other major concerns bedeviling the nation beyond economics. China was growing more regionally assertive, while North Korea responded to famine and economic collapse by developing nuclear weapons and missiles. National fertility declined, meaning that Japan's taxpayers faced a growing burden of paying for older generations—who, having worked hard to achieve a "bright new life," now expected to reap the rewards in retirement. After the postwar boom, Japan had now entered the "lost decades," a less optimistic, more inward-looking, and more anxious era than the one that had preceded it.

The NASDA official's weeping was symptomatic of the gathering despair of the times. The engine failure was just the latest in a seemingly endless stream of mishaps that had bedeviled the H-2 project. In the end, rather than a reliable and cost-effective vehicle, Japan's first effort at producing an entirely domestic MLV would result in engine failures, deaths, and a pair of expensive failed launches. Worse, as the single biggest project failure in Japanese space history—in terms of time, money, and reputational consequence—it affected almost every major stakeholder in the space enterprise. Politicians who had long held dreams of Japan becoming a major player in the commercial space market were disappointed, and concerned by the loss of up to ¥650 billion ($4.7 bn in 2025) in research costs. Scientists, engineers, and technicians suffered lost payloads and scuppered dreams. Corporations saw hopes for a lucrative commercial launch vehicle crushed. Perhaps worst of all, the public came to view the vehicle itself as emblematic of Japan's lost 1990s. The project would, in the end, mark a turning point in Japan's space program: instead of ensuring Japan's ascent to an elite club of space powers, its failure would trigger an end to the decentralized model of space development the country had persisted with since the 1960s.

The Rise and Fall of the H-2 Rocket, 1984–1999

EXPECTATIONS AND AMBITIONS

The H-2 was conceived as the vehicle that would set Japan up alongside the United States, Russia, China, and Europe as one of the "cutting-edge" players in both space exploration and commercial launches.[5] First proposed in 1978, the project was approved in 1984 by the Space Development Commission. As described in the previous chapter, the underlying goal was to break free of the limitations placed on imported US technology, and achieve medium-lift independence.[6] The sheer scale of the project is evident in the fact that it required the involvement not only of NASDA, ISAS, and the NAL, but also seventy-five private corporations, including the usual suspects of MHI, IHI, Kawasaki Heavy Industries, and NEC.[7] Crucially, nearly every single component on board would be domestically manufactured—the most common descriptor attached to it in the press was *junkokusan*, "purely domestic."

In addition to these hefty strategic goals, the project also appealed to the thriftiness of the space programs' sponsors. NASDA's budget was constantly under public scrutiny, and leftist politicians in particular took every opportunity to question its expenditure.[8] As a result, after some dramatic growth, the institution's budget increased only modestly in the early 1980s, meaning that any new project would have to compete for financing with other efforts such as the JEM (as Kibō was known while under development). Indeed, the H-2 project experienced the potential hazards of this firsthand in 1985, when the Ministry of Finance diverted around ¥960 million ($4 million) of funding from it, to fuel ongoing investigations into whether or not Japan should participate in ISS.[9] As a result, cost-effectiveness was a critical element of the H-2 project's institutional appeal. Project leaders consistently spoke about how much it would cost in comparison to similar projects in other countries.[10] Their arguments were only slightly tempered as the price tag for the project increased between 1985 and its first launch in 1994. In the end, development costs for the rocket alone came to around ¥285 billion ($2.5 billion) in 1992; for comparison, the equivalent Ariane-5 cost around $5 billion, while the American Atlas V clocked in at $6 billion.[11]

Yet it was not enough for H-2 to be cheap—it also aimed to be profitable. Japanese corporations had long dreamed of producing commercially viable launch vehicles. In the early 1960s, for example, Prince Motors (later Nissan) and MHI had parlayed MITI funding into the development of small-scale weather-observation sounding rockets as part of an "IX" ("Inexpensive") rocket program.[12] These same companies would have been encouraged when, in 1985, Koji Shibato, head of NASDA's public relations division, stated at a symposium in Tokyo that launches on the H-2 would be "cheaper than on America's Space Shuttle or Europe's Ariane." Chief project engineer Tomifumi Godai echoed this by floating a figure of $24,000 for 2.2 lb (1 kg) to GEO, two-thirds of Ariane-4's costs of about $36,000.[13] The rocket would also be flexible enough to deploy multiple variations, allowing it to tap into demand from diverse domestic and overseas customers. As a result, NASDA anticipated the need to launch several times a year.[14]

A lucrative launch system became even more attractive as the country slipped into an era of low growth in the 1990s. The activation of Super-301 against Japan's satellite industry, the main source of commercial income from space technology in the country, had been a further motivator to find other ways of making money in space.[15] Thus in 1990, even before the rocket was complete, the Rocket System Corporation was established in Japan to handle sales of payload launch services.[16] The company set an initial goal of 25 percent of the global launch market, and wasted no time in bidding for a 1994 INMARSAT launch, marking Japan's first major foray into the commercial launch market. Eventually, in November 1996, Hughes Space and Communication signed a ten-launch deal for $830 million on the planned spin-off H-2A rocket.[17] That the deal was struck soon after one of H-2's rivals, the new Ariane-5, suffered a major failure, was surely a bonus.

In keeping with these expectations, NASDA committed to raising the public profile of Japan's glorious new spacecraft. Even by the standards of the Japanese space programs' extensive outreach, H-2 received VIP treatment. In 1990, four years before its first launch, Kakuda Space Center erected a full-scale model of the vehicle on its grounds. Another model was featured at the YES 89 Space Hall at the Yokohama Expo, and was later moved to the Tsukuba Expo Center.[18] Various parts of the rocket made appearances in places as far-flung as Higashiura in Aichi, Tochigi, Tokyo, Chiba, Sendai (here, displayed with mock-ups of the HOPE-X space plane); on boards in train stations; as soft toys; and on TV.[19] In 1999, Tanegashima museum even added an "H-2 Launch Experience Theatre," so visitors could experience "the launch of the H-2 rocket on a large screen and with loud sound" in order "to create a more powerful experience" of launches "than ever before."[20]

THE PROBLEMATIC DEVELOPMENT OF THE H-2 ROCKET

Despite these high expectations, however, there were early signs that the road to medium-lift independence would be bumpy. First, the greater size and power of the rocket necessitated the creation of swathes of entirely new infrastructure. Some of these costs were defrayed through private companies such as Mitsubishi, which had invested in an 830,000 sq ft (82,000 sq m) expansion of its rocketry production facilities in Nagoya.[21] Meanwhile, NASDA's budget was increased so as to expand Tanegashima launch center to accommodate H-2's requirements with an entirely new launch pad, a vehicle assembly building, and a mobile launcher capable of moving the 250,000 lb (115,000 kg) rocket 1,650 ft (500 m) to the pad's all-new service tower.[22] Meanwhile, creating a system to handle its liquid hydrogen fuel was so complex it had been the subject of studies since at least 1977.[23] Liquid hydrogen has a nasty tendency to soak through a variety of materials and evaporate, so hydrogen disposal facilities were needed to prevent roving clouds of the stuff from wafting around the facility, looking for a flame. Thus in 1983, MHI joined with the Iwatani Manufacturing Group to create a new commercial entity, the Japan Hydrogen Fuel Corporation, for the sole purpose of developing and building the systems that would be required to handle the finicky fuels.[24] They joined the construction boom at Tanegashima, contributing to a 16,000 sq ft (4,800 m^2) storage yard that featured two (later three) huge 1,800 cu ft (540 cu m) hydrogen storage tanks.

Beyond this expense, technical issues bedeviled the project from the very beginning.[25] Some had been anticipated; Tomifumi Godai had observed in 1985 that the rocket posed the fearsome challenge of developing "a number of difficult technologies in tandem."[26] Godai would certainly have known—he'd previously worked on ISAS's successful Kappa, Sigma, and Lambda rockets.[27] He thus understood that the seven-year development period for H-2's laser guidance system, for example, was par for the course for such efforts.[28] Nor would he have been particularly concerned about the development of the second-stage LE-5A engine (modified from the H-1 rocket's LE-5 engine).[29] What did end up taxing his fortitude—and that of nearly everyone involved—was the 3,400 lb (550 kg) first-stage LE-7.[30] Like Godai, many of the engineers who worked on it had experience with other Japanese launch vehicles, and so went in confident that this new device would not be a major challenge, only to find themselves "utterly knocked back" (*migoto ni hanekaesareta*) by persistent problems.[31] By far the most difficult component of the engine proved to be its turbopump. LFRs work by igniting and blasting out thousands of liters of fuel to provide thrust; turbopumps are required to ensure that enough fuel is in the combustion chamber to move the rocket along. In the case of H-2, turbopumps had to move oxygen and hydrogen, both at high temperatures, and the blades within spin many hundreds of times per second. The slightest flaw can cause a malfunction. Given that each LE-7 had to provide ten times more thrust than any predecessors, malfunction it did—repeatedly.[32] In 1989 and 1990 alone, multiple fires occurred during testing, and the speed of the blades was reduced by 4,000 rpm, cutting thrust, and lowering payload capacity.[33] In 1992, just one year before H-2's planned first launch, more problems in the component necessitated even more changes.[34] Adding to all the difficulty was the fact that the turbopump wasn't the only part of H-2 with a tendency to buckle; the very next month, cracks in the nozzles of two engines delayed the rocket's first launch by a year.[35] By this point, concerns had begun to build in political circles. In May 1992, members of the Diet started to grumble about the rising cost of the project, and missed deadlines.[36] The LE-7 engine's failure came the very next month. Yet again, the culprit was its "weak heart," the turbopumps.[37]

To make matters worse, the H-2 project also faced circumstantial issues that slowed its development. First was the fact that any sort of test in Tanegashima involving liquid hydrogen required a 1.85 mi (3 km) radius exclusion zone be initiated up to four hours prior to the actual test.[38] Such activities thus fell firmly under the agreement with fishermen in the area, which permitted launch activities for only two 90-day slots per year. If issues were not resolved by the end of this window, developers had to wait months to test their alterations.[39] In fact, even launches of the completed H-2 were limited to these two six-week slots until 1997, when the windows were extended to 130 days each, in return for various concessions to the fishermen.[40] These delays were particularly unwelcome to budget-minded members of the Diet, some of whom even suggested that Japan should buy access to overseas launch pads so as to guarantee they could launch H-2 often enough to make money.[41] Meanwhile, other issues cropped up that seemed to be less problems of policy than bad maintenance or bad luck. In May 1991, an explosion at Kakuda Space Center caused by rocket fuel blew the slate roof off a facility, alarming nearby residents.[42] Then, on August 9, 1991, twenty-three-year-old technician Arihiro Kanaya died at Mitsubishi's Nagoya Guidance and Propulsion Systems after an injector pipe burst due to defective welding. His was the first death in Japan's civilian space programs, and it led to an investigation into potential violations of safety law by MHI—although in the end, it was let off the hook.[43]

THE FAILURE OF THE H-2

Despite all these travails, H-2's first launch went off without a hitch on February 3, 1994.[44] Instead of drama, the launch team celebrated not only a successful test, but also the fact that Mariko Okuda, wife of the LE-7's combustion test leader Ta-

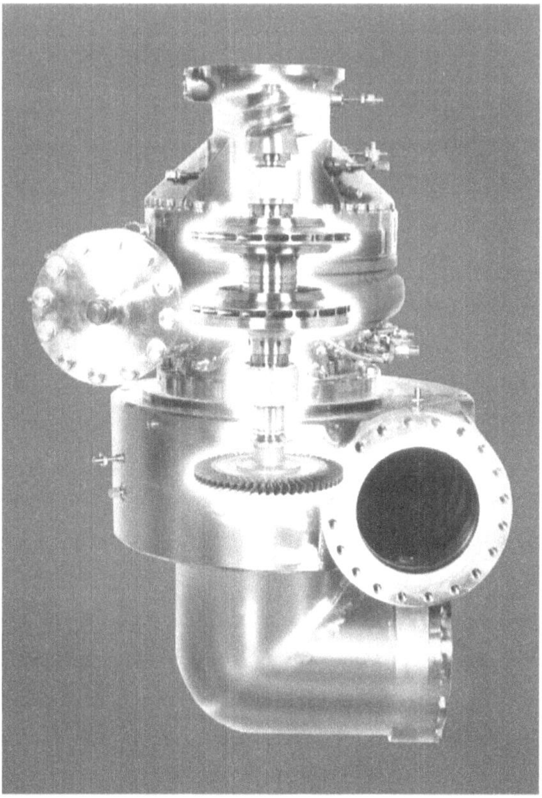

FIGURES 8.1 A AND B. Turbopumps for the LE-7 engine.
Courtesy of JAXA.

FIGURE 8.2. H-2A rocket prepped for launch. Note the large Mitsubishi emblem on its side.

Source: Wikimedia Commons.

dashi Okuda, had given birth during the launch. There had to be some connection, they argued—so perhaps Okuda could name his boychild "H-2 Tarō" in honor of his timing?[45] Meanwhile, outside the control room, residents of Tanegashima celebrated as well; *ryōkan* owners were pleased that their hotels were full, and local bento makers sold four times the usual amount of their products.[46] Even Shima Hideo, now ninety-three, tuned in to watch the launch—although, alas for him, his satellite feed was so bad he had to rewatch the whole thing on video later.[47]

Unfortunately for all concerned, however, the good times did not last. It quickly became apparent that even when everything was working as intended, serious problems remained that would prevent H-2 from living up to the high—arguably too high—expectations of everyone involved. First, the rocket remained persistently pricey, meaning it had no future as a source of profit. This issue was, in some ways, beyond the control of the rocket's developers. When H-2 was under development in the late 1970s and early 1980s, the yen had settled at between ¥270 and ¥230 to the dollar; NASDA had used this exchange rate to calculate the project's budget in

1984. The very next year, however, the Plaza Accord forced a revaluation. By 1987, the exchange rate was down to ¥130 to the dollar, and by 1995 it had reached ¥81.[48] For the H-2, this meant that the 1984 price tag of around ¥25 billion was the equivalent of $100 million per launch—generally regarded as a base competitive price. By 1992, however, this same price in yen had ballooned to $190 million. This was devastating to commercial prospects; there was simply no way overseas customers would choose to pay double to fly on an unproven new rocket with a troubled development.[49] And, indeed, they didn't: by H-2's first launch in 1994, Rocket Systems had made and lost four overseas bids.[50] Alarmed that one of the chief promises of H-2's developers now appeared to be a nonstarter, lawmakers demanded solutions.[51] One suggestion was to explore the possibility of creating a separate, cheaper vehicle, primarily for the purpose of commercial launches. NASDA's first attempt at doing so came in 1991, when it collaborated with ISAS to create the J-I, an SFR made from preexisting parts. This project failed on account of "severe schedule and budget constraints" and was canceled after just one launch.[52] Another approach was to pursue a program of rigorous cost-cutting. In 1995, Rocket Systems managing director Hidetaka Kawashima assured lawmakers and future customers that "by changing the design and importing parts," prices could be reduced, by the rocket's twelfth launch, to roughly half of 1994 levels.[53] However, the prospect of such stringent measures alarmed NASDA's private sector partners like IHI, who pointed out that such economies wouldn't be possible without producing parts for at least fifty H-2s.[54] There were also consequences for the hundreds of subcontractors who produced specialist parts. Some, such as Isomura Manufacturing, a fifty-person enterprise located in Nagoya, had invested heavily in new equipment and training after receiving contracts for the project. Although its involvement in the project had certainly raised its profile, and hence brought in some new business, concerns that high launch costs might scupper the whole effort were a source of great anxiety to managers and employees alike. The company's president put it bluntly: "Dreams don't pay the bills."[55] Such companies also felt trickle-down pressure to reduce production costs, and understood that if this wasn't possible, the next generation of Japanese rockets would almost certainly import their components instead. In 1997, NASDA decided to pursue both cost-cutting and the development of a cheaper rocket concurrently. The first step was simplifying the LE-7 into a cheaper and simpler variant, the LE-7A.[56] This engine would be part of a cheaper spinoff from H-2, dubbed H-2A, which from 1997 offered launches for as low as $100 million—but at the cost of being built with some imported components.[57] The rocket eventually secured a contract from Hughes for ten launches, though in the event it only ever carried one commercial payload, a South Korean satellite in 2012.[58] This is, to date, Japan's only such launch.

Second, even if costs were lowered, the global orbital launch market in the 1990s proved to be both more competitive, and smaller, than expected, meaning the Japanese faced extreme difficulty in establishing a place within it. First, Japan's economic crunch of the 1990s put severe constraints on the space plans not only of official institutions like NASDA, but of private corporations as well. There were, thus, far fewer domestic launches in the 1990s than had been anticipated. Concurrently, the satellite launch market outside Japan had also failed to grow as much as expected. Early projections had suggested that by the time H-2 was active, there would be international demand for up to twenty-five launches per year. Yet by 1993, developments such as globe-spanning fiber optic cable networks had left that figure closer to an anemic fifteen.[59] On top of this, post-Soviet Russia entered the launch market as a particularly fierce competitor, offering launches for between \$15 million and \$30 million a go aboard its Proton MLVs.[60] Meanwhile, while LE-7 was in flaming limbo, ESA's Ariane-4 established itself as capable of reliably launching ten times a year, and had more than forty launches booked between 1992 and 1997. These included Japanese companies that were either unwilling to use, or unable to afford, domestic launches.[61]

The final nail in the H-2 project's coffin was the rocket itself, which proved to be a fickle and unreliable beast. In 1995, NASDA president Masato Yamano had claimed an estimated reliability rating of 97.44 percent for the vehicle—confidence that, in hindsight, seems comically misplaced, given that the rocket's career ended with two major launch failures in 1998 and 1999.[62] Its first major failure came on February 21, 1998, when the LE-5 engine in its second stage failed to fire properly.[63] Given the engine's usual reliability, this was both unexpected and deeply worrying. As a result of the mishap, the H-2 put its payload, the pricey COMETS/Kakehashi satellite, in the wrong orbit. The resulting "dark cloud . . . over Japanese space development" provoked apologies from the director general of the Science and Technology Agency, the minister of Posts and Telecommunications, and the president of NASDA, the last of whom expressed his "deepest apologies for being unable to live up to the expectations of the public and related organizations."[64] Yet the run of bad luck continued. In March, forty-nine-year-old Kiyotaka Fukamachi, a construction worker at the new launch pad for the H-2A rocket, died after an accident while moving a ditch covering.[65] Preparations for the next H-2 launch, scheduled for September, were beset with problems, including damage to the rocket's liquid hydrogen detection sensor. Resolving the situation required the removal of the first stage liquid hydrogen tank, and the formation of a "special on-site inspection team" to improve "the reliability and awareness of on-site work by on-site engineers."[66] When the rocket finally launched on November 15, it sailed upward for four minutes, and then tumbled into the sea, along with the meteorological MTSAT-1 satellite it was carrying. While Kakehashi had been an experimental satellite, MTSAT-1 had been intended to replace Japan's fleet of aging weather satellites; the failure meant that when the GMS-5 satellite finally ceased functioning in 2003, Japan would be without its own weather satellite system for the first time in decades. Determined to find out the cause, NASDA organized an extensive deep-sea search (itself a remarkable technological accomplishment) and recovered the wreckage from waters 10,000 ft (3,000 m) deep, just off the Ogasawara islands.[67] To its dismay, yet again, the crash had been caused by the LE-7's turbopumps.[68]

The 1999 launch was H-2's last outing. Not only was it unlikely anyone would want to launch on it, but even if they did, it would be nearly impossible to insure payloads, reducing its potential customer base to effectively zero.[69] Furthermore, it would take many years of expensive revision for the vehicle to reach the standards required. The press understood what the crash signified: "Longtime dreams of 'getting fully into the satellite launch business,'" *Asahi Shimbun* observed, "have fizzled out."[70]

The institutional fallout from the crash was dramatic. Within a couple of days, Minister of Transport Toshihiro Nikai had visited Tanegashima to meet with NASDA representatives to discuss the failure. Within weeks, the vice minister for science and technology, Toshi Okazaki, had resigned, taking responsibility for the project's demise. The same day, Prime Minister Keizō Obuchi spoke of the necessity to "fundamentally rebuild science and technology administration" in Japan.[71] By the end of the year, Japan's "purely domestic," high-tech, profit-generating rocket project was dead.[72] Still, the aftershocks continued. Within a year, both Isao Uchida, the head of NASDA, and Eiji Sogame, the agency's director of rocket development, had also resigned. In a final blow, in May 2000, a spooked Hughes pulled the plug on its $830 million deal for launches on the H-2–derived H-2A. Japan's hopes of becoming a competitive commercial space launch provider had been well and truly snuffed out.[73]

PROFESSIONAL FRUSTRATION AND PUBLIC DISAPPOINTMENT

The extent and the extreme nature of the political reaction to H-2's failure becomes easier to understand when considering that nearly every major stakeholding group within Japan's civilian space programs came away from the project with a profound sense of failure. First, the sheer difficulty of the development work had come as a shock to satellite designers and industry workers, for whom H-2 had provided a glimmer of hope at a time when the space industry was reeling from the consequences of Super-301.[74] Senior figures responsible for the project at IHI and MHI had despaired of the whole thing by the early 1990s, having seen for themselves, as the director of Mitsubishi's Space Systems Department, Saito Hiroshi, put it, "how tough it is to complete something so big." Yet their frustrations paled in comparison to the pressures employees of NASDA were under. The inability to get basic elements of the project right were particularly galling—the chairman of the H-2 project board, Masato Yamano, recalled thinking after the June 1992 (largely Mitsubishi-manufactured) LE-7 failure that "it was just a ten second burn test. I was sure it would be a success."[75] Compounding this stress was the fact that Japan's space programs found themselves buffeted by negative press coverage. Looking back on the period, Yasunori Matogawa, head of ISAS at the time, would complain about what he saw as the media's selective vision in emphasizing the space programs' failures, without highlighting the positives, or understanding the issues properly.[76] Meanwhile NASDA's staff had become so sensitized to criticism by the early 1990s that when technical problems delayed the second H-2 launch in 1994, they were keen to emphasize that it was "not a failure" of the rocket, but of ground equipment. *Asahi Shimbun* obligingly reported their comments in a piece snarkily titled "Not a Failure" ("Shippai de nai").[77] In the end, the staff working on the project had to develop thick skins. Tomifumi Godai himself later observed that they learned to become "unfazed by things" in order to get the job done.[78] Despite this, they still had to face the slew of procedural work that resulted from every failure. Though intended to ensure that components work, these processes could be onerous—using the LE-7 engine alone had come to require some 103 coordination meetings between different groups involved in building and designing.[79] Prior to the final H-2 launch, specialists had carried out four test firings and conducted other major hardware evaluations as little as five weeks before launch.[80] Thus engineers now found themselves having to deal with more paperwork and procedure, while also dealing with pressure from deadlines and technological challenges.

One response was to lean further into outreach, at least partly to show the public that the failures weren't complete write-offs of technology and effort. In May 1999, a report emphasized that the public should be informed clearly that space technology is a cutting-edge and therefore high-risk field of study, and liable to produce mishaps.[81] Specialists pointed out that prior to the Space Shuttle's success, there had been up to twenty engine failures, while NASDA released analyses of rocket mishaps in the United States, such as Titan 4A and Delta 3 failures in August 1998. Though technical documents, such reports served to emphasize that Japanese space vehicles were not the only ones that could occasionally blow up.[82] Those who had worked on the project began their analyses with graphs positioning the failure rate of H-2 (10 percent) within the context of other, similar vehicles like the US Atlas (also 10 percent), and China's Long March (11 percent).[83] Yet despite their diligence, they still found themselves on the receiving end of intense criticism from both lawmakers, and even their own institution: a 2000 NASDA report, for example, cited the "work mistakes of skilled workers" as a key element in H-2's failure.[84] The death of H-2 not only cast a shadow on their past work, but on potential future projects as well—plans for a new LE-5B engine, for example, were delayed, and the H-2A project was put on hold while it was thoroughly inspected and overhauled.[85] For the residents of Tanegashima, the idea that regular space launches would turn their village into a future transport hub lay in ashes; a local man, expressing his disappointment, commented that he had "wanted [H-2] to succeed for the sake of Tanegashima's future."[86] Others lamented that despite all the money spent, it hadn't, in the end, been enough. These included a Nagoya woman whose family owned a *ryōkan* in Tanegashima, who opined that "the rudimentary mistakes reported" about the rocket's launch delays were "related to the reduced budget." "It would have been nice," she went on to comment, "to have thought of [these problems] before the launch failure."[87] When Minister of Transport Toshihiro Nikai deployed the argument that H-2's failures were no worse than those of equivalent rockets in other countries, critics argued that Japan could ill afford such waste nevertheless.[88] For others, a Japanese failure so close to the launch of North Korea's Taepodong and the success of China's Shenzhou I in December 1999 was a huge blow to national pride. This anger was reflected in the press: "Which is the technological empire now?" demanded *Asahi Shimbun*.[89] H-2 even came to be emblematic of Japan's broader woes in the late 1990s. When the country's tenuous coalition government looked set to collapse in 1999, an anonymous poem appeared in a newspaper, written from the perspective of an H-2 sympathizing with beleaguered Prime Minister Ichiro Ozawa:

"Mr. Ozawa Ichiro," it says, "all your launches are failures too!" (*sochira no uchigete ha / shippai bakari shiteiru ne / —H2 roketto / Ozawa Ichiro san*).[90]

Such disappointment was compounded by the fact that the H-2 was not the only high-profile failure of Japan's civilian space programs of the 1990s. In fact, NASDA had suffered from a series of issues from 1994, when the ETS-VI engineering satellite failed to enter the proper orbit, and was abandoned—the first such failure in fourteen years. The mishap was later attributed to a lapse in ground testing of the device.[91] In 1996, the $1 billion ADEOS 1/Midori, which carried equipment from NASA and France's CNES, suffered from observation equipment failures.[92] The next year it failed altogether, due to a solar panel malfunction.[93] ISAS was having a rough time of it, too. In 1994, an M-3 rocket failed to put the Express satellite, a joint German-Japanese project, into the correct orbit; it eventually crashed in, and had to be recovered from, Ghana.[94] Then, in December 1998, its ambitious Nozomi Mars orbiter mission went off course, meaning not only that it would reach Mars four years late, but be incapable of entering orbit when it got there.[95] Aside from the loss of at least $88 million, and the data the probe would have carried, ISAS also lost face—the institution had encouraged ordinary people to put their name on a postcard and send it in, so that it could be carved onto a plate, attached to the probe, and sent to Mars. Some 270,000 people had taken up the offer. Their names would now never reach the Red Planet; in 2003, the probe was declared lost.[96] Project science director (and future director of ISAS) Koichiro Tsuruta later wrote of the "exhaustion" and "shock" his team felt after the incident.[97] Despite these protests, however, blame was laid squarely on them, and the mission's failure was again attributed to a quality assurance oversight. In comparison to that debacle, the failure that caused an ISAS M-V to drop the ASTRO-E earth observation satellite into the Indian Ocean on February 10, 2000, would have appeared paltry, if the payload hadn't been so valuable.[98] The gloom these failures provoked fed into a broader sense of the "Lost Decade."[99] After the economic crunch, there now seemed to be a rash of fundamental industrial issues bedeviling a country that had become rich building functional, reliable, high-tech products. In 1999, the year of H-2's failures, a lethal incident involving radioactive material occurred at the Tokaimura nuclear plant, and in Fukuoka, part of a tunnel collapsed onto a Shinkansen. The trio of mishaps seemed a suitable end to a dismal decade.

The H-2 rocket project was a defining experience of Japan's development of space technologies, and one of its greatest failures. Born of a desire to both break free of the United States and establish the country as a major commercial launch provider, the project received the support of nearly every major stakeholding group within the space enterprise. For politicians and the public, it meant national power and pride; for private corporations, profits and exports; for engineers and specialists, the fulfillment of long-held technical ambitions. Unfortunately for all of them, it proved to be an unworthy vessel for their dreams. Problems undermined the project from the beginning, extending its timelines and bloating its budget. Changing economic circumstances scuppered its potential as a source of income. And, when it was finally complete, it proved to be unreliable and expensive. Even before it had taken its first flight, some within the space industry were beginning to look at cheaper alternatives, but the price they would pay for this economy was abandoning the goal of creating a "purely domestic" rocket. Japan would, in the end, never become a major commercial launcher of payloads for overseas customers. Despite the reliability of the H-2–derived H-2A and H-2B, their customers would remain Japanese—with the sole exception of a South Korean satellite launched in the early 2000s. This is not to say that Japan has given up on its ambitions of cracking the market; more than two decades after the failure of H-2, longtime observers of the Japanese space industry might have felt something akin to déjà vu when a newly developed H-3 rocket suffered an even more inauspicious start to its career than H-2 in 2023. On March 7 of that year, due to a second stage failure, the vehicle's self-destruct sequence was initiated. Debris rained down into the sea. Japan's travails in space seemed to have come full circle.

Yet the failure of the H-3 is quite unlike that of the H-2, if only because the H-2 was the leading cause of the single biggest restructuring of Japan's civilian space programs in its history. Much of this came down to the shift in power within the debate about centralization—a talismanic goal lawmakers had been pursuing since the 1960s. When lawmakers set out to find answers as to why H-2 had failed, what they found caused them to redouble their criticism of the status quo. Reeling from the consequences of the dismal 1990s, those who had fought to defend ISAS, NASDA, and other entities from outside interference found that they'd run out of time. By 2003, Japan's space programs would finally have been integrated into one agency: the Japan Astronautical Exploration Agency, better known as JAXA.

NINE

Reform and the Creation of JAXA

THE FAILURE OF THE H-2 project hit one group of stakeholders in Japan's space programs particularly hard: the lawmakers who had been responsible for its funding and oversight. Enthused by the technonationalist implications of the project, they had been among H-2's most vocal supporters—a particularly passionate group went to the 1992 final fight of H-2's predecessor, the part-Japanese H-1, simply because it was expected to be the last Japanese launch reliant on foreign technology.[1] Others were more appreciative of the fact that the MLV would mean Japan could finally launch its own spy satellites whenever it wanted.[2] To still others, it was a "wonderful" (*subarashii*), "great" (*taihen*), and "top level" (*toppu reberu*) project, simply in terms of its ambition. One of the few complaints came in 1989, from LDP lawmaker, and future director of the Taisho Japanese zither (*kotō*) association, Kazuhiko Kimiya. "H-2 rocket is a bad name," he grumbled in the Diet. "Maybe we should change the name to something a little bit more romantic."[3]

In return, as agencies in receipt of funds from the state, both NASDA and ISAS were expected to keep various Diet committees abreast of their activities, and listen to their concerns. Diet members regularly conducted piercing examinations of their work; in 1994, for example, the director general of the Science and Technology Agency (which oversaw NASDA), Makiko Tanaka, was forced to stage a robust defense of the H-2 in the Diet's oversight body, the Science and Technology Committee.[4] Similarly, the rocket's chief designer, Tomifumi Godai, was interrogated in depth about the actual mechanism of H-2's dastardly turbopump in a March 2000 meeting of the same committee.[5] Failures in Japan's civilian space programs thus had very loud, very political repercussions, beyond the clamor of exploding rockets. When investigations into the H-2's failure placed blame squarely on the inefficiencies created by this decentralized system, even those bureaucrats and managers who had defended the status quo had to concede that reform was critical. The result was that in three short years, from 2000 to 2003, Japan's government brought an end to the country's decentralized space programs. The future would belong to a new, centralized agency, JAXA.

Political Concerns around Japan's Civilian Space Programs

Japan's lawmakers had always been dissatisfied with the organization of space R&D in the country. As early as 1967, the issue of how to better organize ISAS had already produced two subcommittees with the astonishingly anodyne names of "Comprehensive Subcommittee" and "Technical Subcommittee."[6] Lawmakers were already overwhelmed by the sheer number of space research bodies that had seemingly popped up overnight; as one complained in the Diet:

> The University of Tokyo currently has the Institute of Space and Astronautical Science. Furthermore, the Science and Technology Agency has the Aerospace Technology Research Institute and the Space Development Promotion Headquarters . . . Furthermore . . . the Ministry of Posts and Telecommunications has created a budget for communication satellites, [and] under the same supervision of the Ministry of Posts and Telecommunications, the Japan Broadcasting Corporation, Nippon Telegraph and Telephone Public Corporation, Kokusai Denden, and others all set their own budgets for broadcasting satellites and communication satellites. In addition to this, for example, the Japan Meteorological Agency is also considering meteorological satellites. In that case, over the next one or two years, government ministries and agencies will see a dramatic increase in science and technology-related budgets . . . it would be a waste of national funds.[7]

Others simply couldn't keep up; in 1965 Kagoshima representative Ki'ichi Murayama had complained that he could only make sense of what was happening at the Uchinoura space site through articles in the *Nihon Keizai Shimbun*. Thirty-four years later, in a 1999 meeting, some lawmakers sometimes had to be reminded precisely what the difference between ISAS and NASDA was.[8]

Those who did manage to get to grips with Japanese space organization didn't much like what they saw.[9] Take the division between NASDA and ISAS, for example. NASDA had been founded in 1969 as a "special corporation," according to director general of the Science and Technology Agency, Shiro Kiuchi, "to carry out space development comprehensively, systematically and efficiently" (*sōgō-teki, keikaku-teki katsu kōritsu-teki ni jisshi shiyō to suru*).[10] The organization was certainly presented to the public as an attempt at undertaking satellite and rocketry production in a more unified (*ichigenteki*) manner.[11] Yet, as shown in previous chapters, NASDA's reach and power did not extend even as far as ISAS. Lawmakers' confusion only increased when, in the early 1980s, the latter institution stopped developing its own rockets to focus on designing space probes.[12] If there was no clash in the field of rocketry development, wondered some, why then did ISAS need to remain independent of NASDA? Yet independent it remained; by 1994, ISAS was putting its satellites in orbit on rockets built by MHI, Nissan, and IHI—at least two of which were also main contractors for NASDA.[13] Even more confusing, ISAS was also actively involved in NASDA projects, such as the development of the HOPE-X space plane intended for use with the H-2 and, indeed, the H-2 itself.[14] This was just the tip of the iceberg. Japan's multitudinous space agencies frequently stepped on each other's toes.

This complexity was known, even before H-2's failure, to have two major consequences. The first was the potential for budgetary inefficiencies. Keeping the civilian space programs cheap had been a stated goal of their political overseers. Lawmakers had tolerated some growth in budgets over the years as Japan's activities in the field had been seen to bring tangible economic and political benefits, but still balked at spending too much.[15] Then, as projects like H-2 and Kibō gathered steam, Japan's overall expenditure on space research enjoyed increases of several percent a year from 1988, including an 11 percent leap for fiscal year 1989.[16] Space officials were repeatedly grilled over the use of these funds, particularly by leftists with an eye for detail; in 1988, for example, NASDA chief Uchida Isao was cross-examined by Japan Socialist Party councilor Jin'ichi Katayama over the relatively small failure of two antennas on a payload.[17] By 1994, after the failure of that year's ETS-6/Kiku satellite, members of the Science and Technology Committee pointedly observed that "space development [was] carried out using taxpayers' money," and hence such fail-

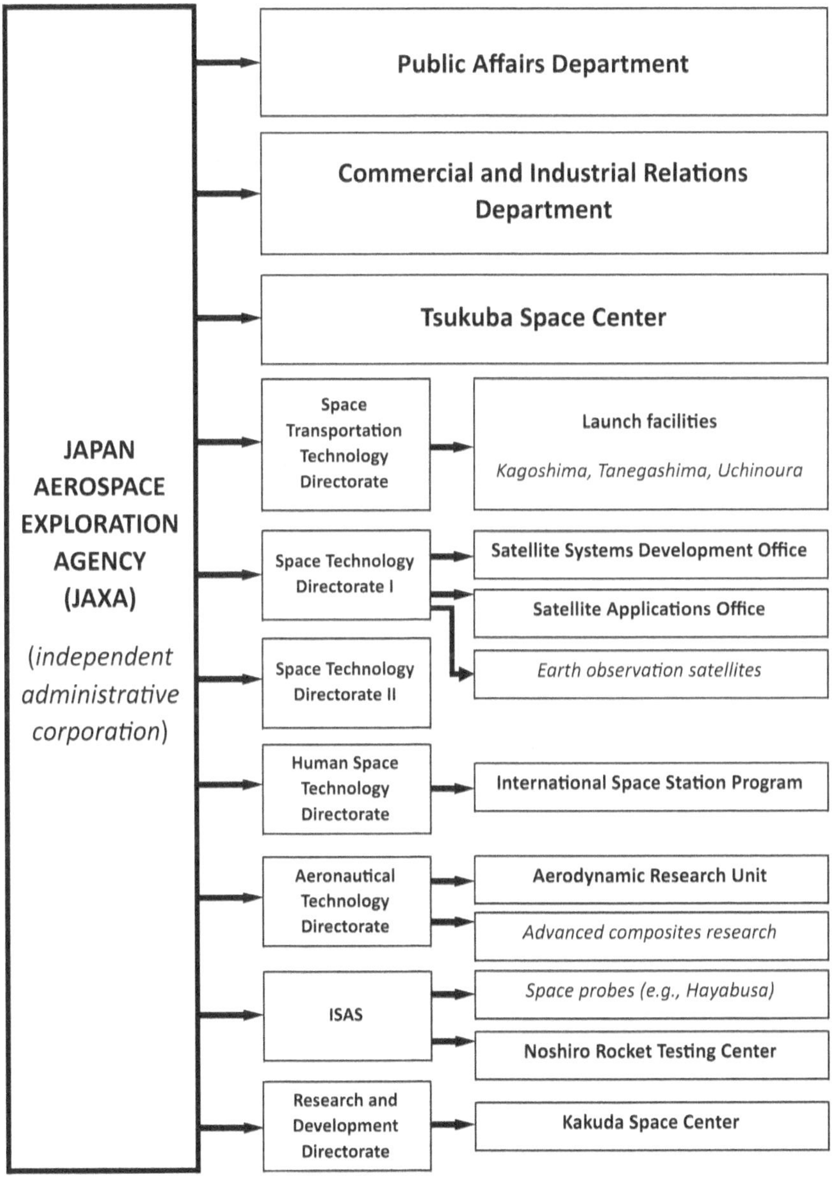

FIGURE 9.1. Japanese space programs after the creation of JAXA.

Chart created by author.

ures could not be treated lightly.[18] A particularly fierce critic of the money spent on H-2 emerged in the late 1990s in the form of former musical theater actress, and Diet member from the center-right New Kōmeitō party, Akira Matsu.[19] As a member of the Science and Technology Committee, Matsu repeatedly pointed out Japan's dire economic straits, and questioned how expenditure on rockets could, in that context, be justified.[20] After H-2's failure in 1998, she wondered aloud if maybe Japan would be better off just launching its payloads abroad.[21] Her questioning of officials shows just how much ire representatives felt about the rocket's troubles:

> The failure of the launch of the H-2 rocket has wiped out ¥685 billion in development costs . . . Failures are bound to occur in research and development, but ¥685 billion is a large amount of money. How do you intend to recover this amount? Or, who will be responsible for this failure, and how will it be handled?[22]

Given this sort of pressure, it was to be expected that after the 1998 launch failure of H-2, one of the first reactions of the director general of the Science and Technology Agency, Hiroshi Hase, was to initiate an investigation into NASDA's budget.[23]

The second consequence of this decentralization was that it rendered oversight and generalized decisionmaking essentially impossible, as lawmakers were forced to navigate difficult bureaucratic terrain every time they wished to exercise their powers. Simple procedures could cause tremendous headaches. Take, for example, the tendency of the Science and Technology Committee to discuss projects that had technical commonalities together, particularly if they were suffering from similar problems. This was based on the sensible understanding that if space projects relied on a shared network of R&D and infrastructure, there would be common solutions to similar problems. Hence joint discussions occurred in the Science and Technology Committee in February 1995 of both ISAS's loss of its Express satellite, and NASDA's loss of ETS-VI/Kiku.[24] Yet it proved impossible to institute broad-ranging solutions, because these projects came under different agencies that devised and implemented their own in-house solutions, resulting in multiple, agency-specific methods for handling identical problems.[25] Hence after the loss of ADEOS I/Midori in 1997, NASDA handled the inquiry, and suggested changes.[26] ISAS did the same in 2000 for the failure of an M-V rocket launch.[27] For lawmakers, it was baffling that so much time and effort was being expended to generate so many solutions to so many problems, and yet none of these seemed to be working.

Further complicating their ability to effectively control the space program was the fact that these various agencies were more than a little prone to squabbling, as when the decision to participate in ISS in 1985 "touched off fierce competition between government ministries and amongst business groups for the leading role in administering the program."[28] A particularly egregious case of bureaucratic squabbling occurred when, in 1995, the Ministry of Transportation, the Ministry of Education, Science, Sports, and Culture, and MITI collaborated on a dual launch on H-2.[29] A leak on one of the payloads, an ISAS/Ministry of Education/MITI probe dubbed the Space Flyer Unit, delayed the launch. Ministry officials delved into the schematics of each other's projects, demanding to know who had designed the faulty valve responsible for the delay, and forcing a series of all-night discussions about the matter. An intense period of finger-pointing ensued, culminating in the Ministry of Transportation demanding a letter of apology from the Ministry of Education for the delay caused to *its* payload. An even worse falling out between ministries came after H-2's final failure in 1999, when the sponsors of its lost payload, the Ministry of Transportation and the Japan Meteorological Agency, refused to complete payment for the launch to the Space Activities Agency, arguing that it had not fulfilled its end of the launch deal. In response, the agency filed a civil suit against them in the Tokyo District Court in 2000 for the sum of ¥3.5 billion ($58 million in 2024).[30] Though they were eventually resolved, such quarrels not only wasted time and funds, but were terrible PR for the civilian space programs, as the press took an eagle-eyed interest in the budget overruns and failures.[31]

The result of all this, as historian William Wray observed in 1992, was that in Japan, multiple "agencies and institutions have a role in policy making, but there [was] no single government agency with comprehensive decision making authority over current issues."[32] Nor was it a model conducive to cooperation; Japanese space policy was "not governed by consensus, for there are splits between the agencies involved; rather it is implemented when alliances among them form a majority opinion, though these shifts depend on the issue."[33] Hence, from the late 1980s, lawmakers were often playing catchup with space policy, rather than determining it. Although the Space Activities Commission only planned to issue a "Fundamental Policy" document every five to ten years, by the late 1970s it had begun doing so yearly, in order to account for changes that had already occurred within ministries and agencies. Reflecting the sense of frustration, following the failure of the November 1999 H-2 launch, Science and Technology Committee member Kiyomi

Tsujimoto glumly suggested that perhaps the group should "now be renamed the Accident Management Committee."[34]

Reform

THE AFTERMATH OF H-2

It was in the context of these ongoing anxieties that investigations into H-2's failure laid the blame squarely at the door of this institutional disorganization and lack of effective oversight. First, it became clear that there was a "normalization of deviance" of the same sort identified by sociologist Dianne Vaughn in her excellent study of the Challenger disaster.[35] Causing this was the sheer number of Japanese space research institutions issuing their own guidance, overwhelming workers on joint projects like H-2, which had 162 quality assurance points *beyond* those required for individual components (recall that the LE-7 engine alone had required some 103 assessments)—leaving plenty of room for human error.[36] Furthermore, not only were these policies inconsistent across the various institutions, but new policies and systems were continuously being implemented, meaning that workers (particularly those moving from one institution to another, or from the private sector to the state) were too busy learning how things were *supposed* to be done to actually do them properly. This issue was so severe that it had, in fact, been raised in the Diet just a few months before the November 1999 H-2 launch.[37] On top of this, the agencies had also failed to notice that their vigorous cost-cutting efforts had caused short-staffing, which meant those who worked on their projects were expected to do more, with less time and support.[38] In a 2000 consultation with the new president of NASDA, long-time satellite specialist Fugono Nobuyoshi pointed out that many within the industry had been complaining about insufficient manpower and overwork for quite some time.[39] The result was that across Japan's space program, "quality control [had become] a mere shell."[40] These pressures had also created a toxic work culture; the Ministry of Education, Culture, Sports, Science and Technology's special Failure Knowledge Utilization Study Group, for example, had begun studying Japan's various R&D failures in the 1990s, and in an August 2001 report identified a "focus on status [which does] not allow for the acknowledgment of failures" as a major problem in the civilian space programs.[41]

The space agencies were also now understood to have mishandled the way in which they liaised with their private sector partners; there was, as one report put it, a "lack of communication within and outside the organization of manufacturers and business associations."[42] This was nothing new either; in 1998, tensions between NASDA and its private sector partners had reached enough of a pitch that the matter was discussed in the Diet.[43] In the aftermath of H-2's failure, however, observers found multiple points of failure, starting with a fundamental clash of objectives between space agencies and corporations. A Space Activities Commisssion report pointed out that, on one hand, NASDA independently funded and developed its own technologies, while on the other, it relied on a diverse set of manufacturers to produce components and hardware for its rockets. Hence sometimes, while NASDA pushed for more sophisticated capabilities, corporations were more keen to finish developing what they were working on first, in order to make them cost-effective. Nor was it just technical issues that had slipped past NASDA—in 1999, the agency discovered that its long-term partner NEC had been using double-bookkeeping methods to swell the cost of contracts between them. By the time of H-2's final failure, it was planning to sue the corporation.[44]

Economic pressures were also found to have had a deleterious effect on the space programs' private sector partners. NEC's skulduggery may well have been inspired by the fact that companies manufacturing rocket and missile components in the post–Cold War world found that the market for their products had shrunk. Nagoya Guidance & Propulsion Systems Works, a division within MHI responsible for producing missile and aerospace engines, was a major victim of this trend. It was spun off from the Nagoya Aircraft Works in 1989 in anticipation of surging profits, and indeed had achieved profitability in 1991. By 1993, however, it was cutting costs in the face of declining demand for the sort of rocket-based weaponry (such as SAM batteries) it produced.[45] MHI also posed problems. A NASDA report of May 2000 on the H-2 failure of 1999 was particularly damning of its quality control, stating that "various operational errors that led to the postponement of the launch occurred in the manufacturing and launch preparation phases. These stages are the responsibility of the manufacturer."[46] Furthermore, the idea that the failure was the result of a misfit between IHI's turbopump and Mitsubishi's components gained public credence on the back of analyses in the press and the Diet.[47] *Asahi Shimbun* opined that "the names of the manufacturers do not appear anywhere in the report. However, if one reads it, one can see that both IHI and Mitsubishi failed in their parts of the project."[48] Still, under this system, it was NASDA's responsibility as prime contractor to ensure the parts worked together, rather than relying on the producers independently to meet these requirements; here, the institution had clearly failed.[49]

Lastly, a series of reports revealed that the Science and Technology Agency itself, responsible for overseeing several space bodies, including NASDA, had failed in multiple areas of oversight. Lawmaker Shigefumi Matsuzawa bluntly stated a few days after the rocket's failure: "I believe that the responsibility for these two accidents is partly due to the lack of supervision of the Science and Technology Agency, which is the supervisory authority."[50] A May 2000 report published by a special committee of the Space Development Agency (populated, it should be noted, by a large number of outside investigators) concluded that the agency had failed to "accurately understand the reliability of the National Space Development Agency of Japan." The report identified multiple minor concerns, including a need for greater theoretical input in technical development, the necessity for improvements to the institution's information-sharing apparatuses, and improved day-to-day management.[51]

By this point, even the ministries that had defended their own institutions had accepted the need for reform, and conceded that insular culture and intensely hierarchical management had undermined their ability to sustain reliable innovation. Among those managers, however, there was one prophetic figure who would have been pleased that change was coming. Yasunori Matogawa had long supported the idea of a centralized space program; in 1991, he had suggested amalgamating ISAS and NASDA as a way of not falling behind in rocketry development.[52] Now his voice joined a chorus of others in arguing for reform.[53] Over the next three years, they'd get a large measure of what they asked for.

THE ROAD TO, AND FROM, JAXA

Reform of Japan's space agencies began months after H-2's final failure, starting with the appointment of Shūichirō Yamanouchi as NASDA chief in early 2000. Yamanouchi had a background in Japan Railways, much like Hideo Shima; unlike Shima, however, his major contribution—as vice president, and then chairman, of JR East—had come after the company's privatization in the early 1980s. His private sector nous, it was hoped, would help cut NASDA's costs, and improve efficiency. One of his first steps was to axe the HOPE-X space plane.[54]

Meanwhile, deeply concerned by public criticism, senior managers in the space field felt sure that "a failure to launch such a rocket again could even undermine the public support for Japan's space development." NASDA concurred: "It is also important to clearly demonstrate that space development is an important national project . . . to gain public understanding of the need to invest sufficient resources to enable reliable space development."[55] Thus when the Space Activities Commission's Basic Strategy Subcommittee began considering a complete overhaul of Japan's mid- and long-term space objectives in late 2000, it posted its report online, with the specific intention of soliciting input from the public.[56] Steps were also taken to mitigate the impact of reforms on Japan's small manufacturers, which had suffered under stringent cost-cutting during the H-2 project. In March 2000, the government launched a Space Development Venture High-Tech Development System (Uchū kaihatsu benchā haitekku kaihatsu seido), later known as the Space Venture Office, with a budget of ¥120 million ($1.8 million in 2024). The institution was charged with encouraging "attractive ideas and unique technologies possessed by domestic small and medium-sized venture companies."[57]

Then came major institutional reforms designed to finally integrate Japan's space programs under one administrative entity. In February 2001, the Science and Technology Agency; the Ministry of Education, Culture, Sports, Science and Technology; ISAS; NASDA; and the NAL began consultations on further coordinating their space research activities.[58] On April 6, 2001, a temporary arrangement coordinating their work was formalized under a joint Operational Headquarters, while the new minister of Education, Culture, Sports, Science and Technology, Takashi Aoyama, created a task force on integrating them on a more permanent basis.[59] Finally, on August 31, 2001, it was announced that Japan's three biggest space agencies—NASDA, ISAS, and NAL—would be amalgamated into one. This new agency would be charged with raising "the level of space science and technology and aeronautical science and technology," and promoting the "development and utilization of space."[60] After some debate over a name, a committee headed by Yasunori Matogawa eventually settled on the title of Nihon uchū kūkō kenkyū kaihatsu kikō. In English, this translated roughly as the Japan Aerospace Agency, but the contraction, JAA, left the committee less than satisfied. In the end, it inserted "exploration" into the title; "Japan Aerospace Exploration Agency" could be contracted to JAXA. Yasunori Matogawa would serve as an associate executive director and head of outreach; Shūichirō Yamanouchi would be the entity's first president.[61]

These reforms had considerable impact beyond the cancellation of major projects such as HOPE-X and the reshuffling of senior figures. For starters, all agencies underwent restructuring to centralize key functions. Management was consolidated into a number of directorates responsible for overseeing particular areas of activity across *all* of Japan's space institutions, such as the Space Transportation Mission Directorate (overseeing rocket launches) and the Space Applications Mission Directorate (overseeing satellites).[62] NASDA ceased to be; in 2002, its project management department, project support department, and planning department were integrated, and moved into JAXA's headquarters at Tsukuba Space Center. ISAS endured, but would need to go through a centralized materials procurement system henceforth. Other, smaller space institutions were now slotted into a hierarchical system designed to centralize coordination, generalize policy, and maintain interdepartmental transparency. What pockets of independent space development in Japan remained afterward were largely small projects undertaken at university level; Kyushu Institute of Technology's Space Systems Laboratory, for example, continued to work on real-time guidance systems and carbon fiber reinforced tanks, as well as reusable winged rockets, into the 2010s, while Wakayama University continued research into balloon-deployed "balloonsats."[63] Yet the new agency did not merely subsume earlier ones; it would also differ markedly in its functioning and objectives. It would be run as a *dokuritsu gyōsei hōjin*, that is, an independent agency, overseen by MEXT and the Ministry of Internal Affairs and Communications (MIC). This granted JAXA a considerable degree of independence, something bolstered by the fact that Japan's chief space policy determining bodies were also gradually removed from ministerial control. The Space Activities Commission, responsible to MEXT, was abolished, and replaced in 2009 by the Strategic Headquarters for Space Policy. By 2012, the Japanese government had "elevated and institutionalised space policy-making directly under the Prime Minister in the Cabinet Office," meaning it could now, theoretically, circumvent the ministries altogether.[64] Competition between Japan's space institutions was diminished, though ISAS, stubborn and persistent, did manage to maintain some leeway in its functions.

The period after 2000, but particularly after the foundation of JAXA in 2003, also saw several profound shifts in relations between Japan's space program and its industrial base. Two case studies are particularly revealing for the scale and significance to Japan's space industry of the companies that formed its manufacturing base. Nissan, by 1986, produced the bulk of SFR motors used in Japan's civilian space programs—in fact, the Lambda 4S-5 which had put Ohsumi in orbit had largely been a Nissan vehicle. By 2000, however, the company was running at an annual loss of ¥590 billion ($5.5 billion). Four days after an M-V rocket it had built for ISAS failed to put a satellite into orbit in February 2000, it sold its aerospace division as part of a wholescale restructuring.[65] IHI stepped up to purchase the assets, securing the technology, and contracts to provide missiles to all three of Japan's self-defense forces—Ground, Air, and Maritime. Thus a major consolidation occurred in the production of Japanese space hardware, meaning that henceforth, the field would be dominated by IHI and MHI. Software and electronics underwent a similar process. Hitherto, Melco, Toshiba, and NEC had enjoyed a "revolving carousel" of government largesse, with each company "taking turns to be the prime integrator on a NASDA contract, while the other two companies would receive important systems contracts." The NEC overcharging scandal, and errors in the integration of satellites such as ETS-VI/Kiku and ETS-VII, resulted in a realignment here, as well. NEC merged its satellite activities with Toshiba, but in the long term, both began to lose market share to MELCO, which became the dominant player; it not only expanded its dominance of civilian satellite production, but successfully lobbied to be the prime producer of military satellites in Japan as well.[66]

Lastly, as MELCO's expansion shows, the early 2000s also marked the beginning of a profound shift in the explicitly stated objectives of Japan's activities in space—namely, the gradual legitimization and integration of military activities. In fact, JAXA would, in time, abandon even the semblance of "peaceful purposes." Until the agency's foundation, military involvement in Japan's space institutions was almost entirely mediated through private corporations, or through the movement of specialized engineering personnel from civilian to military activities or vice versa (some early members of NASDA, for example, had been seconded from the JSDF, partly because they were the only people with relevant experience in the field).[67] The rise of tensions in East Asia in the late 1990s had expanded demand in Japan for space-based military assets, such as the (MELCO-manufactured) QZSS. Just as pressure from these corporations in the 1960s and 1970s had led to the creation of NASDA, they now promoted projects such as the development of reconnaissance and early-warning satellites, as well as missiles, via groups such as the Society of Japanese Aerospace Companies. Combined with a gathering sense of regional threat, this led to the gradual alteration of Japan's long-held commitment to "peaceful usage" of space only. In 2008, the new Basic Space Law officially overturned this commitment; within two years, JAXA would become responsible for the organization and running of Japan's growing fleet of spy satellites.[68] MELCO, meanwhile, has at various times in the twenty-first century made the bulk of its income from satellite manufacturing by producing military application devices for the state.[69] Starting from 2021, military expenditure was integrated into JAXA's budget; national security is now one of the entity's central responsibilities, alongside supporting industry and technological independence. Japan's purposes in space were no longer just peaceful.

The period from 1980 to 2000 included a great deal of instability, contestation, and disappointment in the Japanese civilian space programs. The H-2 project's failure was a profound disappointment to the engineers, specialists, and private sector researchers who worked on its development. Coming at the end of a decade marred by numerous failures in the civilian space programs and broader economic malaise, it became emblematic of a sense that something had gone profoundly wrong in Japan's civilian space programs. Such views tallied with those of the program's political overseers, who through the 1990s had grown increasingly anxious about costs, inefficiencies, and organizational complexity. These circumstances concatenated in the period from 2000 to 2003 to produce a consensus that Japan's civilian space programs needed better coordination both between their various component agencies, and between these agencies and their private partners. Thus Japan moved between 2001 and 2003 to overturn nearly half a century's worth of entrenched decentralization within the civilian space programs, and integrate under a singular agency, the Japan Aerospace Exploration Agency.

Despite this disruption, Japan is no stranger to new starts, and even this dismal period in its spacefaring history would provide a seedbed for future successes. First, a derivative of the H-2, the H-2A, enjoyed a career that saw it launch forty-five times, with one single failure. Beyond that, the travails of the 1990s and early 2000s proved that despite complaints about spending, organization, and usefulness, Japan's presence in space was going nowhere. Rather, the disappointment of stakeholders stemmed precisely from the fact that so many were invested in the space programs' success. Nor did any more large corporations follow Nissan out of the space launch game. Public interest in space exploration held steady—the Japan Astronaut Club, for example, grew in size throughout this period, while some members of the public argued that Japan was better off spending money on rockets than on huge projects like the Tokuyama Dam, with its woeful environmental consequences.[70]

And belief that Japan had a future in space remained entrenched within political circles. Within months of the eighth H-2's failure, Japan's annual budget actually increased the amount of funding that would be available for the H-2A.[71] Minister of Science and Technology Policy Kōji Omi summarized the enduring government attitude toward space exploration in 2001 thus: "[it] is a dream of the people, and we will actively work to promote it."[72] That responsibility would come to rest on the shoulders of JAXA, as shown by the full-size replica of the H-2 at the entrance to the Tsukuba facility—testament to an ambitious vision of Japan in space that, for all its failures, has never truly faded.

Conclusion

Billions of years from now our sun, then a distended red giant star, will have reduced Earth to a charred cinder. But the *Voyager* record will still be largely intact, in some other remote region of the Milky Way galaxy, preserving a murmur of an ancient civilization that once flourished—perhaps before moving on to greater deeds and other worlds—on the distant planet Earth.

—Carl Sagan[1]

MODERN SPACE RESEARCH IN JAPAN began its hundred-year-long history, as with so many things, during the country's early twentieth-century attempt to build an empire. Its genesis lay in the period between roughly 1920 and 1960. In the first half of this period, riven by geopolitical and cultural anxieties, the country invested in industrial and R&D systems that produced a cohort of excellent aeronautical engineers. During World War II, this network was deployed to develop the most advanced aeronautical industry outside of the West—one that included rocket-based weapons. In the postwar period, many specialists who worked on these contributed to the country's recovery in a variety of industrial settings. Others laid the narrative, organizational, and engineering foundations for resuming work on rocketry—for the sake of peacetime prosperity, and not martial success.

The country's journey to space began with a six-inch-long rocket dubbed the Pencil in 1954. By 1970, Japan had put a satellite in orbit. Between the 1960s and 1980s, space research in Japan became increasingly institutionalized, but in a fashion quite unlike that of the centralized US system. Rather, private corporations, state entities, and bureaucrats competed to have their voices heard and their objectives realized within a constellation of administrative bodies that were responsible for diverse areas of space technological research. Concurrently, engineer-managers established themselves as a powerful voice in the direction of the space programs, and citizens living near their facilities fought to ensure that rapid technological development did not occur at the cost of their safety and prosperity. These conflicting interests bred multiple space programs, with two particularly important institutions—ISAS and NASDA.

Despite this organizational complexity, the 1970s and 1980s were a period of great success for Japan's civilian space programs. The 1970s saw acquisition and domestication of US LFR technology under NASDA's guidance. By the 1980s, Japanese space agencies and their researchers were designing world-class devices like the comet-explorer Sakigake, and carrying out world-class research in such areas as ion thrusters. Riding the momentum of Japan's rise to being the world's second biggest economy, this bred high hopes for various ambitious space projects. Few, however, would come to fruition. Amid a general downturn in the 1990s, Japan's space programs suffered a series of major technical failures, ranging from lost Mars probes, to the high-profile collapse of the H-2 project. These let-downs finally tipped the balance in favor of those who had long argued that Japan's decentralized space exploratory system needed radical reform. Thus, in 2003, the era of multiple independent space agencies and projects came to an end. Henceforth, a new, central entity would govern Japan's civilian space program: JAXA.

Japan Still Dreams of Space

Just as a person flying in a plane sees no national borders on the face of the world, observers of the past will find no clear-cut moments where one "period" ends and another begins. Academics are forced—for the sake of time, publishing, and sanity—to choose a start and end point for our inquiries. But no historian worth their salt genuinely thinks that the stories they seek to tell end on a given date. Where we choose to end should be defensible, no doubt; but in the end, it is a choice. Thus, although this book ends with the formation of JAXA in 2003, the truth is that many of the trends and processes described over the preceding pages have continued, in one form or the other, into the twenty-first century. In some cases, these give every sign of continuing to develop into the foreseeable future.

Perhaps the most surprising survival is ISAS. Although it has been integrated into JAXA, it continues to endure as an administrative entity *within* that organization, albeit one that is now limited to the design and coordination of specific scientific payloads. In fact, one of JAXA's first big successes, the 2003–2010 asteroid sample return mission Hayabusa, was largely an ISAS production. Other missions, such as NASA's Galileo, had visited asteroids before, but none had landed on one, let alone brought back a sample.[2] Hayabusa not only intended to change that, but to deploy a small lander, Minerva, to the surface of the object it was visiting. The probe was developed within ISAS, run by academics who worked there, and was even part-built by ISAS's old collaborator, Hitachi.[3] Launched on May 9, 2003, from the (ISAS-built) Uchinoura Space Center, the probe scooted along on four (ISAS-designed) experimental xenon ion engines for two years before reaching its objective.[4] It was launched aboard an ISAS-commissioned SFR, the M-V, and its management was handled by an ISAS team.[5]

The narratives that emerged in the mid- to late twentieth century also endure, particularly in the notion of Hideo Itokawa as the "father" of Japanese rocketry, and the idea that success in space is reflective of Japan's status as a "first rate" technological power. Again, the example of Hayabusa is salutary. The asteroid it visited was called 25143 Itokawa; originally discovered in New Mexico by a team from Massachusetts, it was renamed after Dr. Rocket by Japanese request.[6] It is no coincidence that ISAS chose his name for its objective. Nor is it any coincidence, despite protestations to the contrary, that the probe bears the name of one of Itokawa's most famous wartime projects, the Nakajima Ki-43 Hayabusa fighter. Narratives and legacies from as far back as World War II linger in the civilian space programs even in the twenty-first century.

There are other continuities, perhaps less welcome, such as a tendency for Japanese probes to succeed in their missions despite, and not because of, their complexity. Take Hayabusa — as a contemporary *Gizmodo* headline put it, "Everything That Could Go Wrong for Hayabusa Did, and Yet It Still Succeeded."[7] After reaching the asteroid Itokawa, three separate attempts to collect samples from the asteroid's surface met with varying degrees of failure. When finally the sample canister was sealed, Hayabusa suffered a power outage. Then on December 8, 2005, it suddenly changed orientation, and disappeared from JAXA's communications system.[8] Four months later, on March 7, came a breakthrough: Hayabusa was back under control, and on its way home. By November, after some technical chicanery, the ISAS team finally got it under full control.[9] It returned to Earth, sample intact, in 2010. Its trove of asteroid dust hit the ground unscathed, and the particles it contains have become the basis for a slew of ground-breaking research papers on the origins of the Solar System.[10]

Still, Hayabusa's story was transformed into an uplifting tale of success against the odds by JAXA's skilled outreach machine. During and after the project, press coverage emphasized the heroics of the engineers and technical specialists involved in running and saving Hayabusa. JAXA composed press releases and staged press conferences to satisfy a curious public; *Asahi Shimbun* dedicated an entire webpage to the probe and its travails, while the foreign press was intrigued by Japan's ambitious mission to a space rock 3 billion mi (5 billion km) away, and the remarkable effort to get the spacecraft back under control. The story even got the movie treatment, with two feature films coming out in 2012. One, *Okaeri, Hayabusa* (Welcome Home, Hayabusa; dir. Katsuhide Motoki, Japan, 2012), had as its main protagonist an engine specialist who had helped design the probe's ion engines; notably, his father had worked on the failed Nozomi probe of the 1990s. Meanwhile, *Hayabusa: Harukanaru kikan* (Hayabusa: The Long Voyage Home; dir. Tomoyuki Takimoto, Japan, 2012), was something of a flop, despite starring Ken Watanabe.

Hayabusa's end also points to the way Japan continues to have to accommodate its own particular geography in conducting space research. For the same reason that Japan was forced to situate its testing ranges and launch facilities close to villages and small towns, the country found it difficult to locate a suitable area for the returned samples to land. Hence, just before midnight on June 13, 2010, people on the ground near the Woomera Prohibited Area in South Australia were treated to an extraordinary light show. What began as a single, tremulous light in the clear night sky proliferated into multiple phosphorescent streaks, each blazing for a short while before disappearing. One speck alone, riding out in the front of the others, continued on and hit the ground. Eyewitnesses reported that it was silent, except for the faintest of rumbles murmuring in the outback night.[11] The light was, of course, Hayabusa.

The dreams and objectives of the 1980s and 1990s also survive in various projects carried out under JAXA's purview. Japan's space goals may no longer involve reusable space planes, Moon bases, and space stations, but it has undertaken ambitious missions in other directions. For example, on January 3, 2014, just over a decade after Hayabusa, JAXA launched a follow-up mission, Hayabusa 2. The mission bears the hallmarks of the organizational changes that have occurred within Japan's civilian space program under the aegis of JAXA. Its management reflects a greater integration of ISAS into JAXA's superstructure; it launched from Tanegashima; and its constructor was the JAXA-chosen NEC, as opposed to long-time ISAS collaborator Hitachi.[12] At the same time, it was considerably more complex than its predecessor. The probe featured two small rovers that landed on its target, the asteroid 162173 Ryūgu, and successfully transmitted data, including video.[13] Hayabusa 2 itself not only collected dust from the surface of the asteroid, but also managed to acquire material from beneath the surface. It returned to Earth without a hitch, dropped its samples off at Woomera, and set off again for distant parts.[14] Nor has JAXA given up on the quest for a commercially viable launch vehicle. Despite its long and successful career, the H-2A, child of the monstrous H-2, failed to garner any business outside of Japan. JAXA, however, initiated work on the third vehicle in the series, the H-3. Though this vehicle too has faced some difficulties, its production and launch costs are far lower than H-2's. Rather than being "purely domestic," H-3 relies on a number of components from overseas; the rocket's fairings, for example, are constructed by Switzerland-based Beyond Gravity.[15] Such measures have enabled it to provide launches for around $50 million a pop, about a quarter of H-2's prices.

As Carl Sagan observed, beyond a thin layer of carbon, plastics, and radioactivity in the planetary crust, humanity's most enduring legacy will be the probes we've cast off into the outer dark in search of knowledge about our place in the universe. Perhaps more than anything, this sense of aspiration is the most important quality that survived the tumult and reorganization in Japan's civilian space programs in the late 1990s. If there is one thing this book has sought to emphasize, it is that a space program is the sum total of the dreams of the people who form it. If engineers, bureaucrats, politicians, fishermen, technicians, journalists, explosives experts, and others had not dreamed of reaching for the stars, the institutions and funding for them to do so would never have come together. None of this motivation has gone away.

Hence, it behooves us to end with a story illuminating what this looks like on the ground. On March 7, 2023, the aforementioned H-3 rocket suffered an engine failure on its first launch. Coming not long after the failure of a new-generation Epsilon SFR in 2022, the affair had a dreadful sense of déjà vu about it. Perhaps Japan was right back where it had been in the 1990s. Perhaps there were some things the country simply did not have the wherewithal to do. Then, about three days later, a small delegation arrived at JAXA's headquarters in Tokyo. They had come all the way from Kimotsuki, home of the Uchinoura Space Center, led by its mayor. They presented a group from JAXA, including Yasunori Matogawa, with five banners, around 5 ft (1.5 m) by 3.2 ft (1 m) each, containing messages of support from more than eight hundred association members, schoolchildren, and other inhabitants of the area. In the aftermath of the many failures, all had rallied to send a message of support to JAXA. "Flowers will surely bloom after hardship," one message read. Another urged them to "Give light to Japan's future." A third simply said "Never give up."[16] JAXA didn't—and since then, the H-3 rocket has flown four more times, without a single hitch. One can only speculate how Eiichi Iwaya would have felt, watching these launches.

GLOSSARY

Asuka A space observatory developed in collaboration with NASA, measuring approximately 8 ft (2.4 m) in length and weighing about 3,750 lb (1,700 kg). It carried four X-ray telescopes and a gas imaging spectrum, aiming to study celestial X-ray sources. The Asuka probe was launched on February 20, 1993, marking Japan's first dedicated X-ray astronomy satellite mission.

B-29 The "Superfortress," a very large and long-range strategic bomber aircraft developed by Boeing during World War II.

Baby 47 in (120 cm) long, 22 lb (10 kg) rocket, tested 1956.

"Baika" The Mitsubishi J8M, a rocket-powered interceptor aircraft developed during World War II. It was intended for suicide missions against American bomber formations.

Cassini A collaboration between NASA, the European Space Agency (ESA), and the Italian Space Agency (ASI), Cassini-Huygens was a spacecraft designed to study the planet Saturn, its rings, and its moons.

Epsilon Rocket class introduced in 2016. Half the price of the Mu class, with a slightly lower payload mass.

Hayabusa Asteroid sample retrieval probe, launched 2003, successfully returned in 2010

Hayabusa 2 Asteroid sample retrieval probe, launched 2014, successfully returned in 2020

H-1 A two-stage liquid-fuel rocket measuring approximately 89 ft (27 m) in length with a payload capacity of around 3,300 lb (1,500 kg) to low Earth orbit, making its first launch in August 1986.

H-2 A two-stage liquid-fuel rocket measuring approximately 160 ft (49 m) in length and capable of carrying payloads up to 9,000 lb (4,000 kg) to low Earth orbit, with its inaugural launch taking place in February 1994.

H-2A A two-stage liquid-fuel rocket measuring approximately 174 ft (53 m) in length with a payload capacity of around 9,900 lb (4,500 kg) to geostationary transfer orbit (GTO), commencing its first launch in August 2001.

H-2B A two-stage liquid-fuel rocket measuring approximately 207 ft (63 m) in length and capable of carrying payloads up to 18,000 lb (8,000 kg) to GTO, conducting its maiden launch in September 2009.

H3/H-3 A two-stage liquid-fuel rocket measuring approximately 207 ft (63 m) in length with a payload capacity of around 14,000 lb (6,500 kg) to GTO, making its debut launch in May 2022.

Kappa Rocket class tested from 1958 onwards, up to 500 lb (255 kg) in weight, an altitude of up to 37 mi (60 km).

Keidanren Japan Business Federation, one of Japan's most powerful industrial bodies.

Kibō The Japanese Experiment Module (JEM), a component of the International Space Station (ISS).

Lambda Rocket class used from 1960 onwards. A four-stage L-4S-5 put Ohsumi in orbit on February 11, 1970.

Me-163 Komet Messerschmitt-developed rocket plane. Around 350 were used in battle after 1944.

Medium lift A rocket capable of up to 20,000 kg into orbit.

MEXT The Ministry of Education, Culture, Sports, Science and Technology of Japan.

Mu Rocket class used from 1966–2006. The Mu 5 (M-V) has a launch mass of up to 306,000 lb (137,000 kg) and placed payloads of up to (4,000 lb) 1,800 kg into orbit.

Nakajima A prominent Japanese aircraft manufacturer that played a significant role in the aviation industry during the first half of the 20th century, established in 1917.

N-I A two-stage solid-fuel rocket measuring approximately 63 ft (19.2 m) in length and capable of carrying payloads up to 2,000 kg to low Earth orbit, with its first launch occurring in August 1975.

N-2 A two-stage liquid-fuel rocket measuring approximately 91 ft (28 m) in length and capable of carrying payloads up to 4,400 lb (2,000 kg) to low Earth orbit, with its first launch occurring in February 1981.

New Horizons Mission launched by NASA on January 19, 2006, with the primary objective of studying the dwarf planet Pluto and its moons in the outer regions of the solar system.

Nozomi Japanese Mars orbiter, launched on July 3, 1998. Propulsion failures led to the probe's eventual loss.

Ohsumi First Japanese satellite, launched atop a Japanese rocket, the Lambda, on the February 11, 1970

Ōka The Yokosuka MXY-7, a Japanese suicide air torpedo developed during World War II. It was designed to be carried to its target by a larger aircraft, typically a Mitsubishi G4M "Betty" bomber, and then released to glide toward its target before activating its rocket engine for the final attack.

Operation Paperclip A covert program conducted by the United States Office of Strategic Services (OSS) and later by the Joint Intelligence Objectives Agency (JIOA) after World War II which aimed to recruit German scientists, engineers, and technicians, particularly those involved in Nazi Germany's rocket development program, to work for the United States.

Pencil Japan's first postwar rocket. 9 in (23 cm) long, 0.44 lb (200 g) in weight, first tested in 1955.

Roscosmos Russian federal state corporation responsible for space flights, cosmonautics programs, and aerospace research.

Sakigake First Japanese interplanetary probe, launched on January 7, 1985, to conduct a flyby of Halley's Comet.

Shūsui The Mitsubishi J8M, a World War II–era Japanese rocket-powered interceptor aircraft. It was based on the German Messerschmitt Me 163 Komet.

Suisei Sakigake's sister probe, launched on August 18, 1985.

Sounding rocket A rocket capable of reaching high altitude, but not orbit. Usually used to carry out upper atmospheric experimentation.

Space Activities Commission An advisory body established to provide recommendations and guidance to the Japanese government on matters related to space policy and activities. Founded in 1969.

Thor-Delta US LFR system, first launched in 1960, and retired by 1962. Its technology was transferred to Japan in a 1969 agreement.

V-2 World's first long-range guided ballistic missile. More than 3,000 V-2s were launched by Nazi Germany at Allied targets in the dying months of World War II.

NOTES

Introduction

1. Kenneth M. Swope, "Crouching Tigers, Secret Weapons: Military Technology Employed during the Sino-Japanese-Korean War, 1592–1598," *The Journal of Military History* 69, no. 1 (January 2005): 27.

2. Yasunori Matogawa, "Another Destiny of Rocketry in Japan—Festival Rockets in Japanese Shrines," *History of Rocketry and Astronautics: Proceedings of the History Symposium of the International Academy of Astronautics* 23 (1995): 315.

3. Robert H. Goddard, *Rockets—Two Classic Papers* (Mineola, NY: Dover, 2002); Marcin Pomarański, "Science and Technology in Russian Cosmic Utopias from the Beginning of the Twentieth Century: Konstantin Tsiolkovsky and Alexander Bogdanov," *Utopian Studies* 33, no. 1 (2022): 42–46.

4. Solid-fuel rockets run on a combustible sludge that is relatively stable and easy to handle—but that, once ignited, cannot be halted. Liquid-fuel rockets, in contrast, run on liquefied sources of energy that can be more volatile (and in some cases, such as oxygen, require intense cooling), but can be ignited and then extinguished multiple times. Liquid-fuel rockets are hence easier to control and have higher functionality than solid-fuel rockets.

5. University of Tokyo, "Ohsumi, Japan's First Satellite," http://www.u-tokyo.ac.jp/en/whyutokyo/hongo_hi_009.html.

6. Ryoko Dateki, "Space Industry Promotion in Japan—The Role of JAXA," JAXA, July 22, 2024, https://earth.jaxa.jp/conseo/news/20240618-1/02_JAXA_Business_Development%20and_Industrial_Relations_Dept.pdf; JAXA, "JAXA President Regular Monthly Press Conference," July 18, 2013, http://global.jaxa.jp/about/president/presslec/201307_e.html; World Economic Forum, "Which Countries Spend the Most on Space Exploration?"

January 11, 2016, https://www.weforum.org/agenda/2016/01/which-countries-spend-the-most-on-space-exploration.

7. JAXA, "Transition of Number of Staff and Budget," https://global.jaxa.jp/about/transition/index.html.

8. JAXA, "Space Transportation Systems: H-IIB Launch Vehicle in Operation" (2003), http://global.jaxa.jp/projects/rockets/h2b; Kari A Bingen and Makena Young, *From Earth to Uchū: The Evolution of Japan's Space Security Policy and a Blueprint for Strengthening the U.S.–Japan Space Security Partnership* (Washington, DC: Center for Strategic and International Studies, 2024): 25.

9. Christine Y. L. Luk and Subodhana Wijeyeratne, "Alternatives to GPS: Space Infrastructure in China and Japan," *Roadsides* 3 (August 2020): 57–62.

10. JAXA, "MMX—Martian Moons Explorer," https://www.mmx.jaxa.jp/en/#:~:text=Approximately%20one%20year%20after%20leaving,formation%20in%20the%20Solar%20System; JAXA, "JAXA President Regular Monthly Press Conference," January 13, 2017, http://global.jaxa.jp/about/president/presslec/201701.html; Planetary Society, "The Japan Aerospace Exploration Agency (JAXA)," https://www.planetary.org/the-japan-aerospace-exploration-agency-jaxa#:~:text=There%20are%20plans%20for%20JAXA,robotic%20lunar%20lander%20and%20rover.

11. Richard J. Samuels, *Rich Nation, Strong Army: National Security and the Technological Transformation of Japan* (Ithaca: Cornell University Press, 1996).

12. Bingen and Young, *From Earth to Uchū*, 14.

13. Igor Oganesoff, "Japan Builds Rocket Industry, Emphasizes Peaceful Use, Exports," *Wall Street Journal*, October 22, 1962.

14. George A. Eberstadt, "Japan's High Frontier: Viewing the Civilian Space Programs As Industrial Policy," M.A. thesis, Harvard College, 1989, 2.

15. See, for example, "Japan Tests Rocket," *New York Times*, October 30, 1957; Arthur J. Dommen, "Japanese Rocketeers Have Come a Long Way," *Los Angeles Times*, October 3, 1965; "Much Talk, Little New: U.S–Japan Talk," *Los Angeles Times*, July 18, 1966.

16. ISAS, "Hayabusa," https://www.isas.jaxa.jp/en/missions/spacecraft/past/hayabusa.html; JAXA, "Hayabusa2 Project," 2013, http://www.hayabusa2.jaxa.jp; Eberstadt, "Japan's High Frontier," 2.

17. Walter McDougall's most extensive discussion of Japan's space development is relegated to a note, and reads as follows: "Japan initially confided rocketry and space science to an academic group at the University of Tokyo, which also succeeded in orbiting a satellite, on a pencil-shaped Vanguard-type rocket, in 1970. But Japanese industry prevailed on the government to inaugurate a commercially motivated space program. Thanks to the purchase of American Delta technology and the fastest-growing space budget in the world in the 1970s, Japan has emerged as a growing competitor in space applications." McDougall, *The Heavens and the Earth: A Political History of the Space Age* (New York: Basic Books, 1985), 528. The

work of Stephen J. Dick and Roger D. Launius, meanwhile, reserves its most detailed mentions of Japan for ithe development of photo reconnaissance and spy satellites in the early 2000s; see Steven J. Dick and Roger D. Launius, eds., *Critical Issues in the History of Spaceflight* (Washington, DC: NASA, 2006), 513–514, 519, 531.

18. Yasushi Sato, "A Contested Gift of Power: American Assistance to Japan's Space Launch Vehicle Technology, 1965–1975," *Historia Scientiarum* 11, no. 2 (2001): 176–204; Hitoshi Yoshioka, "Uchū kaihatsu kyūsei no kakuritsu," in *Nihon no kagaku gijutsushō dai-3 ki*, ed. Nakayama Shigeru (Tokyo: Gakuyō Shobō, 1995), 172–183; Bingen and Young, *From Earth to Uchū*, 25–28.

19. Yasunori Matogawa, "Yasunori Matogawa hakase no uchū yomoyama hanashi: jimae roketto kaihatsu wa kaku ketsudan sareta," *Sekai shūhō* 79, no. 38 (October 20, 1998): 56.

20. Junji Matsui and Masakane Shimizu, "Uchū kaihatsu jigyōdan no sesturitsu," *Toki no hōrei* 692 (October 1969): 24–33.

21. John O'Sullivan, *Japanese Missions to the International Space Station: Hope from the East* (Cham, Switzerland: Springer Praxis, 2019), 54, 72, 187.

22. Sogame Eiji, "H-I Roketto," *Nihon kōkū uchū gakkaishi* 36, no. 418 (November 1988): 492.

23. Brian Harvey, *Japan in Space: Past, Present, and Future* (Chichester, UK: Springer Praxis, 2023); John O'Sullivan, *Japanese Missions to the International Space Station: Hope from the East* (Cham, Switzerland: Springer Praxis, 2019).

24. Asif Siddiqi, "Spaceflight in the National Imagination," in *Remembering the Space Age*, ed. Steven J. Dick (Washington, DC: NASA Office of External Relations History Division, 2008), 28.

25. Uchū kaihatsu jigyōdan hō (1969), JAXA, https://www.jaxa.jp/library/space_law/chapter_1/1-1-2-10/index_j.html.

26. Michel Callon, "Some Elements of a Sociology of Translation: Domestication of the Scallops and the Fisherman of St. Brieuc Bay," in *Power, Action, and Belief: A Sociology of a New Knowledge?* ed. J. Law (London: Routledge, 1986), 196–223. His work is particularly salient here as he acknowledges the narrative role of inanimate and nonhuman historical quantities (in the case of his seminal work on the subject, scallops harvested off the coast of France) on a par with the humans who interact with them. For an analysis of a field concerned with objects that are often ascribed value as the offspring of national capacity and inventiveness, whose success is couched as a national success, and whose failures are often understood as national tragedies, a framework that can provide some sense of the narratives generated by these inanimate objects is crucial.

27. Peter Fritzsche, *A Nation of Flyers: German Aviation and the Popular Imagination* (Cambridge, MA: Harvard University Press, 1992), 4–5.

28. James E. Webb, *Space Age Management: A Large Scale Approach* (New York: McGraw Hill, 1969), 89.

29. Matthew Hersch, *Inventing the American Astronaut* (New York: Palgrave Macmillan, 2012), 29–34.

30. Anne Walthall, *The Weak Body of a Useless Woman* (Chicago: University of Chicago Press, 1998); John Dower, *Embracing Defeat: Japan in the Wake of World War II* (New York: W. W. Norton, 2000); Timothy S. George, *Minamata: Pollution and the Struggle for Democracy in Postwar Japan* (Cambridge, MA: Harvard University Asia Center, 2002); Martin Dusinberre, *Hard Times in the Hometown: A History of Community Survival in Modern Japan* (Honolulu: University of Hawaii Press, 2012); Amy Stanley, *Stranger in the Shogun's City: A Japanese Woman and Her World* (New York: Scribner, 2020).

31. Asif Siddiqi, *The Challenge to Apollo: The Soviet Union and the Space Race, 1945–1974* (Washington, DC: NASA, 2000).

32. McDougall, *The Heavens and the Earth,* 74–97. Malina was a famously grouchy member of Caltech's pioneering "Suicide Squad" of rocketeers who experimented with the technology in the 1930s. The group was so named for the recklessness of some of their experiments.

33. Asif Siddiqi, "Fighting Each Other: The N-1, Soviet Big Science, and the Cold War at Home," in *Science and Technology in the Global Cold War,* ed. Naomi Oreskes and John Krige (Cambridge, MA: MIT Press, 2014), 189–225.

34. . Indian Space Research Organization, "Archive of Demands for Grants," April 3, 2025, https://www.unoosa.org/oosa/en/ourwork/spacelaw/nationalspacelaw/japan/nasda_1969E.html.

35. European Space Agency, *Convention for the Establishment of a European Space Agency* (Noordwick: ESA Publications Division, 2007), 11.

36. Hyoung Joon An, "South Korea's Space Program: Activities and Ambitions," *Asia Policy* 15 no. 2 (2020): 34–42; Livia Peres Milani. "Brazil's Space Program: Finally Taking Off?" The Wilson Center, October 24, 2019, https://www.wilsoncenter.org/blog-post/brazils-space-program-finally-taking.

37. "Cha no ma ni chikyū no kao Sōren no jinkō eisei kara TV chūkei NHK—uchūchūkei," *Asahi Shimbun,* June 9, 1967.

38. United Nations Office for Outer Space Affairs, "Law Concerning the National Space Development Agency of Japan," June 23, 1969, https://www.unoosa.org/oosa/en/ourwork/spacelaw/nationalspacelaw/japan/nasda_1969E.html.

39. National Diet of Japan, NASDA Foundation Law (Uchū kaihatsu jigyōdan hō), June 23, 1969, https://www.shugiin.go.jp/internet/itdb_housei.nsf/html/houritsu/06119690623050.htm.

40. Bin Umino, Kyo Kageura, and Shinichi Toda, "A Sixty Year History and Analysis of the Japanese Publishing Industry: A Statistical Analysis of Circulation," *Publishing Research Quarterly* 26, no. 4 (December 2010): 278–279.

41. "Matogawa hakase no uchū yomoyama hanashi: jinrui no shōrai no tenbōshita Itokawa Hideo," *Sekai jōhō* 85, no. 17 (May 4, 2004): 62–63.

42. R. W. Home and Morris F. Low, "Postwar Scientific Intelligence Missions to Japan," *Isis* 84 (September 1993): 537.

43. Andrew Gordon, *A Modern History of Japan: From Tokugawa Times to the Present* (Oxford: Oxford University Press, 202), 273–275.

44. Takashi Nishiyama, "Swords into Plowshares: Civilian Application of Wartime Military Technology in Modern Japan, 1945–1964," PhD diss., Ohio State University, 2005, 15.

45. Pluto will always be a planet to the author of this book.

46. Ling Xin, "China's Moon Atlas Is the Most Detailed Ever Made," *Nature*, April 25, 2024, https://www.nature.com/articles/d41586-024-01223-0.

47. Josh Smith, "North Korea's First Spy Satellite Is 'Alive,' Can Manoeuvre, Expert Says," *Reuters*, February 28, 2024, https://www.reuters.com/technology/space/north-koreas-first-spy-satellite-is-alive-can-manoeuvre-expert-says-2024-02-28/.

Chapter 1

1. Robert A. Pape, "Why Japan Surrendered," *International Security* 18, no. 2 (Fall, 1993): 154.

2. Hisamitsu Matsuoka, *Nihonsho no roketto sentōki Shūsui* (Tokyo: Miki Shobō, 2004), 49.

3. Jun Okamura, *Kōkū gijutsu no zenbō (jō)* (Tokyo: Shippitsusha shōkai, 1976), 3.

4. Masuo Mori, "Kaiso dampen," in *Nippon kaigun nenryōshi (ge)* (Tokyo: Hara Shobō, 1973), 831. One reason that Japanese technology wasn't regarded as being worth seizing at the end of World War II was that much of it was designed to take account of issues in Japan that weren't considerations in the United States. For example, by war's end, with oil imports running dry, the Japanese turned to sources as diverse as pine roots, soybean oil, orange peel, birch bark, sweet potatoes, and rubber to devise substitutes. The United States, meanwhile, had a surfeit of petroleum, and had no interest in these innovations. For more details, see NavTech, "Japanese Fuels and Lubricants, Article 1—Fuel and Lubricant Technology," in *Report of the US Naval Technical Mission to Japan, Ship and Related Targets—Japanese Navy Diesel Engines* (San Francisco: US Naval Technical Mission to Japan, February 21, 1946), National Diet Library, Tokyo, 8, 11.

5. Matsuoka, *Sentōki Shūsui*, 47; Eiichi Iwaya, "Roketto ki Shūsui ni tsuite," *Sekai no kōkū* 22, no. 9 (January 1952): 82–88; Hitoshi Fujihira, ed., *Kimitsu heiki no zenbō toshō shippitsu henshū* (Tokyo: Hara Shobō, 1976), 30, 34; Tamechika Yamamoto, "Roketto tsui ni hasshin sezu," in *Nihon kaigun nenryō sho (jō), nenryō konwakai* (Tokyo: Hara Shobō, 1972), 1069.

6. Matsuoka, *Sentōki Shūsui*, 55.

7. Karl T. Compton, "Mission to Tokyo," Office of the President Records: Jonathan Dick-

inson to Harold W. Dodds Subgroup, Box 205, Folder 11. Princeton University Archives, Department of Rare Books and Special Collections, Princeton University Library, 8.

8. NavTech, "Enclosure (A): Historical Narrative of the US Naval Technical Mission to Japan," in *Report of the US Naval Technical Mission to Japan, History of Mission*, A, January 1946, 10–12.

9. John Dower, *Japan in War and Peace* (New York: New Press, 1993), 60.

10. NavTech, 'Enclosure (A)," 532; Compton, "Mission to Tokyo," 1, 9.

11. Compton, "Mission to Tokyo," 1.

12. Members of this mission included David Griggs and Andrew Longacre, members of the Radiation Laboratory at MIT, and H. Kirk Stephenson of the Office of Field Service of Office of Scientific Research and Development (OSRD). Military members included Major William R. Hewitt of the New Developments Division of the General Staff, and Chemical Warfare Service personnel including Lieutenant Colonel Murray Sanders (Special Projects Division), Major Howard E. Skipper (Medical Division), and Lieutenant Harry H. Youngs (Special Projects Division). Though not officially assigned to the project, Lieutenant Gordon T. Walls, Chemical Warfare Officer in the Far East Air Forces (FEAF), assisted them in their research. GHQ, *Report on Scientific Intelligence Survey in Japan*, 1, 9.

13. US Air Force—Biographies, "General H. H. Arnold," http://www.af.mil/AboutUs/Biographies/Display/tabid/225/Article/107811/general-henry-h-arnold.aspx. Arnold was a pioneer of military aviation and a student of the Wright Brothers themselves.

14. Michael H. Gorn, *The Universal Man: Theodore von Kármán's Life in Aeronautics* (Washington, DC: Smithsonian University Press, 1992), 4, 13, 34–35.

15. Eto Jun, *Senryō shiroku 4—Nihon hontō shinchū* (Tokyo: Kodansha, 1989), 48.

16. General Headquarters, Supreme Commander for the Allied Powers, *History of the Nonmilitary Activities of the Occupation of Japan*, vol. 9: *Reparations and Property Administration—Part A: Reparations* (1945–January 1951), National Diet Library, Tokyo, 5–6; NavTech, "Enclosure (A)," 13.

17. General Headquarters, Supreme Commander for the Allied Powers, *History of the Nonmilitary Activities of the Occupation of Japan*, vol. 9: *Reparations and Property Administration—Part A: Reparations*, 5–6; NavTech, "Enclosure (A)," 13.

18. Fritz Zwicky "Remarks on the Japanese War Technical Report," December 1945, Archives of the Secretariat of the US Air Force Scientific Advisory Board, Washington, DC, 3–4.

19. , Yasunori Matogawa, "Japan Liquid Rockets in the Second World War," in *History of Rocketry and Astronautics—AAS History Series 26*, ed. Donald C. Elder and George C. Jams (San Diego: AAS, 2005), 123.

20. Martin Caidin, "Japanese Guided Missiles in World War II," *Journal of Jet Propulsion* 26, no. 8 (1956): 691–694.

21. NavTech, "Japanese Naval Rockets," December 1945, National Diet Library, Tokyo, 5.

22. NavTech, "Shipboard Rocket Launchers," February 1946, National Diet Library, Tokyo, 7.

23. NavTech, "Japanese Guided Missiles," December 1945, National Diet Library, Tokyo, 1.

24. NavTech, "Japanese Naval Rockets," December 1945, National Diet Library, Tokyo, 8.

25. John Dower, *War without Mercy* (New York: Pantheon, 1986), 28, 53.

26. Robin Reilly, *Kamikaze Attacks of World War II: A Complete History of Japanese Suicide Strikes on American Ships, by Aircraft and Other Means* (London: McFarland, 2010), 40–42.

27. Because certain people were, in Zwicky's words, bastards no matter which angle they were observed from.

28. Fritz Zwicky, "Remarks on the Japanese War Technical Report," December 1945, Archives of the Secretariat of the US Air Force Scientific Advisory Board, Washington, DC, 1–2.

29. There was, for example, a wartime bullet train project in Japan. For details, see Takanori Maema, *Dangan Ressha* (Tokyo: Jitsugyō no nihonsha, 1994).

30. Andrew Gordon, *A Modern History of Japan: From Tokugawa Times to the Present* (Oxford: Oxford University Press, 2020), 97–99.

31. Masai Hashimoto, "Tenbun kyūta no seimitsu koioseizō no kyūmu," *Tenbun Geppō* (Astronomical Herald) 2, no. 12 (March 1910): 135–136.

32. Aaron Moore, *Constructing East Asia Technology, Ideology, and Empire in Japan's Wartime Era, 1931–1945* (Stanford: Stanford University Press, 2013), 4.

33. Dong-Won Kim, "Nishina Yoshio and the Two Cyclotrons," *Historical Studies in the Physical and Biological Sciences* 36, no. 2 (March 2006): 246; Andrew John Robertson, "Mobilizing for War, Engineering the Peace: The State, the Shop Floor, and the Engineer, 1965–1960," Phd diss., Harvard University, 2000, 40–41; James R. Bartholomew, *The Formation of Science in Japan* (New Haven: Yale University Press, 1989), 199–201.

34. Jürgen P. Melzer, *Wings for the Rising Sun: A Transnational History of Japanese Aviation* (Cambridge, MA: Harvard University Press, 2020), 28–30.

35. Kim, "Nishina Yoshio," 246.

36. Bartholomew, *The Formation of Science in Japan*, 249–50.

37. Robertson, "Mobilizing for War," 42.

38. Bartholomew, *The Formation of Science in Japan*, 231.

39. Kim, "Nishina Yoshio," 248, 260.

40. Seiji Takata, "Keisokugaku Genron Nōto: Keisokugaku no Ryōbun (6)," *Keisei* 12, no. 5 (May 1969): 71–73.

41. Robert Wohl, *A Passion for Wings: Aviation and the Western Imagination, 1908–1918* (New Haven: Yale University Press, 1994); Melzer, *Wings for the Rising Sun*, 24–36, 40–65; "Hikōki kenbutsu-ki Yoyogi Hara no shiyō Hino taii Tokugawa taii (Yoyogi renpeiba de

shiken hikō)," *Asahi Shimbun*, December 15, 1910; "Tokugawa taii sōjū no fuaruman-shiki fukuyō hikōki," *Asahi Shimbun*, December 16, 1910, 5; "Abura nure no hikō-fuku no naka ni hana no Hyōdō-san jū ni-ki tsubasa o nabete Shimoshidzu no kyōgi taikai kowaretaru do kei hikō-ka ni fuhei ga ōi Negishi-kun wa yarinaoshi/ hikō kyōgi taikai," *Asahi Shimbun*, June 3, 1922; "Kyōdan kara sora e Ferisu-kō jokyōyu Mabuchi Chōko jō," *Asahi Shimbun*, July 10, 1933.

42. Gorō Suzuki, "Jinbutsu fairu koksuan roketto no oto: Itokawa Hideo hakase to nihon no uchū kaihatsu," *Maru/Chō shobō* 65, no. 12 (December 2012): 152.

43. Hiromi Mizuno, *Science for the Empire: Scientific Nationalism in Modern Japan* (Stanford: Stanford University Press, 2009); Janis Mimura, *Planning for Empire: Reform Bureaucrats and the Japanese Wartime State* (Ithaca: Cornell University Press, 2011), 3, 195–199.

44. Shigeru Nakayama, "The Central Laboratory Boom and the Rise of Corporate R&D," in Shigeru Nakayama and Kunio Gotō, eds., *A Social History of Science and Technology in Contemporary Japan* ,vol. 3: *High Economic Growth Period* (Melbourne: Trans Pacific Press, 2006), 67.

45. Bōeichō hōseikenshūjōū senshishitsu, *Rikugun kōkū no gunbi to unyō—Shōwa jūsan nen shūki made* (Tokyo: Chōun shimbunsha, 1971), 6–10.

46. Not to be confused with noted World War II admiral, Isoroku Yamamoto.

47. Bōeichō hōseikenshūjō senshishitsu, *Kaigun kōkū gaishi* (Tokyo: Chōun shimbunsha, 1976), 1.

48. Takehiko Hashimoto, *Hikōki no tanjō to kūkiryoku gaku no keisei: okkateki kenkyū kaihatsu no kigenomotomete* (Tokyo: Tokyo Daigaku Shuppankai, 2012), 248; Bōeichō hōseikenshūjō senshishitsu, *Kaigun kōkū gaishi*, 1–2.

49. Hirosige Tetu, *Kagaku no shakaishi—kindai nihon no kagaku taisei* (Tokyo: Chūō Kōron, 1973), 154, 176–200, 207–209; Tokyo University, "Daigaku sōchō kaigi ni okeru sankō shiryō," August 25, 1943, Uchida Yoshikazu Papers, Tokyo University archives (sorting no. 4–1), 64.

50. Ben-Ami Shillony, "Universities and Students in Wartime Japan," *The Journal of Asian Studies* 45, no. 4 (August 1986), 769.

51. Shillony, "Universities and Students in Wartime Japan," 773–774.

52. Chalmers Johnson, *MITI and the Japanese Miracle* (Stanford: Stanford University Press, 1982), 133; Hirosige, *Kagaku no shakaishi* (1973), 156, 162, 200.

53. Ichirō Tani, "Kenkyū Kaihatsu to Gakkai—kaigun kōkū to no sōgū," in Nihon kaikūkai kaigun kōkūgaishi kankōkai, ed., *Kaiwashi no kōseki—Nihon kaigun kenkyū gaishi* (Tokyo: Hara shobō, 1982), 38–45; Hashimoto, *Hikōki no tanjō*, 249–268, 285.

54. Williams, "Japanese Aeronautical Research Program and Achievements," 154; Hashimoto, *Hikōki no tanjō*, 261–285.

55. Ichiro Tani, "Kenkyū no kaiko," *Tōkyō daigaku uchū kōkū kenkyūjo hōkoku* 4, no. 2 (1968), 163–178.

56. *Nihon kaigun kōkū shi dai 3,* 372.

57. General Headquarters, United States Army Forces, Pacific, Scientific and Advisory Section, *Report on Scientific Intelligence Survey in Japan, September and October 1945,* November 1, 1945, National Diet Library, Tokyo, 15.

58. Tetu Hirosige, ed., *Nihon shihonshugi to kagaku gijutsu* (Tokyo: Sanichi Shobō, 1962), 119; Yamazaki Toshio, *Gijutsushi* (Tokyo: Tōyō keizai shimpōsha, 1961), 227; Fritz Zwicky, "Remarks on the Japanese War Technical Report," December 1945, Archives of the Secretariat of the US Air Force Scientific Advisory Board, Washington, DC, 4.

59. GHQ, *Report on Scientific Intelligence Survey in Japan, September and October 1945,* 2.

60. Zwicky, "Remarks on the Japanese War Technical Report," 2.

61. Shigetaka Onda, *Tokkō* (Tokyo: Kodansha, 1991), 403; "Kaigun kōkūgijutsushō-chō ninmei Wada Misao shōshō / Wada shōshō ryakureki," *Asahi Shimbun,* May 24, 1940, 2.

62. Kumao Hino, *Theory and Practice of Blasting* (Asa: Nippon Kayaku Co., Ltd., 1959); Nippon kayaku kabushiki kaisha, email to author, October 3, 2016.

63. McDougall, *The Heavens and the Earth,* 9–31; Walter E. Grunden, Yutaka Kawamura, Eduard Kolchinsky, Helmut Maier, and Masakatsu Yamazaki, "Laying the Foundation for Wartime Research: A Comparative Overview of Science Mobilization in National Socialist Germany, Japan, and the Soviet Union," *Osiris* 20 (2005): 79–106.

64. GHQ, *Report on Scientific Intelligence Survey in Japan, September and October 1945,* 5.

65. Matsuoka, *Sentōki Shūsui,* 53–54; Yasunori Matogawa, "Japanese Solid Rockets in the Second World War," in *History of Rocketry and Astronautics—AAS History Series 25,* ed. Herve Moulin and Donald C. Elder (San Diego: AAS, 2003), 124.

66. Matogawa, "Solid Rockets," 126.

67. Testing facilities and machinery were among some of the most important materials developed in this period. Rockets required specific technologies, such as improved altimeters, to track and process their performance. In testing the Ichi-go 1A, for example, Japanese engineers developed a system to measure trajectory and performance. See Matogawa, "Solid Rockets,"124.

68. Yasunori Matogawa, "Japan Liquid Rockets in the Second World War," in *History of Rocketry and Astronautics—AAS History Series 26,* ed. Donald C. Elder and George C. Jams (San Diego: AAS, 2005), 117.

69. Akira Takamoto, "Heiki toshite roketto," *Gunji to Gijutsu* 111 (March 1936), 51; Zwicky, "Remarks on the Japanese War Technical Report," 3–4; Bōeichō hōseikenshūjō senshishitsu, *Rikugun kōkū no gunbi to unyō,* 452–454; *Nihon kaigun kōkū shi 3—seido gijutsu hen,* ed. Chikao Yamamoto (Tokyo: Jiji Press, 1969), 40–43.

70. Matogawa, "Solid Rockets," 125.

71. Eiffel-type wind tunnels are open-ended structures with little to no internal wind circulation, and low power. The design was named for its inventor, Gustave Eiffel, of Eiffel Tower fame. Göttingen-type apparatuses, in contrast, are closed-circuit structures.

72. Jun Okamura, *Kōkū gijutsu no zenbō (ge)* (Tokyo: Hara Shobō, 1976), 126, 268–272.

73. Jun Okamura, *Kōkū gijutsu (jō)*, 18–21.

74. Masao Suzuki, "Dai Tōa Sensō to dai ichi kaigun nenryō shō," in *Nippon kaigun nenryōshi (jō)* (Tokyo: Hara Shobō, 1973), 680–683; Shigeaki Nabeshima, "Kaigun nenryōshō chōkyū kaijukyoku chō jidai no omoide," in *Nippon kaigun nenryōshi (ge)* (Tokyo: Hara Shobō, 1973), 8, 4. Nabeshima was later the head of the No. 3 Fuel Depot in Yamaguchi; see NavTech, "Japanese Fuels and Lubricants—Article 5. Research on Rocket Fuels of the Hydrogen Peroxide-Hydrazine Hydrate Type" (February 1946), National Diet Library, Tokyo, 14.

75. Suzuki, "Dai Tōa sensō," 680–689.

76. Bōeichō hōseikenshūjō senshishitsu, *Kaigun kōkū gaishi*, 34–35.

77. Bōeichō hōseikenshūjō senshishitsu, *Rikugun kōkū no gunbi to unyō*, 160–161.

78. Okamura, *Kōkū gijutsu no zenpō (jō)*, 28.

79. Johnson, *MITI and the Japanese Miracle*, 23.

80. Okamura, *Kōkū gijutsu no zenpō (jō)*, 2, 6; Bōeichō hōseikenshūjō senshishitsu, *Kaigun kōkū gaishi*, 35–26, 126–130.

81. Bōeichō hōseikenshūjō senshishitsu, *Rikugun kōkū no gunbi to unyō*, 404.

82. Iwaya, "Roketto ki Shūsui ni tsuite," 85.

83. Mitsubishi Heavy Industries, *Mitsubishi jukōgyō kabushikigaisha shi* (Tokyo: Mitsubishi jūkōgyō kabushiki kaisha, 1953), 294.

84. Tetsuji Okazaki, "The Supplier Network and Aircraft Production in Wartime Japan," *Economic History Review* 54, no. 3 (August 2011): 987.

85. NavTech, *Report of the US Naval Technical Mission to Japan*, "Japanese Propellants—Article 1. Use and Manufacture of Ortho Tolyl Urethane for Stabilizing Rocket and Gun Propellants," Report of the US Naval Technical Mission to Japan, "Intelligence Targets Japan," 0-10-1 (November 1945), 7; NavTech, *Report of the US Naval Technical Mission to Japan*, "Japanese Propellants—Article 2. Rocket and Gun Propellants—General," "Intelligence Targets Japan," 0-10-2 (November, 1945), 3.

86. Records of the U.S. Strategic Bombing Survey, *2 USSBS YD-208* (1947), microfilm, National Diet Library, Tokyo, 23.

87. Hodogaya Chemical Company, "History of Hodogaya Chemical Group," https://www.hodogaya.co.jp/english/company/history.

88. Arnold Krammer, "Transfer as War Booty: The US Technical Oil Mission to Europe, 1945," *Technology and Culture* 22, no. 1 (January 1981): 68; Tetu Hirosige, *Kagaku no shakaishi (ka)* (Tokyo: Iwanami Shōten, 2013), 61.

89. Francis K. Danquah, "Japan's Food Farming Policies in Wartime Southeast Asia: The Philippine Example, 1942–1944," *Agricultural History* 64, no. 3 (Summer 1990): 62.

90. Pape, "Why Japan Surrendered," 159; William M. Tsutsui, "Landscapes in the Dark

Valley: Toward an Environmental History of Wartime Japan," *Environmental History* 8, no. 2 (April 2003): 305.

91. Tsutsui, "Landscapes in the Dark Valley," 295.

92. GHQ, *Report on Scientific Intelligence Survey in Japan, September and October 1945,* 15.

93. Tsutsui, "Landscapes in the Dark Valley," 295; NavTech, "Japanese Fuels and Lubricants, Article 1—Fuel and Lubricant Technology," 15.

94. Theodore Cohen, *Remaking Japan* (New York: Free Press, 1987), 6.

95. Compton, "Mission to Tokyo," 14; John J. Geoghegan, *Operation Storm: Japan's Top Secret Submarines and Its Plan to Change the Course of World War II* (New York: Crown, 2013), 110–130.

96. Frank Williams, "Japanese Aeronautical Research Program and Achievements," in *Towards New Horizons: Technical Intelligence Supplement: A report prepared for the AAF Scientific Advisory Group* 3 (Dayton: Army Air Force Headquarters Air Materiel Command, May 1946), 157; Cohen, *Remaking Japan,* 3–5.

97. Takashi Hosokawa, ed., *Rekishi gunzō shiriiizu [Ketteihan] Taiheyō sensō 3: "Nanpō shigen" to Ranin shisen* (Tokyo: Gakken, 2009), 13.

98. Laurie M. Brown and Yoichiro Nambu, "Physicists in Wartime Japan," *Scientific American* 279, no. 6 (December 1998): 100.

99. Grunden et al., "Laying the Foundation for Wartime Research," 107–130; NavTech, *Pharmacology and Malariology in Japan: Civilian and Naval,* February 21, 1946, National Diet Library, Tokyo. For a detailed description of the development of the death ray, see Walter Grunden's excellent *Secret Weapons and the Second World War* (Lawrence: University Press of Kansas, 2005), 112–116; Melzer, *Wings for the Rising Sun,* 234; Robertson, "Mobilizing for War," 253–256.

100. Michael Gorn, *Harnessing the Genie: Science and Technology Forecasting for the Air Force, 1944–1986* (Washington, DC: Office of Air Force History, United States Air Force, 1988), 17.

101. Melzer, *Wings for the Rising Sun,* 236.

102. NavTech, "Japanese Guided Missiles," in *Report of the US Naval Technical Mission to Japan* (San Francisco: US Naval Technical Mission to Japan, December 1945), National Diet Library, Tokyo, 11.

103. Matogawa, "Japan Liquid Rockets in the Second World War," 116–117; NavTech, "Japanese Guided Missiles," December 1945, National Diet Library, Tokyo, 21.

104. Bōeichō hōsei kenshūjō senshishitsu, *Sōsho senshi rikugun kōkū heiki: seisan, hokyū no kaihatsu* (Tokyo: Tokyo Publishing, 1975), 440–441, 458–459; NavTech, "Japanese Guided Missiles," December 1945, National Diet Library, Tokyo, 21–22, 87.

105. *Nihon kaigun kōkū shi dai 3,* 541.

106. Jun Okamura, *Kōkū gijutsu (ge),* 244–257; Jun Okamura and Eiichi Iwaya, *Nihon no kōkūki kaigunki hen* (Tokyo: Shuppan kyōdō-sha, 1960), 259–261.

107. Eiichi Iwaya, "Ichi—arishi nichi no waga kaigun kōkūki no zenbō to sono jittai," in *Kōkū gijutsu no zenbō (ge)*, ed. Jun Okamura (Tokyo: Hara Shobō, 1976), 231–246; *Nihon kaigun kōkū shi dai 3*, 547; Okamura and Iwaya, *Nihon no kōkūki kaigunki*, 259–261, 237.

108. Matsuoka, *Sentōki Shūsui*, 56.

109. Reilly, *Kamikaze Attacks of the Second World War*, 40–42.

110. Iwaya, "Roketto ki Shūsui ni tsuite," 82.

111. Matogawa, "Japan Liquid Rockets in the Second World War," 112–116; Yamamoto, "Roketto tsui ni hasshin sezu," 1079.

112. Iwaya, "Kaigun kōkūki no zenbō," 231–246.

113. GHQ, *Report on Scientific Intelligence Survey in Japan, September and October 1945*, 8.

114. NavTech, "Japanese Naval Rockets," December 1945, National Diet Library, Tokyo, 5; *Nihon kaigun kōkū shi dai 3*, 781; NavTech, "Japanese Bombs," September 4, 1945, 34; NavTech, "Japanese Torpedoes and Tubes, Article 2: Aircraft Torpedoes," March 1946, 1; *Nihon kaigun kōkū shi dai 3*, 668.

115. Dai-ichi rikugun gijutsu kenkyūjō, "Ichi giken san kenhou dai kyū ban (hō 101) shisei yon shiki ni ju shiki senchi funshinhō kenkyū hōkoku" (March 1945), https://www.digital.archives.go.jp/das/image/F0000000000000218330.

116. Dai-ichi rikugun gijutsu kenkyūjō, "Shisei yon shiki yonjū sencha mokusei funshinhō, shisei yon shiki yonjū sencha funshin ryūdan setsumeisho" (March 1945), https://www.digital.archives.go.jp/das/image/F0000000000000218310; Matogawa, "Solid Rockets," 126–129.

117. United States Pacific Fleet and Pacific Ocean Areas, "Japanese Artillery Weapons," CINCPAC-CINCPOC Bulletin no. 152, vol. 45 (July 1, 1945), 80, https://www.bulletpicker.com/pdf/CINCPAC-Bulletin-152-45-Japanese-Artillery-Weapons.pdf.

118. Matsuoka, *Sentōki Shūsui*, 57, 60–63; Yasunori Matogawa, *Nippon uchū kaihatsu hishi—ganso tori ningen kara minkan roketto e* (Tokyo: NHK Shuppan, 2017), 53.

119. Jun Okamura, *Kōkū gijutsu no zenpō (jō)*, 481–483.

120. Shigeo Itani, "Ro-go yaku no omoide," in *Nihon kaigun nenryō shi (ge)* (Tokyo: Hara Shobō, 1972), 1084.

121. *Nihon kaigun kōkū shi dai 3*, 643.

122. Fujihira, *Kimitsu heiki no zenbō*, 6.

123. Yamamoto, "Roketto tsui ni hasshin sezu," 1069–1070.

124. Matogawa, "The Shūsui Japanese Rocket Fighter," 177.

125. *Nihon kaigun kōkū shi dai 3*, 604.

126. NavTech, "Japanese Propellants—Article 2. Rocket and Gun Propellants—General," 2–15.

127. Toshikazu Uchida, "Kaigun nenryō shō to watashi," in *Nihon kaigun nenryō sho (jō), nenryō konwakai* (Tokyo: Hara Shobō, 1972), 812–814.

128. Yamamoto, "Roketto tsui ni hasshin sezu," 1076.

129. Matogawa, "Japan Liquid Rockets in the Second World War," 113–117.

130. NavTech, "Japanese Fuels and Lubricants—Article 1, Fuel and Lubricant Technology," 14; NavTech, "Japanese Fuels and Lubricants—Article 5. Research on Rocket Fuels of the Hydrogen Peroxide-Hydrazine Hydrate Type," Report of the US Naval Technical Mission to Japan, *Intelligence Targets Japan*, February 1946, 11–13; No. 1 Naval Fuel Depot, Ōfuna, "Asetoarudihido kōsei ni kansuru kenkyū," *Kenkyū jikken seiseki hōkoku* 158 (Yokohama: No. 1 Naval Fuel Depot, Ōfuna, November 9, 1943); No. 1 Naval Fuel Depot, Ōfuna, "Asetoarudihaido yori kurtonaruderu no kōei ne kansuru kenkyū," *Kenkyū jikken seiseki hōkoku* 159 (Yokohama: No. 1 Naval Fuel Depot, Ōfuna, November 11, 1943).

131. Hydrogen peroxide was produced through an electrochemical process using platinum wires—which have very high resistance—to oxidize a solution through which they ran. This process has long been superseded in the industrial production of the chemical by more efficient methods.

132. *Nihon kaigun kōkū shi dai 3*, 605; Yamamoto, "Roketto tsui ni hasshin sezu," 1074.

133. Iwaya, "Kaigun kōkūki no zenbō," 231–246.

134. NavTech, "Japanese Fuels and Lubricants—Article 1, Fuel and Lubricant Technology," 14; NavTech, "Japanese Fuels and Lubricants—Article 5. Research on Rocket Fuels of the Hydrogen Peroxide-Hydrazine Hydrate Type," 11–13.

135. Matogawa, "The Shūsui Japanese Rocket Fighter," 178–184; Sukezō Atachi, "Rō go yaku," in *Nihon kaigun nenryō sho (jō)* (Tokyo: Hara Shobō, 1972), 1068.

136. NavTech, "Japanese Fuels and Lubricants—Article 5. Research on Rocket Fuels of the Hydrogen Peroxide-Hydrazine Hydrate Type," 11–13; Atachi, "Rō go yaku," 1067.

137. Jun Okamura, *Kōkū gijutsu no zenpō (jō)*, 483.

138. Iwaya, "Roketto ki Shūsui ni tsuite," 84–85.

139. Matsuoka, *Sentōki Shūsui*, 85, 91; Iwaya, "Roketto ki Shūsui ni tsuite," 85; Kamura and Iwaya, *Nihon no kōkūki kaigunki*, 232; Jun Okamura, *Nihonsho roketto sui 'Shūsui': Ekitai rokettoenjin-ki no tanjō* (Tokyo: Miki Shobō, 2004), 85.

140. Okamura and Iwaya, *Nihon no kōkūki kaigunki*, 232.

141. Matsuoka, *Sentōki Shūsui*, 55.

142. Matogawa, "The Shūsui Japanese Rocket Fighter," 177–187; Nishiyama, "Swords into Plowshares," 2. It should be noted that this does not mean there were no female aviators in Japan in the prewar period. Indeed, two—Chōko Mabuchi and Kiku Nishizaki—have been the subjects of TV adaptations of their life stories and remain well regarded in Japan. However, in terms of specialist theoreticians and engineers working on aircraft, Japan's universities and armed forces remained entirely male domains.

143. Fujio Nakanishi, "On the Yield Point of Mild Steel," PhD diss., Tokyo Imperial University (University of Tokyo), 1931; Fujio Nakanishi, "Hikōki no roketto suishin ni tsuite," *Nihon koukū gakkaishi* 5, no. 39 (July 1938): 669–678; Fujio Nakanishi and Kazuo Sato, "Ōryoku jōtai to sōsei henkei no katachi," *Nihon kikaigaku hōron bunshū* 26, no. 170: 1327–

1332; Ichirō Tani, "Kenkyū Kaihatsu to Gakkai—kaigun kōkū to no sōgū," in Nihon kaikūkai kaigun kōkūgaishi kankōkai, ed., *Kaiwashi no kōseki—Nihon kaigun kenkyū gaishi* (Tokyo: Hara shobō, 1982), 43; Hidemasa Kimura, Kazuo Ochiai, and Shinaki Tsukuda, "Shokyū guraidā no chijō shōgeki sokutei kekka," *Nihon kōkū gakkaishi* 5, no. 45 (1957): 29–30; Hidemasa Kimura, "N-58 Shigunetto keihikōki," *Nihon kōkū gakkaishi* 9, no. 84 (1961): 29–30. There was, in fact, a concerted effort on the part of Japan National Railways to recruit former aviation engineers, specifically in order to preserve their knowhow and provide them with a means of living in the tumultuous postwar period, details of which can be found in Minoru Sawai, "Gijutsu-sha no gunmin tenkan to tetsudō gijutsu kenkyūjo," *Osaka daigaku keizaigaku* 59, no. 1 (2009): 1–19.

144. Atsushi Shibuya, *Hino Kumazō den: Nihonkōkū shoki no shinsō* (Tokyo: Aoshio, 1977), 225–226, 292.

145. Hino was, incidentally, the naval specialist on these subjects. The army had its own research facilities in the form of the Seventh and Eighth Military Technical Laboratories, headed by Major General Masahiko Nomura and Major General Takeo Nagao, respectively. Army development of rocketry appears to have been entirely separate, with fundamental research happening at the Seventh, and further testing happening at the 1st Military Laboratory, while propellant manufacturing took place in Tokyo and Osaka—and not at the navy's Funaoka facility, which will be discussed below. See Records of the US Strategic Bombing Survey, "Principal Army Ground Force Laboratories," *Report on Scientific Intelligence Survey in Japan* II (1946), National Diet Library, Tokyo, frame 98; W. W. Burgess, "Solid Rocket Propellants Research," 1945, National Diet Library, Tokyo.

146. NavTech, "Enclosure (A)," 8.

147. NavTech, "Enclosure (A)," 8.

148. Compton, "Mission to Tokyo," 19–22.

149. Kumao Hino, Letter from Hino Kumao to Dr. F. Zwicky, *Records of the US Strategic Bombing Survey, Extracts from Intelligence Bulletin pertaining to the Bakam Report* (November 21, 1945), National Diet Library, Tokyo.

150. Kumao Hino, "On the Combustion and Detonation Velocities of Solid Explosives (including the Applications of the Principles)," Records of General Headquarters Far East Command, Assistant Chief of Staff, November 22, 1945.

151. Kumao Hino, *Kayakurui no bakusoku ni kansuru kenkyū*, PhD diss., Tokyo Imperial University (University of Tokyo), 1945; Nippon Kayaku kabushiki kaisha, email to author, October 3, 2016.

152. Shigeru Nakayama, ed., *A Social History of Science and Technology in Contemporary Japan*, vol. 1: *The Occupation Period, 1945–1952* (Melbourne: Pan-Pacific Press, 2006), 29.

153. General Headquarters, Supreme Commander for the Allied Powers, SCAPIN-301 (AG 360 [18 Nov 45])—Commercial and Civil Aviation, https://en.wikisource.org/wiki/SCAPIN-301.

154. Samuels, *Rich Nation, Strong Army,* 197–200.

155. Kumao Hino, "Kayōsei enrui no kyōshū sokudōshiki," *Kōgyō kayaku kyōkaishi (Journal of the Industrial Explosives Society)* 10, no. 4 (March 1950): 142–145; "Theory of the Burning Rate of Industrial Safety Fuse," *Kōgyō kayaku kyōkaishi (Journal of the Industrial Explosives Society)* 15, no. 1 (March 30, 1954): 12–21; "Theory of Blasting with Concentrated Charge," *Kōgyō kayaku kyōkaishi (Journal of the Industrial Explosives Society)* 14, no. 4 (December 1954): 233–250; "Concentrated Type of No-cut Round of Blasting," *Kōgyō kayaku kyōkaishi (Journal of the Industrial Explosives Society)* 16, no. 3 (September 1955): 166–179; "Fragmentation of Rock through Blasting," *Kōgyō kayaku kyōkaishi (Journal of the Industrial Explosives Society)* 1 (March 1956): 2–11.

156. Kumao Hino, "Shock Wave Theory of Blasting with Cylindrical Charge," *Kōgyō kayaku kyōkaishi (Journal of the Industrial Explosives Society)* 18, no. 1 (March 1957): 2–21; "Theory of Relations between the Detonation Velocity of Solid Explosives and the Thickness of Cases or the Diameter of the Charge," *Kōgyō kayaku kyōkaishi (Journal of the Industrial Explosives Society)* 19, no. 3 (August 1958): 169–180.

157. Kumao Hino, *Theory and Practice of Blasting* (Asa: Nippon Kayaku Co., Ltd., 1959); Nippon kayaku kabushiki kaisha, email to author, October 3, 2016.

158. Nippon kayaku kabushiki kaisha, "A Cap Sensitive Ammonium Nitrate-Fuel Oil Explosive and a Method of Manufacturing the Same," US patent no. 3111437, filed September 6, 1960, issued November 19, 1963.

159. Nippon kayaku kabushiki kaisha, email to author, October 3, 2016.

160. Yasunori Matogawa, *Gyakuten no Tsubasa—Penshiru roketto monogatari* (Tokyo: Shin Nippon Shuppansha, 2005), 66.

161. Hashimoto, *Hikōki no tanjō,* 289.

162. Daizō Kusayanagi, "'Suzō sangyō' no sutā Itokawa Hideo," *Bungei shunju* 47, no. 7 (July 1969): 260–273.

163. Bartholomew, *The Formation of Science in Japan,* 199–201; Hashimoto, *Hikōki no tanjō,* 250–254; *Nihon kaigun kōkū shi dai 3,* 259.

164. Okamura, *Kōkū gijutsu no zenpō (jō),* 18.

165. Gorō Suzuki, "Itokawa Hideo hakase," 152–153; Takanori Maema, "Maema ga iku! Kokusanki o tsukuru otokotachi dai 9 kai—Itokawa, Hideo (Nihon uchū kaihatsu, roketto no oto) (jō) roketto hakase no kōkū gijutsusha jidai," *Kōkū jōhō* 61, 4, no. 811 (April 2011): 83–88.

166. Yasunori Matogawa, *Gyakuten no Tsubasa: Penshiru roketto monogatari* (Tokyo: Shin Nippon Shuppansha, 2005), 65–66; Shichibei Kozawa, "Senji chu no Itokawa sensei," in *Ningen Hideo Itokawa hakase to wa,* ed. Iwao Kanezawa (Tokyo: Hirohide Kōgei, 2003), 3; Gorō Suzuki, "Itokawa Hideo hakase," 153.

167. Between 1935 and 1941, Itokawa published a variety of articles in the *Report,* ranging in length from five pages (Hideo Itokawa, "Tsubasa jōmen ni shōzuru henryū o riyō shita shi

hayakei," *Tōkyō teikoku daigaku kōkū kenkyūjo ihō* 154 [June 1937]: 348–351) to a whopping thirty pages (Hideo Itokawa and Yu Taisaka, "Sō hatsuki no naseru no hanryū ga shōkōda kōka oyobi sōantei ni oyobi su eikyō —[suihei biyoku no jōge ichi kettei shiryō dai ni-pō]," *Tōkyōteikokudaigaku kōkū kenkyūjo ihō* 201 [May 1941]: 103–134).

168. Hideo Itokawa, "Yoko oyobi hōkō no anteisei oyobi sōju sei," *Nakajima kenkyū jōhō* 4, no. 1 (March 1939): 19–32; Hideo Itokawa, "Yoko oyobi hōkō no anteisei oyobi sōju sei dai 2 hō," *Nakajima kenkyū jōhō* 4, no. 3 (September 1939): 163–176.

169. Yasunori Matogawa, "'Pencil' Rocket and Hideo Itokawa: A Pioneering Work of Japanese Rocketry," *History of Rocketry and Astronautics: Proceedings of the History Symposia of the International Academy of Astronautics* 28–29 (1994–1995): 123.

170. Takanori Maema, "Maema ga iku! Kokusanki o tsukuru otokotachi—Itokawa, Hideo (Nihon uchū kaihatsu, roketto no oto) (ge) roketto hakase no kōkū gijutsusha jidai," *Kōkū jōhō* 61, 5, no. 812 (May 2011): 87; Atsushi Shirasawa, "Denkan yōwa: Nihon ni okeru nōhakei no rekishi," *Denkan no sōgō gakkaishi* 14, no.2 (November 2020): 130.

171. Gorō Suzuki, "Itokawa Hideo hakase,"154.

172. Yasunori Matogawa, "Matogawa hakase no uchū yomoyama hanashi: Itogawa Hideo no baiorin dzukuri," *Sekai shūhō* 85, no. 15 (April 20, 2004): 64; Matogawa, *Nippon uchū kaihatsu hishi,* 49.

173. Yasunori Matogawa, "'Pencil' Rocket and Hideo Itokawa: A Pioneering Work of Japanese Rocketry," *History of Rocketry and Astronautics: Proceedings of the History Symposia of the International Academy of Astronautics* 28–29 (1994–1995) 123.

174. Kazuya Toda, "Sengo mo daikū yumemita oto tachi," in *Dai Nippon teikoku rikukai-gun kōkū butai zenshi: Rekishi dokuhon* (Tokyo: Henshūbu hen, 2015), 248–255.

175. "A Passion for Rocketry: Ryojiro Akiba," Japan Aerospace Exploration Agency, http://global.jaxa.jp/article/interview/2013/vol77/index_e.html.

176. Matogawa, "'Pencil' Rocket and Hideo Itokawa," 122.

177. Matogawa, "'Pencil' Rocket and Hideo Itokawa," 122–129.

178. ISAS, "History of Japanese Space Research," 2008, https://www.isas.jaxa.jp/e/japan_s_history/detail/backgr.shtml.

179. JAXA, "A Passion for Rocketry: Ryojiro Akiba."

180. Though direct inflation-based comparison of prices would yield a figure of around $120,000, in this book I have preferred the methodology pioneered by the team at MeasuringWorth, which emphasises *percentage of GDP.* This provides a clearer sense of how much the figure mentioned would have appeared to those spending it at the time. The figure mentioned in the text is based on the prevailing exchange rate of around 360 yen to $1 in the 1950s. For details on their methodology, see Samuel H. Williamson and Louis P. Cain, "Measures of Worth," MeasuringWorth, http://www.measuringworth.com/defining_measures_of_worth.php.

181. Zwicky, "Remarks on the Japanese War Technical Report," 3.

Chapter 2

1. *Space Battleship Yamato,* dir. Takashi Yamazaki, Tōhō, 2010.

2. *Space Battleship Yamato,* dir. Leiji Matsumoto, Yomiuri TV: 1974–1975; *Space Battleship Yamato,* dir. Leiji Matsumoto, 1976.

3. Sato, "Contested Gift," 178.

4. Asif A. Siddiqi, "Spaceflight in the National Imagination," in *Remembering the Space Age,* ed. Steven J. Dick (Washington, DC: NASA Office of External Relations History Division, 2008), 18.

5. Robertson, "Mobilizing for War," 39, 211.

6. McDougall, *The Heavens and the Earth,* 4.

7. "Shinsekkaisha no yume," *Sora* (September 1936), 261; (February 1938), 202; (March 1938), 308; (October 1938), 210; (December 1938), 418; (April 1939), 117; (January 1941), 98.

8. Asif Siddiqi, *Sputnik and the Soviet Space Challenge* (Gainesville: University of Florida Press, 2010), 45, 98–101.

9. *Frau im Mond,* dir. Fritz Lang, UFA, 1929; *Things to Come,* dir. William Cameron Menzies, United Artists, 1936.

10. Kazuo Harada, *Dare ni mo wakaru kagaku zenshū dai kyū ken: saikin hatsumei romansu* (Tokyo: Kokumin Chizu, 1929), 11.

11. *Chūō Kōrōn,* "Gurabia han: saishin kagakkai no kyōi," 46, no. 1 (January 1, 1931): 453.

12. Shinzaburo Hori, *Saki dzu detarame nise bunka o hōmure: Katei keizai jukyū keizai kokka keizai no hatan o sukū ni wa seikō-ai katei-ai kokka ai no fusoku o oginau ni wa* (Tokyo: Nihon sōgō kagaku kenkyujo, 1933), 12–13. This title translates, rather spectacularly, as "Bury the Bullshit and False Culture first: To save the family economy, supply and demand economy, and national economy from bankruptcy, and to make up for the lack of love for sexual intercourse, familial love, and love of nation." One can only assume from the following two decades of Japan's history that Hori's advice did not reach the highest levels of government.

13. Keitarō Kaishima, *Saikin no hikōki to shōrai* (Tokyo: Kaishima Keitarō, 1930), 153–164.

14. *Hatsumei.* "Kanten mo nan'nosono roketto riyō no jinkō kōu," 28, no. 8 (August 1931): 47.

15. Akio Saito, "Roketto o kataru," *Gunji to gijutsu* 96 (December 1936): 38.

16. Yūshichi Nishizawa, *Hanabi no kenkyū* (Tokyo: Uchida rōkakuho, 1924); Yūshichi Nishizawa, "Über die Strukturviskosimetrischen Untersuchungen," PhD diss., Tokyo Imperial University (University of Tokyo), 1932; Yūshichi Nishizawa, *Kōryō oyobi kashōhin* (Tokyo: Kyōritsusha, 1934).

17. Yūshichi Nishizawa, *Kagaku sen heiki* (Tokyo: Sanshōdō, 1939), frontispiece, 173–174, 177.

18. *Shonen gonensei*, 214; Shuichi Kaneyama, *Saikin kodomo kagaku dokuhon* (Tokyo: Bunka shobō, 1931), 277–288.

19. "Yūbin roketto," *Shōnen kurabu* 23, no. 10 (September 1936); "Roketto sono hatsumei ka," *Shonen gonensei* 17, no. 3 (June 1936): 215.

20. Thomas R. H. Havens, *Valley of Darkness: The Japanese People and World War Two* (New York: W. W. Norton, 1978), 25–33, 73–83.

21. Yūshichi Nishizawa, *Kagaku sen heiki* (Tokyo: Sanshōdō, 1939), 178.

22. Mitsuo Harada, *Kodomo rigaku e-banashi. Dai 1-kan (jidōsha no maki)* (Tokyo: Kodomo Rigaku kai, 1932), 54–55.

23. Tarō Sunō, "Roketto," in Nihonhōsōkyōkai Kantō shibu, ed., *Tsūzoku kagaku: JOAK kaki tokubetsu kōza (dai ni hōsō tekisuto)* (Tokyo: Nippon hōsō shuppan kyōkai, 1930), 280–287. Perhaps the only notable exception to this description was that such fancies do not appear to have spread to those who perhaps would be most interested in travel to space—astronomers. Both professional and amateur interest in the stars had reached enough of a critical mass for the creation of the Japanese Astronomical Society in 1908, eighteen years after the creation of its British equivalent in 1890, and less than a decade after the American Astronomical Society's creation in 1899. However, its official journal, *Tenmon Geppō* (The Astronomical Herald), made little to no mention of rockets during the entire prewar period. In fact, the publication included no substantial discussions of rocketry until the mid-1950s.

24. Ito, "Values of 'Pure Science,'" 76.

25. *Momotarō no Umiwashi*, dir. Seo Mitsuyo, Geijutsu Eigasha, 1943; *Momotaro: Umi no Shimpei*, dir. Seo Mitsuyo, Shochiku, 1945, 26:00–34:00.

26. Kenkichiro Koizumi, "In Search of 'Wakon': The Cultural Dynamics of the Rise of Manufacturing Technology in Postwar Japan," *Technology and Culture* 43, no. 1 (January 2002): 29–49.

27. Suzuki, "Itokawa Hideo hakase," 152; Matogawa, *Nippon uchū Kaihatsu*, 46–47.

28. Kenji Ito, "Values of 'Pure Science': Nishina Yoshio's wartime Discourse between Nationalism and Physics, 1940–1945," *Historical Studies in the Physical and Biological Sciences* 33, no. 1 (2002): 68–69.

29. Jixing Pan, "On the Origin of Rockets," *T'oung Pao*, col. 73, livr. 1/3 (1987): 2–15; Joseph Needham, "The Fire-lance, Ancestor of All Gun-barrels," in C. Le Blanc and Susan Blader, eds., *Chinese Ideas about Nature and Society: Studies in Honour of Derk Bodde* (Hong Kong: Hong Kong University Press, 1987), 295–333.

30. Susumu Kichinaga, "Roketto no haitatsu kōshi," *Gunji to gijutsu* 1, no. 121 (January 1937): 95–97; Rajesh Kochhar, "Zinc, Steel and Rocketry: Western Mainstreaming of Empirical Indian Technologies," *Current Science*, 113, no. 9 (November 10, 2017): 1796.

31. Saito, "Roketto o kataru," 38.

32. Kōyaku Matsuda, "Roketto no Shōrai," *Gunji to gijutsu* 5, no. 8 (August 1931): 43–48.

33. *Gunji to gijutsu* 82 (October 1933): frontispiece.

34. "Roketto no kenkyū," *Kagaku Zasshi* 18, no. 4 (April 1933): 33. Japan did not officially approach Germany with suggestions of a pact to counter the activities undertaken to spread communism by the USSR's Communist International (Comintern) organization until 1935.

35. Watarō Tsurihashi, *Norakura shōi roketto tokkan-tai: warai no manga bukku* (Tokyo: Taikodō, 1934).

36. "Roketto heiki no haitatsu," *Shūkan shokokumin* 4, no. 6 (February 1945): 2.

37. Hiroshi Tanaka, "Tokkyō kenkyū: roketto," *Hatsumei* 42, no. 1 (June 1946): 14.

38. Nihon kōtsū kyōkai, *Kōtsū kankei shiryō dai 6 shō—Kagaku no kyōi roketto* (Tokyo: Nihon kōtsū kyōkai, 1936), 17, 43, 45, 47, 50–54, 59, 63–64, 66–67. Ohara's comment on piloted rockets can be found on p. 45.

39. "Roketto hikōki ni tsuite," *Kaigun gijutsu kenkyū shō* (December 17, 1941).

40. Hideo Itokawa, "Gakugei/saikin kōkūki no jūyō kadai (kōkū-bi tokushū)," *Asahi Shimbun*, September 19, 1943.

41. Hideo Itokawa, "Sudeni soku I de shiken-zumi Amerika no roketto hikōki," *Asahi Shimbun*, January 12, 1944.

42. JAXA, "A Passion for Rocketry: Ryojiro Akiba."

43. Dower, *Embracing Defeat*, 492–495.

44. Letter from Yoshio Nishina to Hidehiko Tamaki, August 7, 1945, in *Hiroshima-ken shi: genbaku shiryō hen* (Hiroshima, 1972), 429.

45. John Dower, *Empire and Aftermath: Yoshida Shigeru and the Japanese Experience, 1878–1954* (Cambridge, MA: Harvard University Asia Center,1988), 278.

46. Robertson, "Mobilizing for War," 141–143.

47. Nishiyama, "Swords into Plowshares," 74.

48. "Chōon roketto-ki Tōdai Itokawa kyōju-ra no kenkyū kagaku," *Asahi Shimbun*, December 12, 1955; "Kōha roketto to zengeru fusai Doitsu kagaku no monogatari," *Asahi Shimbun*, June 20, 1958.

49. Melissa Willard-Foster, "Planning the Peace and Enforcing the Surrender: Deterrence in the Allied Occupation of Germany and Japan," *The Journal of Interdisciplinary History* 40, no. 1 (Summer 2009): 43, 50–51; "Shin chōkyori roketto-hō butai Ōshū e rikugun tōkyoku happyō gunji," *Asahi Shimbun*, October 31, 1954.

50. Kihara, "Production of Weapons in Postwar Japan," 4–5.

51. Historian and sociologist Elise K. Tipton has succinctly summarized the concept of "dark valley" as a "common view amongst liberal and 'progressive' (leftist) historians in both Japan and Western countries that the highly authoritarian, expansionist system of wartime Japan was an aberration in Japan's modern development, and that defeat and postwar occupation reforms steered Japan back to the democratization process begun during the Taisho period." This view has been intensively critiqued, not least by those who have pointed out that "mobilization for total war laid the foundation for the postwar democratic system which is characterized by rationalization, mobilization and high levels of social integration and

control." Nevertheless, the period of mobilization for total war "left a lasting memory," and "even more than six decades after its end it continues to haunt Japanese domestic politics and foreign relations," particularly in terms of associating the period with material deprivation, chaos, and violence. See Elise K. Tipton, *Modern Japan: A Social and Political History* (Oxford: Routledge, 2008), 132–152.

52. Nishiyama, *Swords into Plowshares*, 30.

53. NHK purojekutoekkusu seisaku han, *Purojekutoekkusu dai 2-kan fukkatsu e no butai-ura* (Tokyo: NHK Shuppan, 2003), 19. Nakayama was presumably unaware of weapons such as the German 31 in (80 cm) bore Schwerer Gustav railway gun, capable of firing 15,000 lb (7 tonne) projectiles up to 29 mi (47 km).

54. Moore, *Constructing East Asia Technology, Ideology, and Empire in Japan's Wartime Era, 1931–1945*, 226–227.

55. "Sutekina roketto," *Ichinen gakushū* (June 1959), 36–39; "Tsuki e iku roketto basu," *Ichinen gakushū* (February 1960), 2–3.

56. Asahi Shimbun Urawa shikyoku, ed., *Arakawa: Sono tsuchi to kokoro* (Tokyo: Asahi-sonorama, 1972), 293; Yoshidamachi kyōiku iinkai, ed., *Yoshidamachi shi* (Yoshida: Yoshida City, 1982), 1071–1072.

57. Yasunori Matogawa, "Another Destiny of Rocketry in Japan: Festival Rockets in Japanese Shrines," *History of Rocketry and Astronautics: Proceeding of the History Sumposium at the International Academy of Astronautics* 23 (1995), 325.

58. Chichibu kankō kyōkai, "Ryūsei matsuri," http://www.chichibuji.gr.jp/event/ryūsei/.

59. Yoshida Ryūsei hozonkai, "2023 matsuri jōhō," https://ryusei.biz/?page_id=21.

60. Jiro Horikoshi and Eiichi Iwaya, "96 shiki kanjō sentōki," *Kōkū jōhō Rinji zōkan: Nihon Suguru-ki monogatari* (April 1959), 104–115.

61. Jiro Horikoshi and Eiichi Iwaya, "96 shiki kanjō sentōki (jō)," *Kōkū jōhō* 64, no. 2 (February 2014): 60–65; Jiro Horikoshi and Eiichi Iwaya, "96 shiki kanjō sentōki (ge)," *Kōkū jōhō* 64, no. 3 (March 2014): 60–65.

62. Eiichi Iwaya, "Nihon sentōki no ayunda michi," *Kōkū jōhō* (October 1958): 39–49.

63. Unfortunately these, and other sources from Iwaya, have large lacunae in the collections at the National Diet Library and elsewhere.

64. Frank Whittle, *Jets* [*Jetto Furanku Hoittoru*], trans. Eiichi Iwaya (Tokyo: Hitotsuboshi, 1955), 417–419.

65. Ito, "Values of 'Pure Science,'" 71.

66. Eiichi Iwaya, "Taiheiyō sensō mitsuwa kaitei ichi-man go-sen ri no misshi," *Gekkan Yomiuri* (June 1951), 37; Eiichi Iwaya, "Jettoki 'Baika' monogatari," *Sekai no kūkō* 1, no. 2 (December 1951): 10–15.

67. Iwaya, "Roketto ki Shūsui ni tsuite," 82–88.

68. Fujihira, *Kimitsu heiki no zenbō*, 6–46.

69. Matsuoka, *Sentōki Shūsui*, 40.

70. Fujihira, *Kimitsu heiki no zenbō*, 30, 34.

71. Iwaya, "Roketto ki Shūsui ni tsuite," 88; Eiichi Iwaya, "Taiheiyō sensō mitsuwa kaitei ichi-man go-sen ri no misshi," *Gekkan Yomiuri* (June 1951), 37.

72. Iwaya, "Roketto ki Shūsui ni tsuite," 83.

73. "Kokusai kyōsō ni osore? Roketto yosan, sanbai heru," *Asahi Shimbun*, Akita suppl., January 22, 1958.

74. "'Heiki no senshi' ni hatsu no kōshō gijutsu no homare 100 yōmei 'fuhai no sōbi' ni kagayaku rikugun yūkōshō," *Asahi Shimbun*, November 15, 1941; "Jūgo no gunji hatsumei ni hatsu no kōshō senshō in no shukun kō 'rikugun gijutsu yūkō-shō' kagayaku 28-ken," *Yomiuri Shimbun*, November 15, 1941.

75. "Shin eiki no hanashi—Itokawa sensei kakonde," *Shūkan shokokumin* 3, no. 1 (January 1944): 24; Hideo Itokawa, "Shineiki no miwake kata—senzoomotsu mono motanu mono," *Shūkan shokokumin* 3, no. 50 (December 1944): 15.

76. "Sen koto hikkei/bōgyoryoku ni nayami saikin no bakugeki-ki kaisetsu (1)," *Asahi Shimbun*, June 21, 1943; "Sen koto hikkei/kogata de kōsoku-ryoku e saikin no bakugeki-ki kaisetsu (2)," *Asahi Shimbun*, June 22, 1943; "Sen koto hikkei/sokuryoku to tōsai karyoku saikin no bakugeki-ki kaisetsu (3)," *Asahi Shimbun*, July 23, 1943; "Sen koto hikkei/kōzokuryoku no jūyō-sei saikin no bakugeki-ki kaisetsu (4)," *Asahi Shimbun*, July 24, 1943; Hideo Itokawa, "B 24, B 29 no seinō nihon hondo kūshū-yō o kogō," *Yomiuri Shimbun*, June 18, 1944; Hideo Itokawa, *Kōkū rikugaku no kiso to ōyō* (Tokyo: Kyōritsu shuppan, 1943); Hideo Itokawa, *Kōkūki no shomondai* (Tokyo: Meiji Shobō, 1944); Hideo Itokawa, "V 2-gō seisōken kara sakaotoshi tekichide wa kakkū suru kai 'denshinbashira'—Hideo Itokawa-shidan," *Asahi Shimbun*, November 13, 1944.

77. "Report on Meetings between Chinese and Soviet Representatives on Rocket Production," September 23, 1957, Wilson Center Digital Archive, RGAE f. 8157, op. 1, 1957, d. 1991, l. 77–80; obtained and translated for CWIHP by Austin Jersild, https://digitalarchive.wilsoncenter.org/document/116821.

78. "Chūkyō ki, roketto shiyō ka," *Mainichi Shimbun*, October 4, 1958; "Gaiden memo—roketto seisaku de Seidoku ga Chūgoku o kyōso," *Mainichi Shimbun*, January 10, 1969.

79. 58th Diet, House of Representatives, Budget Committee, No. 2, February 6, 1968, https://kokkai.ndl.go.jp/simple/detail?minId=105805261X00219680206&spkNum=158&current=324#s158.

80. 40th NDHR E&C, No. 8, February 28, 1962, https://kokkai.ndl.go.jp/simple/detail?minId=104005077X00819620228&spkNum=99&current=28#s99.

81. Matogawa, "Japan Liquid Rockets in the Second World War," 71.

82. Moore, *Constructing East Asia Technology, Ideology, and Empire in Japan's Wartime Era, 1931–1945*, 226–239.

83. David Aldritch, *Site Fights: Divisive Facilities and Civil Society in Japan and the West* (Ithaca: Cornell University Press, 2010), 2, 119–136.

84. Martin Dusinberre, *Hard Times in the Hometown: A History of Community Survival in Modern Japan* (Honolulu: University of Hawaii Press, 2012), 149.

85. Particularly salient among the many analyses of these figures are Iris Chang, *Thread of the Silkworm* (New York: Basic Books, 1995); Alexander C. T. Geppert, "Space *Personae:* Cosmopolitan Networks of Peripheral Knowledge, 1927–1957," *Journal of Modern European History* 6, no. 2 (September 2008): 262–286; Michael J. Neufeld, "Three Heroes of Spaceflight: The Rise of the Tsiolkovskii-Goddard-Oberth Interpretation and Its Current Validity," *Quest: The History of Spaceflight Quarterly* 19, no. 4 (2012): 4–14; and Asif Siddiqi, *The Red Rockets' Glare: Spaceflight and the Soviet Imagination, 1857–1957* (Cambridge: Cambridge University Press, 2010), 43–73.

86. Hideo Itokawa, *Roketto* (Tokyo: NHK Books, 1965), 103–143.

87. Nihon kagakushi gakkai, *Nihon kagaku gijutsushi taikei dai 14 ken* (Tokyo: Dai ippō ki shuppan, 1964): frontispiece.

88. Itokawa, *Nihon sōseiron* (Tokyo: Yoshimasado Insatsu, 1990), 82. The first three countries to launch satellites on their own rockets were the Soviet Union (Sputnik, 1957); the United States (Explorer I, 1958), and France (Astérix, 1965). Between 1958 and 1964 the United Kingdom, Canada, and Italy all had satellites launched aboard US rockets, while between Astérix's launch in 1965 and Ohsumi's in 1970, Australia, West Germany, and an alliance of Belgium, Denmark, France, West Germany, Italy, the Netherlands, Spain, Sweden, Switzerland, and the United Kingdom all launched satellites on American rockets.

89. "Shin eiki no hanashi—Itokawa sensei kakonde," *Shūkan shokokumin* 3, no. 1 (January 1944): 24–25.

90. Hideo Itokawa, "Penshiru roketto Kara kappa 8 kei made," *Seisan kenkyū* 12, no. 12 (December 1960): 14–15.

91. Hideo Itokawa, *Nihon sōseiron*, 82.

92. Shōzō Matsukawa, Hideo Itokawa, and Hiroshi Miyazaki, "Purantoru tiichensu ryūtai rikigaku," *Kikai gakkaishi* 252, no. 41 (April 1938): 178.

93. 27th NDHR SST No. 3, November 9, 1957, https://kokkai.ndl.go.jp/simple/detail?minId=102703913X00319571109&spkNum=84&current=10#s84. The desire to remain independent is covered in detail in Sato, "Contested Gift," 176–177.

94. Hideo Itokawa, *Uchū o sampo suru: roketto zuihitsu* (Tokyo: Ryūnan Shobō, 1957), 12, 28.

95. Hideo Itokawa, *Roketto* (Tokyo: NHK Books, 1965), 145.

96. Itokawa, *Nihon sōseiron*, 82.

97. Oganesoff, "Japan Builds Rocket Industry."

98. Itokawa, *Uchū o sampo suru*, 16.

99. Oganesoff, "Japan Builds Rocket Industry."

100. "Eisei yōwa. moto tsūshin sōgō kenkyūsho-chō Fugono Nobuyoshi 1," *Space Japan Review* 82 (February–May 2013): 2–4.

101. Hideki Mizuno, "Rensai tokushū. Eisei yōwa," *Space Japan Review* 27 (February–March 2003): 1–2.

102. United Nations Office for Outer Space Affairs, "Law concerning the National Space Development Agency of Japan," June 23, 1969, https://www.unoosa.org/oosa/en/ourwork/spacelaw/nationalspacelaw/japan/nasda_1969E.html.

103. Hideo Itokawa, "Ni juppun de Taiheiyō kōdan," *Mainichi Shimbun*, January 3, 1955.

104. *Asahi Shimbun*, December 12, 1955; "Roketto—watashi no hanashi," *Asahi Shimbun*, August 19, 1955.

105. Arthur J. Dommen, "Japanese Rocketeers Have Come a Long Way," *Los Angeles Times*, October 3, 1965.

106. Hajime Tawara, "Itokawa sensei to Isuraeru," in *Ningen Hideo Itokawa hakase to wa*, ed. Iwao Kanezawa (Tokyo: Hirohide Kōgei, 2003), 131.

107. Reinhard Drifte, *Arms Production in Japan: The Military Applications of Civilian Technology* (New York: Taylor and Francis, 1986), 65–70.

108. Yasushi Sato, "Managing the Interface between Politics and Technology: Hideo Itokawa, Shima Hideo, and the Early Japanese Civilian Space Programs," *Historia Scientiarum* 21, no. 3 (2012): 193–210.

109. "Kaku hōmen kara giwaku ni memo," *Asahi Shimbun*, March 21, 1967; "Tōdai uchū ken—giwaku wa fumon—Itokawa kyōjū no jinin ga moto kimari," *Asahi Shimbun*, March 28, 1967.

Chapter 3

1. Akira Meguro, "Eisei Yōwa—subarashiki eisei jinsei," *Space Japan Review* 42 (August–September 2005): 2–3.

2. Matogawa, "Jimae roketto kaihatsu wa kaku ketsudan sareta," 56.

3. ISAS, *Uchū kūkan kansoku 30 nen shi* (Tokyo: Uchū kūkan kansoku 30nen shi henshū iinkai, 1987), 8; ISAS, "History of Japanese Space Research—Pencil at Michikawa," http://www.isas.jaxa.jp/e/japan_s_history/profito/pencil.shtml.

4. Officially, 62 mi (100 km) above the surface of the Earth. Technically, therefore, the first country in space was Germany, whose V-2s almost certainly broke that barrier at some point between 1943 and 1945.

5. ISAS, *Uchū kūkan kansoku 30 nen shi*, 24.

6. Yasunori Matogawa, "A Survey of Rocketry for Space Science in Japan," in *History of Rocketry and Astronautics: Proceedings of the History Symposia of the International Academy of Astronautics* 22 (San Diego: American Astronautical Society, 1993), 208–214; Nakabe et al., "Uchū kaihatsu 60 nen shi," 19–21; ISAS, *Uchū kūkan kansoku 30 nen shi*, 28.

7. Nishiyama, "Swords into Plowshares," ii, 5.

8. Bartholomew, *The Formation of Science in Japan*, 238–239.

9. Robertson, "Mobilizing for War," 30–35.

10. Nishiyama, "Swords into Plowshares," 27.

11. Nihon kaigun kōkū shi hensan iinkai, *Nihon kaigun kōkū shi dai 2—gunbi hen*, ed. Chikao Yamamoto (Tokyo: Jiji Press, 1969), 337, 643–776.

12. Nobuhei Nakahara, "Kaigun nenryō shō ni kanrensuru omoide," in *Nippon kaigun nenryōshi (ge)* (Tokyo: Hara Shobō, 1973), 814; Hideo Okuda, *Nakahara Nobuhei den* (Tokyo: Toua Nenryō Kōgyō, 1982), 105–176, 374–448.

13. Robertson, "Mobilizing for War," 2.

14. "Haisen chokugo no kagaku gijutsukai no jittai," in *Nihon no kagaku gijutsu*, ed. Shigeru Nakayama (Tokyo: Gakuyo Shobō, 1995), 183–185.

15. Sawai Minoru, "Gijutsu-sha no gunmin tenkan to tetsudō gijutsu kenkyūjo," *Osaka daigaku keizaigaku* 59, no. 1 (2009): 11.

16. Robertson, "Mobilizing for War," 3.

17. Robertson, "Mobilizing for War," 187.

18. Pronounced "Tetsu Hiroshige."

19. Hirosige, *Kagaku to Shakaishi*, 24–25, 36–38.

20. Nishiyama, "Swords into Plowshares," 37–45, 135–144, 154.

21. Nishiyama, "Swords into Plowshares," 27.

22. Matogawa, "A Survey of Rocketry for Space Science in Japan," 203.

23. Hirosige, *Kagaku to Shakaishi (ka)*, 96.

24. Nishiyama, "Swords into Plowshares," 89–92, 180–194.

25. Sawai Minoru, "Gijutsu-sha no gunmin tenkan to tetsudō gijutsu kenkyūjo." Though there are numerous other pieces on this topic, Sawai's has the bonus of meticulously listing the origins, experiences, and overall careers of a great many of them.

26. Nishiyama, "Swords into Plowshares," 27, 32–33. Nishiyama describes a series of ill-tempered confrontations between ex-military engineers and their new colleagues in Japan's national railway system, including a widely read letter complaining about the "occupation" of the industry by military engineers, and one ex-army engineer being called a "traitor" in his interview for the company; Tokyo daigaku seisan gijutsu kenkyūjo, *Tokyo daigaku dai ni kagakubushi* (Tokyo: Tokyo daigaku seisan gijutsu kenkyūjo, 1968), 77–80. For a contemporary and insightful take on this process, see S. Sidney Ulmer, "Local Authority in Japan since the Occupation," *The Journal of Politics* 19, no. 1 (February 1957): 46–65. Ulmer points out that in many cases reform by the occupation was perceived as an opportunity to *re-create* structures and organizations with the *participants themselves* purveyors of norms and rules. Alternatively, in many cases, it led to a simple recasting of the nominal relationship between center and periphery—for example, school governors were now decided locally, rather than assigned from Tokyo. On occasion localities voluntarily opted for recentralization, as in the case of the organization of local police forces. The behavior of engineers and academics in creating genuinely independent organizations of the kind mentioned above is, therefore, all the more noteworthy.

27. Koizumi, "In Search of 'Wakon,'" 30.

28. Nobuhei Nakahara, "Kaigun nenryō shō ni kanrensuru omoide," in *Nippon kaigun nenryōshi (ge)* (Tokyo: Hara Shobō, 1973), 814; Hideo Okuda, *Nakahara Nobuhei den* (Tokyo: Toua Nenryō Kōgyō, 1982), 105–176, 374–448.

29. Matogawa Yasunori, *Gyakuten no Tsubasa: Penshiru roketto monogatari* (Tokyo: Shin Nippon Shuppansha, 2005), 66.

30. Matogawa Yasunori, *Nippon uchū kaihatsu hishi—ganso tori ningen kara minkan roketto e* (Tokyo: NHK Shuppan, 2017), 57.

31. Matogawa, *Uchū ni ichiban chikai machi*, 53–61.

32. A Japanese spirit similar in alcohol level to brandy.

33. Onoda Junjiro, "Uchinoura to watashi—Uchinoura 50-shūnen ni yosete," in ISAS, *Uchinoura uchū kūkan kansokujo no 50 nen 1962–2012* (Uchinoura uchū kūkan kansokujo no 50-nen' kinen-shi henshū iinkai, November 2011), 2; Oyama Koichiro, "Omoide no Uchinoura," in *ISAS, Uchinoura uchū kūkan kansokujo no 50 nen 1962–2012* (Uchinoura uchū kūkan kansokujo no 50-nen' kinen-shi henshū iinkai, November 2011), 10.

34. Asif Siddiqi, "Soviet Secrecy: Toward a Social Map of Knowledge," *American Historical Review* 126, no. 3 (September 2021): 1054–1057, 1061–1062.

35. My thanks to Matthew Hersch for this pithy and insightful explanation.

36. Yoshioka, "Uchū kaihatsu kyūsei no kakuritsu," 172.

37. Sato, "Managing the Interface," 181.

38. Sato, "Contested Gift," 179–180.

39. Watanabe, "Japan–US Space Relations during the 1960s," 1–14.

40. "'Bebii-s' kei mo jōjō—kokusan roketto tesuto," *Asahi Shimbun,* Akita suppl., August 24, 1955; "Kōsō no kiryū—kinō Michikawa de Kappa uchiage," *Asahi Shimbun,* Akita suppl., December 24, 1953; "Roketto jikken ni yosete," *Asahi Shimbun,* Akita suppl., September 23, 1956.

41. "Ichigatsu ni 'Kappa' jikken," *Asahi Shimbun,* Akita suppl., December 28, 1955.

42. Sato, "Contested Gift," 180.

43. Saito Shigefumi, *Nihon uchū kaihatsu monogatari — kokusan eisei ni kaketa senku-shatachi no yume* (Tokyo: Mita Shuppankai, 1991), 28–29.

44. See Sato, "Contested Gift."

45. 61st NDHR Bud., Subcommittee No. 3, February 26, 1969, https://kokkai.ndl.go.jp/simple/detail?minId=106105266X00319690226&spkNum=244&current=13#s244.

46. Siddiqi, "Spaceflight in the National Imagination," 21.

47. Hirotaka Watanabe, "The Evolution of Japanese Space Policy," 274–281.

48. Sato, "Managing the Interface," 199.

49. Sato, "Managing the Interface," 202.

50. Matogawa, "Halley's Comet Exploration in Japan," 225–227.

51. Matogawa, "Halley's Comet Exploration in Japan," 238.

52. Kallendar, "Explaining the Logics of Japanese Space Policy Evolution 1969–2016," 124.

53. "Uchū seisaku no shireitō kinō o meguru giron," *Chōsa to jōhō 748* (April 5, 2012), National Diet Library, Tokyo, 8.

54. Yoshioka, "Uchū kaihatsu kyūsei no kakuritsu," 177–178. The four-stage Lambda 4S that put Japan's first satellite, Ohsumi, in orbit in 1970, was 2 ft (0.64 m) in diameter and 54 ft (16.5 m) long. Compare this to the Sputnik 727, which put Sputnik in orbit and which stood 100 ft (30 m) and had a hefty diameter of nearly 10 ft (3 m).

55. 51st Diet NDHC SST No. 7, June 3, 1966, https://kokkai.ndl.go.jp/simple/detail?minId=105113913X00719660603&spkNum=38&current=5#s38.

56. Yasunori Matogawa, email to author, January 24, 2024.

57. 58th NDHC Cab. No. 14, April 25, 1968, https://kokkai.ndl.go.jp/simple/detail?minId=105814889X01419680425&spkNum=182&current=2#s182.

58. Yasunori Matogawa, "Halley's Comet Exploration in Japan—Japan's First Interplanetary Flight," *49th International Astronautical Congress*, 41 (International Astronautical Federation, 2007), 223.

59. Sato, "Contested Gift," 195.

60. "Uchū kaihatsu jishu rosen hashiru," *Nihon Keizai Shimbun*, October 27, 1969.

61. Saito, *Nihon uchū kaihatsu monogatari*, 69–85.

62. Masafumi Miyasawa, "Waga kuni ni okeru jitsuyou roketto no Kaihatsu to gijutsu dōnyū," *Nihon kōkū uchū gakkaishi*, 39, no. 445 (February 1992), 56.

63. Eiji Sogame, "A History of the Tanegashima Space Center," *History of Rocketry and Aeronautics: Proceedings of the Thirty-Ninth History Symposium of the International Academy of Astronautics: Fukuoka, Japan (2005)* (San Diego: American Astronautical Society, 2011), 262.

64. Takenaka, "N-1 roketto kaihatsu no ayumi," 132–133.

65. Sato, "Managing the Interface," 209.

66. Sato, "Contested Gift," 192.

67. Robertson, "Mobilizing for War," 4.

68. Hashimoto, "Aru eisei tsūshin gijutsusha no omoide—dai-3 kai," 1–2; Kazuhiko Hashimoto, "Aru eisei tsūshin gijutsusha no omoide—dai 4-kai," *Space Japan Review* 68 (June–July 2010): 1–2.

69. Takashi Hirose, Yutaka Imaizumi, and Hideyoshi Yoshida, "Higashinihon daishinsai de riyōsareta NTT saigai taisaku eisei tsūshin shisutemu," *Space Japan Review* 76 (October–November 2011).

70. Masahiro Nakao, Tomii Naoya, Takayama Shinichiro, Horishi Tsuyoshi, Horiuchi Takashi, Hakuchi Hiroki, Ikemi Takashi, Ikema Tadashi, Ikekawa Takashi, and Sunagawa Kei, "Higashinihon daishinsai ni okeru 'kizuna' oyobi 'kiku 8-gō' ni yoru jakusa taiō ni tsuite," *Space Japan Review* 77 (December 2011–January 2012): 1–18.

71. Akira Meguro, "Eisei Yōwa—subarashiki eisei jinsei," *Space Japan Review* 42 (August–September 2005): 1–2.

72. "Eisei yōwa: moto tsūshin sōgōkenkyūsho-chō Fugono Nobuyoshi 2," *Space Japan Review* 83 (June–September 2013). It is interesting to note—though beyond the purview of this book and, in any case, expertly covered by Paul Kallendar-Umezu in his 2017 dissertation—that institutional conflict remains a characteristic of Japan's civilian space programs even today. These clashes were further situated within an even broader system of contestation within the Japanese government that intensified in the late 1990s, with attempts initiated by the DPJ under Ryutaro Hashimoto to assert control over the bureaucracy and continued by LDP politicians, including Koizumi Junichiro, under whose aegis JAXA was founded. The foundation of JAXA effectively gave control of 60 percent of the space development budget and institutional control to MEXT. Yet efforts to centralize Japan's space research and rocketry institutions persisted despite the foundation of JAXA, and as late as the DPJ's government of 2009–2012. Even in the period after 2003, some five other government agencies and institutions received budgets for space development, including MITI, the Ministry of Posts, the Ministry of Education, and the Telecommunications Advancement Agency of Japan. Indeed, MITI even had its own space branch, the Institute for Unmanned Space Experiment Free Flyer (USEF)—founded in 1986 and *not* absorbed into JAXA. Even under the nominal aegis of JAXA, there was extensive competition between bureaucratic sectors and groups—one rocket project was scuppered, for example, when MEXT refused to fund the initiative as being too much under MITI's control. In the aftermath of a critical H-2 rocket failure in November 2003, the MEXT minister Takeo Kawamura was unable to identify a single department that was willing to accept responsibility for the failure and not shift it elsewhere. Projects such as the development of a position, navigation, and telemetry constellation floundered for lack of any institution capable of taking the lead on the project or mustering the budget required. See Paul Kallendar-Umezu, "Explaining the Logics of Japanese Space Policy Evolution 1969–2016, Combining Macro- and Microtheories, Notably The Strategic Action Field Framework," PhD diss., Keio University, 2017.

73. Meguro, "Eisei Yōwa—subarashiki eisei jinsei," 1.

74. Andrew Pollack, "Japan Set to Launch Rocket to Match West: Japan Set to Launch Its Own Rocket," *New York Times*, January 25, 1994.

75. 145th NDHR S&T No. 2, February 9, 1999, https://kokkai.ndl.go.jp/simple/detail?minId=114503911X00219990209&spkNum=147&current=3#s193.

76. Susumu Kitazume, "Mitsubishi jūkō meiyū ni okeru H-IIA roketto enjin no kaihatsu," *Space Japan Review* 19 (October–November 2001): 11.

Chapter 4

1. Casey Drier, "An Improved Cost Analysis of the Apollo Program," *Space Policy* 60 (2022): 1, 3. This equates to $257 billion in 2022. Information on the Soviet budget was kindly provided by Asif Siddiqi in an email to the author, December 4, 2022.

2. 31st NDHR Bud., Subcommittee No. 3, February 27, 1959, https://kokkai.ndl.go.jp/simple/detail?minId=103105266X00319590227&spkNum=55&current=11#s55.

3. Yasunori Matogawa, "Sengo Fukkoki no Roketto Gijutsu," in *Nihon Roketto monogatari: noroshi kara uchū kanko made,* ed. Hirokyuki Osawa (Tokyo: Mita Shuppankai, 1996), 76.

4. Prime Minister's Office of Japan, "The Constitution of Japan," https://japan.kantei.go.jp/constitution_and_government_of_japan/constitution_e.html.

5. Hakuo Nakabe, Toshiaki Takesaki, Yuka Ōno, "Uchū Kaihatsu 60 nen shi. 1955 nen—2015 nen," in *Proceedings of Space Transportation Symposium FY2016* (Tokyo: Institute of Space and Astronautical Science, Japan Aerospace Exploration Agency, 2017), 19.

6. 40th NDHR E&C No. 8, February 28, 1962, https://kokkai.ndl.go.jp/simple/detail?minId=104005077X00819620228&spkNum=99&current=28#s99.

7. ISAS, *Uchū kūkan kansoku 30 nen shi,* 51, 72. The lead designers on this project—Kenrō Murata and Kisaburō Nakazawa—were also Tokyo University professors who had moved into Hitachi. Furthermore, the work at Kobe was conducted under Sukejirō Andō, who later moved to Tokyo University—as did Iwao Nakamura of the Prince Motor Company, who worked on motors in the K-9 rocket. This is an example of yet another dimension of the connectivity between these institutional formations.

8. "Mitsubishi denki de jūchū—roketto tsuibi rēdā," *Asahi Shimbun,* November 29, 1960.

9. "Jinkō kōu, kairyū chōsa ni," *Asahi Shimbun,* Akita suppl., February 10, 1956.

10. Yasunori Matogawa, *Uchū u ni ichiban chikai machi—Uchinoura no roketto hasshajo* (Tokyo: Shunendō Shuppan, 1994), 40–41.

11. "Gimon-darake 'hinomaru eisei' gakujutsu kaigi ga shiteki—Tōdai uchū kōkū-ken mondai," *Asahi Shimbun,* March 21, 1967. KDD was privatized in 1985, and is more commonly known now as NTT.

12. Siddiqi, *Red Rockets' Glare,* 249–260.

13. Siddiqi, *Sputnik and the Soviet Space Challenge,* 210.

14. Dennis R. Jenkins, *Space Shuttle: Developing an Icon: 1972–2013,* vol. 1 (Forest Lake, MN: Specialty Press, 2016), 204.

15. Roger D. Launius, *Historical Analogs for the Stimulation of Space Commerce* (Washington, DC: NASA, 2014), 23.

16. 51st NDHC S&T No. 5, February 17, 1966.

17. Hirotaka Watanabe, "Japan–US Space Relations during the 1960s: Dependence or Autonomy?" in *54th International Astronautical Congress: 29 September—2 October, Bremen, Germany* (Paris: International Astronautical Federation, 2003), 2.

18. ISAS, "Ohsumi," https://www.isas.jaxa.jp/en/missions/spacecraft/past/ohsumi.html.

19. 27th NDHR SST, No. 3, November 9, 1957, https://kokkai.ndl.go.jp/simple/detail?minId=102703913X00319571109&spkNum=84&current=10#s84.

20. 32nd NDHR SST, No. 4, August 11, 1952, https://kokkai.ndl.go.jp/simple/detail?minId=103203913X00419590811&spkNum=91&current=4#s91.

21. 27th NDHC Com., No. 2, November 7, 1957, https://kokkai.ndl.go.jp/simple/detail?minId=102714816X00219571107&spkNum=12&current=68#s12.

22. *Uchū kūkan kansoku 30-nenshi,* 3, 8; Hideo Itokawa, "The Japanese Sounding Rockets," *Journal of Jet Propulsion* 27, no. 3: 286.

23. The World Bank, "GDP per capita (current US$) —Japan," https://data.worldbank.org/indicator/NY.GDP.PCAP.CD?locations=JP.

24. M. Beckley, Y. Horiuchi, and J. M. Miller, "America's Role in the Making of Japan's Economic Miracle," *Journal of East Asian Studies* 18, no. 1 (2018): 1–21; Chiaki Moriguchi and Emmanuel Saez, "The Evolution of Income Concentration in Japan, 1885–2002: Evidence from Income Tax Statistics," Econometrics Laboratory, University of California, Berkeley (August 25, 2005), table 1, https://eml.berkeley.edu/~saez/moriguchi-saez05japan.pdf.

25. 34th NDHR Com., No. 3, February 10, 1960, https://kokkai.ndl.go.jp/simple/detail?minId=103404816X00319600210&spkNum=32&current=3#s32.

26. Yuzo Takahashi, "A Network of Tinkerers: The Advent of the Radio and Television Receiver Industry in Japan," *Technology and Culture* 41, no. 3 (July 2000): 460–484. Indeed, as Takahashi notes, one reason for this staggering growth was that, much in the same way they'd constructed radios at home, hobbyists embraced the habit of constructing their own TVs from kits.

27. "Tsūshin eisei to kokusai chūkei—nyūsu no me," *Asahi Shimbun,* February 13, 1962.

28. "Tōkyō orinpikku ni eisei de terebi hōsō o Okazaki taishi enzetsu—taikiken-gai heiwa riyō-i; Tōkyō orinpikku to terebi eisei," *Asahi Shimbun,* March 23, 1962.

29. "Hana hiraku uchū sangyō eisei tsūshin jidai hikaete—sangyō," *Asahi Shimbun,* September 1, 1968; "Mekishiko Orinpikku Nihon ni mo karā de maitsūshin eisei kaisha keikaku," *Asahi Shimbun,* September 4, 1968. In the five years between 1960 and 1965, for example, *Asahi Shimbun* alone featured more than 160 articles on the subject of communications satellites.

30. "Ranibādo 2-gō kinen kitte o hatsubai—Yūseishō," *Asahi Shimbun,* January 20, 1967.

31. Kazuhiko Hashimoto, "Aru eisei tsūshin gijutsusha no omoide—dai-3 kai," *Space Japan Review* 67 (April–May 2010): 1–3.

32. 26th NDHR Com., No. 34, October 11, 1957, https://kokkai.ndl.go.jp/simple/detail?minId=102604816X03419571011&spkNum=34&current=52#s34.

33. 55th NDHR Com., No. 3, March 24, 1967, https://kokkai.ndl.go.jp/simple/detail?minId=105504816X00319670324&spkNum=22&current=54#s22.

34. Ministry of Finance (Japan), *1972 General Account Budget Reference Book Ministry of Posts and Telecommunications* (Tokyo: Ministry of Finance, 1973), 764.

35. "Tsūshin eisei keiyu hayaku naru 'moshimoshi' Yamaguchi ni chijō-kyoku Nihon—Ōshū Chūkintō," *Asahi Shimbun*, November 3, 1967.

36. Hashimoto, "Aru eisei tsūshin gijutsusha no omoide, dai 3-kai," 1.

37. "Eisei tsūshin nōritsu yoku kokusai denden ga shin hōshiki—tsūshin," *Asahi Shimbun*, July 5, 1970.

38. "Eisei chūkei no ryōkin nesage—terebi," *Asahi Shimbun*, September 22, 1973.

39. Japan Ministry of Posts and Telecommunications, "Radio Research Laboratory," pamphlet (Tokyo: Ministry of Posts and Telecommunications, 1985), 7.

40. Kazuhiko Hashimoto, "Aru eisei tsūshin gijutsusha no omoide, dai 2-kai," *Space Japan Review* 66 (February–March 2010): 1.

41. Pekkanen and Kallender-Umezu, *In Defense of Japan*, 79–92.

42. Shigeru Nakayama, "The Central Laboratory Boom and the Rise of Corporate R&D," in Shigeru Nakayama and Kunio Gotō, eds., *A Social History of Science and Technology in Contemporary Japan*, vol. 3: *High Economic Growth Period* (Melbourne: Trans Pacific Press, 2006), 68.

43. Shigeru Nakayama, "The Central Laboratory Boom and the Rise of Corporate R&D," 68–69.

44. "Nihon nado sangoku ni denwa sābisu maitsūshin eisei kaisha—uchū kaihatsu," *Asahi Shimbun*, July 3, 1966; 48th NDHR Com., No. 8, March 12, 1965, https://kokkai.ndl.go.jp/simple/detail?minId=104804816X00819650312&spkNum=64&current=13#s64.

45. "Mitsubishi denki mo gijutsu teikei e tsūshin eisei setsubi o buntan seisan—uchū kaihatsu," *Asahi Shimbun*, May 17, 1966.

46. Kazuhiko Hashimoto, "Aru eisei tsūshin gijutsusha no omoide, dai 4-kai," *Space Japan Review* 68 (June–July 2010): 2. The satellites had been slated to launch on the Space Shuttle until the Challenger disaster forced the Japanese to find alternate launch systems.

47. "Rēzā de kansoku Hitachi seisakusho jinkō eisei tsuiseki ni seikō uchū kaihatsu," *Asahi Shimbun*, June 13, 1969.

48. "Nihondenki, Beisha to gōben tsūshin eisei chijō-kyoku," *Asahi Shimbun*, August 3, 1966.

49. "Roketto yūdō seigyo no gijutsu dōnyū de gōbenkaisha o setsuritsu Nihondenki to Haniueru," *Asahi Shimbun*, June 19, 1970. NEC subsequently expanded this relationship to include marketing its products, such as supercomputers, via Honeywell in the United States. See "Honeywell Link to NEC," *The New York Times*, July 2, 1986.

50. Kallendar-Umezu, "Explaining the Logics of Japanese Space Policy Evolution," 121, 122.

51. "Keidanren ni uchū heiwa riyōi—keizai dantai," *Asahi Shimbun*, May 17, 1961; "Homi uchu kichi keika chosatan hokokusho," *Uchū* no. 23 (Keidanren: Tokyo, Japan, 1985): iii, 76;

"Sangyō-kai, jimae no eisei kōsō sōki uchiage hitsuyō Keidanren hōkoku—tsūshin eisei 1983 nen," *Asahi Shimbun*, January 5, 1983.

52. 48th NDHC Com., No. 12, March 30, 1965, https://kokkai.ndl.go.jp/simple/detail?minId=104814816X01219650330&spkNum=178&current=19#s178.

53. "Ōte denki mēkā, eiseibijinesu o kyū kakudai kaden fushin o oginau nerai mo," *Asahi Shimbun*, November 11, 1992.

54. "Zanbia no tsūshin eisei shisetsu Nihondenki ga juchū—yushutsu," *Asahi Shimbun*, November 19, 1972.

55. "Arabu no hōsō tsūshinmō seibi Nihon no kyōryoku gutai-ka eisei chijō-kyoku shutokan denwa terekkusu Nichiden teian, Ōbei-sha o shirizokeru," *Asahi Shimbun*, March 15, 1974.

56. "Yūgo ni chijō eisei kichi kensetsu Nichiden to hishi-shō ga kyōdō de,' *Asahi Shimbun*, August 19, 1972.

57. "Jinkō eisei no shisei kenshutsu sōchi yūshūna kokusan-hin ga kansei—uchū kaihatsu," *Asahi Shimbun*, August 15, 1969.

58. "Amerika IBM kei no eisei tsūshin kaisha kara juchū Fujitsū ga henfukuchō sōchi—nichibeibōeki," *Asahi Shimbun*, July 12, 1979.

59. 55th NDHR Committee on Accounts, No. 4, March 28, 1967, https://kokkai.ndl.go.jp/simple/detail?minId=105504103X00419670328&spkNum=132&current=17#s132; "Tōdai no jinkō eisei uchiage chūshi shoho-teki misu tsudzuku," *Asahi Shimbun*, January 22, 1969.

60. Sato, "Contested Gift," 192.

61. Shinya Matsuura, *Kokusan roketto hwas naze ochirunoka* (Tokyo: Nikkei BP Shuppansha, 2004), 38–39.

62. 55th NDHR Bud., No. 10, April 3, 1967, https://kokkai.ndl.go.jp/simple/detail?minId=105505261X01019670403&spkNum=181&current=63#s181; Watanabe, "The Evolution of Japanese Space Policy," 282.

63. 58th NDHR SST, No. 11, April 17, 1968, https://kokkai.ndl.go.jp/simple/detail?minId=105803913X01119680417&spkNum=26&current=1#s26.

64. 55th NDHR Committee on Accounts, March 28, 1967.

65. Sato, "Contested Gift," 178.

66. 26th NDHR For., No. 28, October 15, 1957, https://kokkai.ndl.go.jp/simple/detail?minId=102603968X02819571015&spkNum=76&current=53#s76.

67. "Kon go wa tsūshin eisei—uchū danwa-shitsu," *Asahi Shimbun*, April 18, 1960; 26th NDHR Cab., No. 46, October 8, 1957, https://kokkai.ndl.go.jp/simple/detail?minId=102604889X04619571008&spkNum=240&current=49#s240; 55th NDHR For., No. 18, July 13, 1967, https://kokkai.ndl.go.jp/simple/detail?minId=105503968X01819670713&spkNum=67#s67.

68. Ministry of Finance (Japan), *Budget Reference Documents according to Article 28 of the*

Finance Law, Records on National Treasury Debt Obligation Acts since 1971, "(1) General Account" (Tokyo: Ministry of Finance, 1972), 82.

69. ISAS, *Uchū kūkan kansoku 30 nen shi,* 60.

70. Takenaka, "N-1 roketto kaihatsu no ayumi," 132.

71. 51st NDHC S&T No. 5, February 17, 1966.

72. Stuart Griffin, "Japan Asks for Help," *Science News* 97, no. 21 (May 23, 1970): 516.

73. Jenkins, *Space Shuttle: Developing an Icon: 1972–2013,* 204.

74. Hitoshi Yoshioka, "Uchū kaihatsu kyūsei no kakuritsu," 181.

75. 32nd NDHR SST, No. 4, August 11, 1959, https://kokkai.ndl.go.jp/simple/detail?minId=103203913X00419590811&spkNum=91&current=4#s91.

76. 34th NDHR Cab., No. 3, February 12, 1960, https://kokkai.ndl.go.jp/simple/detail?minId=103404889X00319600212&spkNum=12&current=4#s12.

77. Hirotaka Watanabe, "The Evolution of Japanese Space Policy: Autonomy and International Cooperation," in *History of Rocketry and Astronautics: Proceedings of the Thirty-Ninth History Symposium of the International Academy of Astronautics, Fukuoka, Japan (2005)* (San Diego: American Astronautical Society, 2011), 278, 289; Chihiro Hosoya, ed., *Nichibei kankei shiryō hen 1945–1997* (Tokyo: Tokyo Daigaku Shuppatsukai, 1992), 751–755. Paul Kallendar-Umezu characterizes this investment as the United States "entrapping" Japanese resources and attention, though this affords the Japanese less agency in the process than they perhaps had. Kallendar-Umezu, "Explaining the Logics of Japanese Space Policy Evolution," 107, 122, 133.

78. Sato, "Contested Gift," 176–177.

79. Watanabe, "The Evolution of Japanese Space Policy," 278; John M. Logsdon, "U.S.–Japanese Space Relations at a Crossroads," *Science,* n.s. 255, no. 5042 (January 17, 1992): 294–300.

80. "Taibei yushutsu ni noridasu—Mitsubishi jūkō, Purinsu ryōsha kei no roketto," *Asahi Shimbun,* July 22, 1965. This, at least, is how the approach was covered in Japan. The United States did not, however, have a sounding rocket shortage at the time, which leaves the motivations of the US corporations a little murky. My thanks to Matthew Hersch for his insights on this matter.

81. "Cha no ma ni chikyū no kao Sōren no jinkō eisei kara TV chūkei NHK—uchūchūkei," *Asahi Shimbun,* June 9, 1967.

82. Matsuura, *Kokusan roketto hwas naze ochirunoka,* 271.

83. Watanabe, "Japan–US Space Relations," 1–14.

84. Sato, "Contested Gift," 186; "Beichū kankei kaizen mo tōron Miki gaishō to Rasuku chōkan uchūchūkei o tsūjite," *Asahi Shimbun,* March 5, 1967.

85. Watanabe, "The Evolution of Japanese Space Policy," 275.

86. Sato, "Contested Gift," 185, 187.

87. "Lockheed roketto urikomi—Mitsubishi jūkō nado mōshire," *Asahi Shimbun,* November 15, 1968.

88. Watanabe, "The Evolution of Japanese Space Policy," 276–277. It should also be noted that while politicians remained publicly opposed to importing US technology, many diplomats and ministry bureaucrats forged ahead with discussions to import the technology regardless. See Sato, "Contested Gift."

89. Yoshioka, "Uchū kaihatsu kyūsei no kakuritsu," 174. NASDA was in fact one of four "special corporations" that came into being that year, the others being the Japan Institute of International Affairs, the Pollution Relief Fund, and the Urban Development Corporation.

90. Shigefumi Saito, "Japan's Space Policy: Background and Outlooks," *Space News* 5, no. 3 (August 1989): 193.

91. 61st Diet NDHR Com., No. 3, February 27, 1969, https://kokkai.ndl.go.jp/simple/detail?minId=106104816X00319690227&spkNum=237&current=15#s237; 61st NDHC Com., No. 7, March 25, 1969, https://kokkai.ndl.go.jp/simple/detail?minId=106114816X00719690325&spkNum=251&current=29#s251.

92. "Mitsubishi jū ga Amerika-sha to gōi roketto nenryō gijutsu dōnyū—uchū kaihatsu," *Asahi Shimbun*, February 15, 1970; *Asahi Shimbun*, March 8, 1970.

93. *Asahi Shimbun*, September 14, 1978.

94. Leonard Lynn, "Japanese Research and Technology Policy," *Science* 233, no. 4761 (July 18, 1986): 298.

95. "Bōei kiki no kaihatsu wa, minkan to kyōdō sagyō de Nakasone chōkan—kokubō," *Yomiuri Shimbun*, February 13, 1970.

96. "Uchū kaihatsu ni kansuru wareware no kenkai / keizai dantai rengō-kai uchū kaihatsu suishin kaigi," *Keidanren Geppō* 25, no. 2 (February 1977): 43–47.

97. Yukihiko Takenaka, "N-1 roketto kaihatsu no ayumi," *Nihonkōkū uchū gakkaishi* 23, 362 (1984): 132. These overruns were somewhat par for the course for Shima, who had eventually resigned from the Shinkansen project because it had gone wildly over budget under his administration.

98. Hashimoto, "Aru eisei tsūshin gijutsusha no omoide—dai-3 kai," 1–3.

99. "Ensuring a Viable and Productive Future for Japan's Civilian Space Programs," *Report of the Space Activities Council of the Keidanren*, Federation of Economic Organizations [Keidanren], October 25, 1988.

100. Logsdon, "U.S.-Japanese Space Relations," 279, 298.

101. "Uchū kaihatsu chōki keikaku chōsa-dan shoken oyobi Teigen," *Keidanren Geppō* 19, no. 9 (September 1971): 60–61.

102. "Nihon kigyō mo uchū jikken o Keidanren ga setsumeikai—uchūrenrakusen," *Asahi Shimbun*, September 14, 1978.

103. Keidanren, *Space in Japan 1986–1987* (Tokyo: Keidanren, 1987), 5; Yoshioka, "Uchū kaihatsu kyūsei no kakuritsu," 172–173.

104. Yoshioka, "Uchū kaihatsu kyūsei no kakuritsu," 173.

Chapter 5

1. As of 2005, it is part of the unified village division of Kimotsuki.

2. "Yonki mei uchiage chūshi—Uchinoura Tōkyō daigaku roketto Miyazaki no gyosen ga bōgai," *Asahi Shimbun*, September 13, 1968.

3. Ernst Steinhoff, "The Development of the German A-4 Guidance and Control System 1939–1945: A Memoir," in *Essays on the History of Rocketry and Astronautics: Proceedings of the Third through the Sixth History Symposia of the International Academy of Aeronautics* (Washington, DC: International Academy of Aeronautics, 1977), 203–205. Steinhoff was one of Wernher von Braun's core team members and joined his boss in the United States as part of Operation Paperclip in 1945; he continued to work on rocket guidance systems in the US rocketry program until the 1970s.

4. "Kagodai ni kyōryoku tanomu—Hideo Itokawa roketto uchiage de," *Asahi Shimbun*, April 27, 1963.

5. "Uchū kūkō kenkyū kaihatu kikō (JAXA) no Niijima shishō tōmon o ukete," *Bōei-shō gijutsu kenkyū honbu kōhō* 503 (March 8, 1995): 2.

6. "Tanegashima no roketto kichi—uchū kaihatsu," *Asahi Shimbun*, May 7, 1966.

7. 51st Diet NDHC S&T, No. 7, June 3, 1966.

8. Minoru Oda, *Uchū kūkan kanshu 30 nen shi nenhyō* (Tokyo: Uchū kagaku kenkyūjō, 1987), 6.

9. "Kotoshi no hanashi," *Asahi Shimbun*, Akita suppl., December 17, 1955.

10. "Sensei to seito no gassoku—Yokote ni-chū ni yon tsubō no seiza," *Asahi Shimbun*, Akita suppl., November 2, 1956.

11. "Jinkō eisei no X-masu kēki—Akita, onedan wa 20-man en," *Asahi Shimbun*, Akita suppl., December 10, 1957. The image of this object is alas too deteriorated in quality to reproduce here, but can be seen in the newspaper cited here.

12. "Roketto kansoku dōryōkukai tansei," *Asahi Shimbun*, Akita suppl., July 25, 1957; "Shigoto shiyasui yo—roketto kansoku dōryōku kai hassoku," *Asahi Shimbun*, Akita suppl., July 30, 1957.

13. Timothy S. George, *Minamata: Pollution and the Struggle for Democracy in Postwar Japan* (Cambridge, MA: Harvard University Asia Center, 2002).

14. Daniel Aldrich, *Site Fights: Divisive Facilities and Civil Society in Japan and the West* (Ithaca: Cornell University Press, 2010), 1–7.

15. Dusinberre, *Hard Times in the Hometown*, 157–164.

16. Matogawa, "A Survey of Rocketry for Space Science in Japan," 211; Oda, *Uchū kūkan kanshu 30 nen shi nenhyō*, 7.

17. "Kappa-3 ki tetsu kan hikō jikken 'Chōonsoku no hashiri ya'—Michikawa kaigan kenbutsujin mo odoroku," *Asahi Shimbun*, Akita suppl., June 24, 1957; "Iwaki chō Michikawa kaigun de—ken mo kyōryoku jissoku wa sanjuni nen kara," *Asahi Shimbun*, Akita suppl., August 2, 1955; "Isshun ni kumoma ni—'Kappa' roketto jikken," *Asahi Shimbun*, Akita

suppl., December 9, 1956; "24 nichi no Kappa jikken ni sonaete," *Asahi Shimbun*, Akita suppl., September 21, 1956.

18. "Tōbu kaishū testo—17 nichi Kappa go-kei ni sonaete," *Asahi Shimbun*, Akita suppl., September 13, 1957.

19. "Roketto jikken shippai de jimoto no fuan takamaru—michikawa kaigan," *Asahi Shimbun*, Akita suppl., May 26, 1962. Problems with civilians living in the vicinity continued to bedevil the Japanese civilian space programs—in 1991, for example, an explosion during the testing of engines for the H2 rocket in Kakuda caused shattered windows in homes in the vicinity. See "Roketto jikkenjō de bakuhatsu—H2 keikaku jikō tsuzuki," *Asahi Shimbun*, May 17, 1991.

20. "Roketto asobi de jūshō," *Asahi Shimbun*, Akita suppl., March 4, 1958; "Kōkei roketto bakuhatsu jikken—chūgakusei hitori shinu," *Asahi Shimbun*, Akita suppl., September 25, 1957; "Chūgakusei roketto asobi de jūshō," *Asahi Shimbun*, Akita suppl., January 30, 1959; "Sanin ga jūkeishō, Hanawa," *Asahi Shimbun*, Akita suppl., June 26, 1959.

21. "Kansoku jin zokuzoku Michikawa e—roketto 16 nichi, honban no ichiban ki," *Asahi Shimbun*, Akita suppl., September 12, 1957; "Mujōken de kyōryukō—gyogyōsha ga mōshi-awasu," *Asahi Shimbun*, Akita suppl., September 12, 1957; "Noshiro e iten ga ketteiteki—Roketto sentā Tōdai seiken ga setsumei kai," *Asahi Shimbun*, July 12, 1962.

22. "Kokusai chikyū kansoku nen—sho no roketto kansoku," *Asahi Shimbun*, Akita suppl., September 3, 1957; *Asahi Shimbun*, Akita suppl., September 12, 1957.

23. Dusinberre, *Hard Times in the Hometown*, 171–188; Aldrich, *Site Fights*, 70–94.

24. One reason for this is that the Earth, at its equator, is moving 310 mph (500 kmph) faster than at the poles, reducing the energy cost of getting into orbit.

25. 51st NDHR SST, No. 23, June 2, 1966, https://kokkai.ndl.go.jp/simple/detail?minId=105103913X02319660602&spkNum=71&current=2#s71.

26. 51st Diet NDHC S&T, No. 7, June 3, 1966; "17 nichi kara uchiage Kagakugijutsuchō no roketto—Tanegashima no roketto jikken," *Asahi Shimbun*, March 3, 1967.

27. 51st Diet NDHC S&T, June 2, 1966.

28. "Tōdai uchū kūkan kansokujō o kikō—Kagoshima, Uchinoura," *Asahi Shimbun*, February 3, 1962; Nakabe et al., "Uchū kaihatsu 60 nen shi," 19; "Uchinoura roketto kaihatsujō kaisho," *Asahi Shimbun*, December 9, 1963; "Uchinoura roketto yon nen," *Asahi Shimbun*, Kagoshima suppl., February 3, 1966.

29. "Kondo wa Mu da—yume wa jinko eisei o," *Asahi Shimbun*, Kagoshima suppl., January 1, 1965; "Kensetsu susumu Mu roketto daichi—Uchinoura," *Asahi Shimbun*, Kagoshima suppl., April 6, 1966; "Kankō shiki ni fukikin atsume," *Asahi Shimbun*, November 28, 1966; "Roketto memo—jikkenjō," *Asahi Shimbun*, Kagoshima suppl., July 5, 1964.

30. "Ryokō wa roketto ni notte—ōbei wa wazuka ichi jikan," *Asahi Shimbun*, Kagoshima suppl., January 1, 1969.

31. Kukimoto Takashi, "Tsuisō," *ISAS News*, April 1999, 16; Kimi Tanaka, "Omoidegusa

(roketto obasan)," *ISAS News*, August 1985; ISAS, *Nihon no uchū kaihatsu no rekishi dai 2 sho—Uchinoura no tōjō—chōsa no kaihatsu to ohagi*, https://www.isas.jaxa.jp/j/japan_s_history/chapter02/01/05.shtml.

32. Matogawa, *Uchū ni ichiban chikai machi*, 53–61; Matogawa, email to author, March 22, 2024; Subodhana Wijeyeratne, "Between the Rocket and the Deep Blue Sea: Space Facilities and the 'Fishing Problem' in Southern Japan, 1950–1990," *Historia Scientiarum* 31, no. 22 (January 2022); Kimotsuki jōhōkyoku, "Roketto kichi to chiiki no kizuna o musubu Hashimoto-san," July 31, 2015, https://kankou-kimotsuki.net/archives/10494.

33. ISAS, *Nihon no uchū kaihatsu no rekishi dai 10 sho—takumashiki nakamatachi—Oosumi*, https://www.isas.jaxa.jp/j/japan_s_history/chapter10/01/01.shtml.

34. Kimotsuki jōhōkyoku, "Roketto kichi to chiiki no kizuna o musubu Hashimoto-san," July 31, 2015, https://kankou-kimotsuki.net/archives/10494.

35. "Roketto memo—tantō suru hitotachi," *Asahi Shimbun*, Kagoshima suppl., July 7, 1964; "Kenbutsu mo yōryō yoku—roketto nare no machi no hito," *Asahi Shimbun*, Kagoshima suppl., October 30, 1964; "Roketto no machi Uchinoura genkon—'Kotoshi koso' no kitai," *Asahi Shimbun*, Kagoshima suppl., January 2, 1967; *Asahi Shimbun*, February 3, 1962; Oda, *Uchū kūkan kanshu 30 nen shi nenhyō*, 16; Kagoshima Prefectural Office, *Kakukai kokusei chōsa tok ino shishō sonbetsu jinkō no suii*, https://www.pref.kagoshima.jp/ac09/tokei/bunya/kokutyo/h27kokutyo/documents/55416_20161128120608-1.pdf.

36. 51st NDHR Committee on Education, No. 31, June 27, 1966, https://kokkai.ndl.go.jp/simple/detail?minId=105105077X03119660627&spkNum=0&current=53#s0.

37. Matogawa, *Uchū ni ichiban chikai machi*, 69; *Asahi Shimbun*, February 3, 1962; Oda, *Uchū kūkan kanshu 30 nen shi nenhyō*, 17; *Asahi Shimbun*, Kagoshima suppl., July 7, 1964.

38. "Honnen do ni mo kōji susumu—roketto dōro," *Asahi Shimbun*, Kagoshima suppl., May 7, 1964; "Chigen ga gakkuri—Uchinoura, Tanegashima, 'Yota yota roketto,'" *Asahi Shimbun*, Kagoshima suppl., March 28, 1967; "Hasō kōji isogu—roketto dōro," *Asahi Shimbun*, Kagoshima suppl., October 15, 1968.

39. "Boku no yume, watashi no yume," *Asahi Shimbun*, Kagoshima suppl., January 6, 1964.

40. Yoshioka, "Uchū kaihatsu kyūsei no kakuritsu," 176.

41. "Tanegashima wa secchi būmu," *Asahi Shimbun*, Kagoshima suppl., October 1, 1966; "Raigatsu chū ni wa ichibu kansei—'roketto dōro' settei kyū picchi," *Asahi Shimbun*, Kagoshima suppl., February 24, 1967; "Roketto kichi no meian—Tanegashima," *Asahi Shimbun*, Kagoshima suppl., September 8, 1969.

42. "Umetatechi no sōzei hajimaru," *Asahi Shimbun*, Kagoshima suppl., August 13, 1969.

43. Takenaka, "N-1 roketto kaihatsu no ayumi," 127. Tanegashima's previous claims to fame included being the location of the first arrival of guns to Japan, and as one of the areas most frequently visited by Europeans in the seventeenth and eighteenth centuries.

44. *Asahi Shimbun*, March 28, 1967; "Tanegashima roketto uchiage—junbi, honkakuteki ni," *Asahi Shimbun*, Kagoshima suppl., September 7, 1968.

45. "Nagare hōdai no uchū sentā Tanegashima," *Asahi Shimbun*, Kagoshima suppl., July 13, 1967.

46. "Tōmin, minna ga kangei," *Asahi Shimbun*, Kagoshima suppl., May 24, 1966.

47. *Asahi Shimbun*, Kagoshima suppl., January 2, 1967; *Asahi Shimbun*, February 3, 1962.

48. "'Jūbun kenkyū o kasaneta,'—machi wa onegau," *Asahi Shimbun*, Kagoshima suppl., July 5, 1967.

49. 55th NDHR Bud., First Subcommittee, No. 3, April 21, 1967, https://kokkai.ndl.go.jp/simple/detail?minId=105505266X00319670421&spkNum=182&current=64#s182.

50. 55th NDHR Bud., No. 10, April 3, 1967, https://kokkai.ndl.go.jp/simple/detail?minId=105505261X01019670403&spkNum=181&current=63#s181.

51. "Chūkyō ki, roketto shiyō ka," *Mainichi Shimbun*, October 4, 1958; "Gaiden memo—roketto seisaku de Seidoku ga Chūgoku o kyōso," *Mainichi Shimbun*, January 10, 1969.

52. George, *Minamata*, 261–292.

53. Aldrich, *Site Fights*, 95.

54. "Futatsu no uchū genchi o miru," *Asahi Shimbun*, September 30, 1966; Aldrich, *Site Fights*, 70–94.

55. "Mitooshi shita tanu uchiage," *Asahi Shimbun*, Kagoshima suppl., July 15, 1967.

56. 58th Diet NDHC S&T, No. 11, April 17, 1968.

57. "Roketto kichi chūshin ni kankō kaihatsu," *Asahi Shimbun*, Kagoshima suppl., January 1, 1970.

58. Yasunori Matogawa, email to author, February 27, 2024.

59. *Asahi Shimbun*, March 28, 1967.

60. Cabinet Committee, No. 17, March 16, 1965.

61. Nakabe et al., "Uchū kaihatsu 60 nen shi," 30; Oda, *Uchū kūkan kanshu 30 nen shi nenhyō*, 60, 64; 55th NDHR Bud., No. 10, April 3, 1967.

62. "Roketto mondai ni gatsu ni ugoku ka," *Asahi Shimbun*, Kagoshima suppl., January 25, 1968; "Yon ki mei uchiage chūshi—Uchinoura Tōkyō daigaku roketto Miyazaki no gyōsen ga hōgai," *Asahi Shimbun*, September 13, 1968.

63. Matogawa, "A Survey of Rocketry for Space Science in Japan," 221.

64. 52nd NDHC S&T, No. 1 after closing, November 29, 1966, https://kokkai.ndl.go.jp/simple/detail?minId=105213913X00119661129&spkNum=4&current=60#s4; 51st Diet NDHC S&T, No. 7, June 3, 1966, https://kokkai.ndl.go.jp/simple/detail?minId=105113913X00719660603&spkNum=38&current=5#s38.

65. 51st NDHC S&T, June 2, 1966; 51st NDHC S&T, June 3, 1966. Tanegashima's road network had been a source of concern for economic planners for quite some time, as the island was one of Japan's most productive sugarcane-growing areas.

66. "Chikadzuku roketto shiizon—Uchinoura, Tanegashima kichi," *Asahi Shimbun*, Kagoshima suppl., December 23, 1968; *Asahi Shimbun*, September 17, 1968.

67. "Eikyō kensa nado yōbō—gyomin, uchū sentā de hanashiai," *Asahi Shimbun*, Kagoshima suppl., November 17, 1966; *Asahi Shimbun*, Kagoshima suppl., July 13, 1967.

68. Budget Committee, No. 10, April 3, 1967; "Uchū sentā o shisatsu—secchi hantai no Miyazaki gyomin," *Asahi Shimbun*, Kagoshima suppl., November 13, 1966; *Asahi Shimbun*, Kagoshima suppl., November 17, 1966.

69. Sogame, "A History of the Tanegashima Space Center," 255, 258; Oda, *Uchū kūkan kanshu 30 nen shi nenhyō*, 28; *Asahi Shimbun*, Kagoshima suppl., February 8, 1967; Yoshioka, "Uchū kaihatsu kyūsei no kakuritsu," 176; Matogawa, "A Survey of Rocketry for Space Science in Japan," 213; "Ichibu gyomin no hyōjo kataku—uchiage ni yahari fukuzatsu," *Asahi Shimbun*, September 17, 1968.

70. "Roketto 2-ki no tesshū," *Asahi Shimbun*, Kagoshima suppl., September 21, 1968.

71. *Asahi Shimbun*, Kagoshima suppl., March 28, 1967.

72. "Uchinoura—roketto hassha jumbi wa OK," *Asahi Shimbun*, Kagoshima suppl., August 4, 1969.

73. "Chūshi ni gakkari—roketto uchiage ni kengaku jidō," *Asahi Shimbun*, Kagoshima suppl., September 19, 1968.

74. *Asahi Shimbun*, Kagoshima suppl., September 18, 1968.

75. "Roketto taidan—uchiage matsu Uchinoura Tanegashima," *Asahi Shimbun*, Kagoshima suppl., September 1, 1968.

76. Hamada's behavior here can be construed as an example of what Dusinberre has dubbed "status seduction," a process whereby large projects located in far-flung and sparsely populated areas are granted legitimacy by the support of highly ranked, usually male authority figures. Dusinberre, *Hard Times in the Hometown*, 151.

77. *Asahi Shimbun*, September 17, 1968.

78. *Asahi Shimbun*, Kagoshima suppl., September 18, 1968; *Asahi Shimbun*, Kagoshima suppl., January 1, 1970.

79. *Asahi Shimbun*, Kagoshima suppl., September 8, 1969.

80. 55th NDHR Committee on Accounts, No. 4, March 28, 1967; 58th NDHC S&T, No. 11, April 17, 1968.

81. "Roketto jikken no gyogyō hoshō—seifu no taisoku ni manzoku," *Asahi Shimbun*, Kagoshima suppl., August 18, 1968.

82. "Yamakawa minato ni senin no shukusha," *Asahi Shimbun*, January 21, 1969.

83. *Asahi Shimbun*, Kagoshima suppl., December 23, 1968; "Jikō ni sonae Shobōsha mo—Roketto uchiage, hama no ue no kengakusha," *Asahi Shimbun*, Kagoshima suppl., February 7, 1969.

84. "Ayabumareru roketto uchiage—'jitsujō mushi' okoru gyomin," *Asahi Shimbun*, Kagoshima suppl., May 18, 1969.

85. "Gyogyō chōsa oeru kaiketsu o—Tanegashima roketto uchiage ni," *Asahi Shimbun*, Kagoshima suppl., February 8, 1967.

86. "Chū ni uita shukuga hanabi—'yasumi mo muda' ukanu gyomin," *Asahi Shimbun*, Kagoshima suppl., September 23, 1969.

87. Oda, *Uchū kūkan kanshu 30 nen shi nenhyō*, 60; Sogame, "A History of the Tanegashima Space Center," 259.

88. "Kore de jikken mo kidō ni—kakushu shikenshitsu ga totonou," *Asahi Shimbun*, January 14, 1970.

89. Oda, *Uchū kūkan kanshu 30 nen shi nenhyō*, 90–115.

90. Kagoshima Prefectural Office, *Kakukai kokuseichōsaji no shichōsonbetsu jinkō no suii*, https://www.pref.kagoshima.jp/ac09/tokei/bunya/kokutyo/h27kokutyo/documents/55416_20161128120608-1.pdf; Kagoshima Prefectural Office, *Dai-4 no jinkou doukou*, http://www.pref.kagoshima.jp/ap01/chiiki/kumage/chiiki/documents/71700_20190409114440-1.pdf.

91. Kagoshima Prefectural Office, *Kakukai kokusei chōsa toki no shishō sonbetsu jinkō no suii*. The villages of Uchinourachō and Kōyama were merged into a new town, Kimotsuki, in 2005.

92. Sallie Middleton, "Space Rush: Local Impact of Federal Aerospace Programs on Brevard and Surrounding Counties," *The Florida Historical Quarterly* 87, no. 2 (Fall 2008): 261; Asif Siddiqi, *Challenge to Apollo—the Soviet Union and the Space Race, 1945–1974* (Washington, DC: NASA, History Div., Office of Policy and Plans, 2000), 135; Peter Redfield, "Beneath a Modern Sky: Space Technology and Its Place on the Ground," *Science, Technology, and Human Values* 21, no. 3 (Summer 1996): 261.

93. Seiichi Sakamoto, "Ginga Renpō 25-nen kinen furēmu kitte no hakkō," *ISAS News* 378 (September 2012): 8.

Chapter 6

1. "Shinkansen no kotsu ikashite kidō eisei ga uchigeta yo," *Asahi Shimbun*, March 7, 1977.

2. Andrew Gordon, *A Modern History of Japan: From Tokugawa Times to the Present* (Oxford: Oxford University Press, 2020), 254–255.

3. Shintarō Ishihara and Akio Morita, *"No" to ieru Nihon* (Tokyo: Kōbunsha, 1989).

4. 84th National Diet, House of Representatives, Budget Committee, First Subcommittee, No. 3, March 1, 1978, https://kokkai.ndl.go.jp/simple/detail?minId=108405266X00319780301&spkNum=5&current=11#s5.

5. Fumihiko Tomita, "Chūgoku no saikin no uchū kaihatsu kenmonroku," *Space Japan Review* 20 (December 2001–January 2002): 2; "Chūgoku, uchū kara 'Niihao' yūjin hikō madjika? Junbi wa chakuchaku," *Asahi Shimbun*, January 24, 2000; "Kaigai bunka uchū kaihatsu: Chūgoku, 5-nen inai ni ningen eisei!? Amerika gakusha no yosoku," *Mainichi Shim-*

bun, May 7, 1977; "Kaigai bunka uchū kaihatsu: Chūgoku ni mo yūjin eisei uchiage nōryoku," *Mainichi Shimbun*, June 7, 1985.

6. Yūko Yoshikawa, "Eisei tsūshin to watashi: Miryoku o saidaigen ni tsutaeru koto ga kōhō senden no shimei," *Space Japan Review* 26 (December 2002–January 2003): 13–16.

7. "Sekai no eisei kigyō no CEO ni kiku," *Space Japan Review* 22 (April–May 2002): 1–2; "Sekai no eisei kigyō no CEO ni kiku," *Space Japan Review* 26 (December 2002–January 2003): 1–13; "Sekai no eisei kigyō no CEO ni kiku," *Space Japan Review* 23 (June–July 2002): 1–6; Susumu Kitazume, "Nihon no anzen hoshō to uchū kaihatsu," *Space Japan Review* 36 (August–September 2004): 2–4.

8. JAXA, "Humans in Space—Mukai Chiaki," https://humans-in-space.jaxa.jp/en/astronaut/mukai-chiaki/.

9. JAXA, "Humans in Space—Yamazaki Naoko," https://humans-in-space.jaxa.jp/en/astronaut/yamazaki-naoko/.

10. International Astronautical Federation, "Biographies—Kiyoshi Higuchi," https://www.iafastro.org/biographie/kiyoshi-higuchi.html.

11. Siddiqi, "Spaceflight in the National Imagination," 28.

12. Matsui and Shimizu, "Uchū kaihatsu jigyōdan no settei," 25; Logsdon, "U.S.–Japanese Space Relations," 294.

13. Siddiqi, "Spaceflight in the National Imagination," 7.

14. NASA, "Nasa Spinoff," https://spinoff.nasa.gov/. JAXA maintains its own website dedicated to products that have resulted from its research, which include cooling underwear and insulation materials used in housing; see JAXA, "Spinoff from Japan's Aerospace Technology," https://aerospacebiz.jaxa.jp/en/success-story/.

15. Slava Gerovitch, *Soviet Space Mythologies: Public Images, Private Memories, and the Making of a Cultural Identity* (Pittsburgh: University of Pittsburgh Press, 2015), xii.

16. Siddiqi, "Spaceflight in the National Imagination," 19.

17. Ryūho Ōkawa, *Roketto Hakase Hideo Itokawa—dokuseiteki Mirai Kagaku hassōhō* (Tokyo: Kyubunsha, 2014).

18. Ōkawa, *Roketto hakase Hideo Itokawa*, 25.

19. Gorō Suzuki, "Itokawa Hideo hakase," 151–153.

20. Iwao Kanezawa, ed., *Ningen Hideo Itokawa hakase to wa* (Tokyo: Hirohide Kōgei, 2003).

21. Kusayanagi, "'Suzō sangyō' no sutā Itokawa Hideo," 264, 266, 268, 270. Daizō Kusayanagi established a reputation as an analyst of the psychologies and motivations of many of what he called Japan's postwar "new type" of go-getting intellect; his wide-ranging analysis is best represented in his 1985 work, *Jitsuryoku-sha no jōken—kono hito-tachi no essensu* (Conditions for the Talented—Their Essence) (Tokyo: Bungeishunju, 1985).

22. Yasunori Matogawa, "20 seiki no kōkūjin (sono 17) Hideo Itokawa—Nihon no uchū kaihatsu no oto," *Sora to Bunka* 93 (Spring 2006): 30.

23. There are no notes in much of Matogawa's writing, even though he himself was not present at many of these events. Rather, the implicit assumption in the texts is that he either heard the stories from the man himself or learned from others with similarly intimate knowledge.

24. Yasunori Matogawa, "Lessons from Half a Century Experience of Japanese Solid Rocketry since Pencil Rocket," *Acta Astronautica* 61 (2007): 1108.

25. Hideo Itokawa, "Kōkūki no sekkei to minzoku seishin = ka beiei rokotsuna kojin shugi," *Yomiuri Shimbun*, January 1, 1944; "Kōkūki no sekkei to minzoku seishin = ge nichidoku kōgeki ippon yari Hideo Itokawa," *Yomiuri Shimbun*, January 2, 1944.

26. Kusayanagi, "'Suzō sangyō' no sutā Itokawa Hideo," 260.

27. Hideo Itokawa, "Kōkūki no sekkei to minzoku seishin = ka beiei rokotsuna kojin shugi," *Yomiuri Shimbun*, January 1, 1944; "Kōkūki no sekkei to minzoku seishin = ge nichidoku kōgeki ippon yari Hideo Itokawa," *Yomiuri Shimbun*, January 2, 1944.

28. Itokawa, *Nihon sōseiron*, 16, 92, 94.

29. Matogawa, "Itokawa Hideo no baiorin dzukuri," 64; Gorō Suzuki, "Itokawa Hideo hakase," 152.

30. Mayu Nimiya, "Nihon no Da—Winchi Hideo Itokawa hakase o shinobu," in *Ningen Hideo Itokawa hakase to ha*, ed. Iwao Kanezawa (Tokyo: Hirohide Kōgei, 2003), 17.

31. "'Kokoro no fudoki' Kureshi (Hiroshima ken) uchū kōgaku-sha Yasunori Matogawa-san," *Yomiuri Shimbun*, April 7, 2004.

32. Yasunori Matogawa, *Harē suisei no kagaku hoshizora no pasuwāku* (Tokyo: Shinchō Bunko, 1984); Yasunori Matogawa, *Harē ni idomu—suisei ni kakeru roman* (Tokyo: Dōbun shōin, 1985); Yasunori Matogawa, *Harē suisei no nazo* (Tokyo: Popuraa shagakushō bunko, 1986); Tatsuo Yamanaka and Yasunori Matogawa, "Uchū kaihatsu no ohanashi," *Yomiuri Shimbun*, November 4, 1991; Yasunori Matogawa, *Otona no tame no hontō no uchū no hanashi* (Toyosu: PHP Kenkyūjō, 1992); Yasunori Matogawa, *Uchū de kurasu tame no 69 no kiso chishiki* (Tokyo: Yamato shohō, 1999).

33. Yasunori Matogawa, "Uchū gijutsu no gunji riyō," *Nihon kagakusha* 17, no. 7 (July 1982): 41–43.

34. Matogawa, "Uchū kaihatsu no rekishi to Nihon no yakuwari," 1111; Yasunori Matogawa, "Issho ni uchū e—jinrui shinka ni kado ni okeru yūjin hikō," *Kikai no kenkyū* 48, no. 1 (1996), 53; Matogawa, "Nihon uchū kagaku no saikin no kekka," 38.

35. Matogawa, "Shizen roketto kaihatsu wa kaku ketsudan sareta," 56–57; Matogawa, "Jinrui no shōrai no tenbōshita Itokawa Hideo," 62–63; Matogawa, "Itokawa Hideo no baiorin dzukuri," 64–65.

36. Matogawa, "Jimae roketto kaihatsu wa kaku ketsudan sareta," 56. Susanoo, the "impetuous male," is the younger brother and sometimes foe of Amaterasu, Japanese goddess of the Sun and putative ancestor of the imperial family. Izanagi is his father, who, together with Izanami, sired the sacred islands of Japan.

37. Yasunori Matogawa, "Uchū kaihatsu no rekishi to Nihon no yakuwari," *Kikai no kenkyū* 46, no. 11 (1994): 1109–1110.

38. "Wakusei tansa e shin roketto Uchūken ga 150 oku-en tōji 'M 5' kaihatsu e," *Asahi Shimbun*, January 20, 1990.

39. "Yasunori Matogawa-san 'anata no namae o kasei e' no kagaku-sha (hito)," *Asahi Shimbun*, June 20, 1998; "Uchūken no 'M 5' uchiage shippai 1-dan-me ni atsuryoku ijō tenmon eisei, kidō tōnyū dekizu," *Yomiuri Shimbun*, February 10, 2000.

40. "'H2' seikō de hirogaru yume kokusai kyōryoku de kankyō kanshi eisei hōpu no kaihatsu mo," *Asahi Shimbun*, February 7, 1994; "Kokusan shin roketto 'H2A' no kumitate susumu jigyō-dan ga setsumei/ kagoshima," *Asahi Shimbun*, July 18, 2001.

41. These include the Soviet Union's Mars-exploring Phobos craft ("'Kasei no nazo ni semaru' (jō) eisei tansa-ki fobosu 7 tsuki uchiage (rensai)," *Yomiuri Shimbun*, July 4, 1988); experimental US space technologies ("Deruta kurippā retorona sugata ni shingijutsu mansai tan-danshiki 'jisedai shatoru' fujō," *Yomiuri Shimbun*, September 18, 1993); Japan's policy toward information acquired from the International Space Station ("Kokusaiuchū kichi' tanan'na shuppatsu 'yume no nedan' kakkoku no omoni ni 29nichi ni mo kyōtei shomei," *Yomiuri Shimbun*, January 23, 1998); China's burgeoning role in space ("Chūgoku, uchū kara 'niihao' yūjin hikō majika? Junbi wa chaku chaku," *Asahi Shimbun*, January 24, 2000); the deorbiting of the Russian space station Mir ("Uchū sutēshonn 'Miiru' Nihon ni hahen rakka no osore mo saishū funsha shippai no baai," *Yomiuri Shimbun*, March 22, 2001); and the Space Shuttle Columbia disaster ("Uchū kaihatsu ni dai dageki teitai, Amerika ni todomarazu shatoru tsuiraku jikō," *Asahi Shimbun*, February 3, 2002).

42. "Nosutoradamusu no daiyogen' no kenkyū kyōfu no daiō no 'shōtai' o kenshō," *Yomiuri Shimbun*, February 2, 1999.

43. Matogawa, "Uchū kaihatsu no rekishi to Nihon no yakuwari," 1111; "Roshia no uchū gijutsu sokodjikara to kibishii genjitsu kanmin gōdō no Nihon chōsa-dan ga genchi shisatsu," *Asahi Shimbun*, July 20, 1992.

44. Matogawa, "Uchū kaihatsu no rekishi to Nihon no yakuwari," 1109.

45. Yasunori Matogawa, "A Survey of Rocketry for Space Science in Japan," in *History of Rocketry and Astronautics: Proceedings of the History Symposia of the International Academy of Astronautics* 22 (San Diego: American Astronautical Society, 1993), 208–214; Matogawa, "Another Destiny of Rocketry in Japan—Festival Rockets in Japanese Shrines," in *History of Rocketry and Astronautics: Proceedings of the History Symposium of the International Academy of Astronautics* (San Diego: American Astronautical Society, 1994–1995), 315–326; Matogawa, "Japanese Solid Rockets in the Second World War," 123–136; Matogawa, "Japan Liquid Rockets in the Second World War," 111–125.

46. "Harē suisei' no kinen botoru hatsubai (nyūsurain)," *Asahi Shimbun*, September 7, 1995.

47. "Dai 10-kai Asahi yasashii kagaku no kyōshitsu," *Asahi Shimbun*, April 21, 1990;

"Uchū e no kauntodaun fōramu to ronbun kaiga sakuhin boshū (shakoku)," *Yomiuri Shimbun*, May 27, 1991; "Dai 14-kai Asahi yasashii kagaku no kyōshitsu 'koko made kita uchū kaihatsu'," *Asahi Shimbun*, April 28, 1992; "Dai 20-kai Asahi yasashii kagaku no kyōshitsu," *Asahi Shimbun*, May 16, 1995; "Sentan gijutsu no kōkai kōza 26 nichi, Kawasaki de kagaku gijutsu akademii ga kaisai = Kanagawa," *Yomiuri Shimbun*, August 20, 1998; "Bisei de tenmondai kōza 12nichi ni hiraku kōshi wa matogawayasunori uchū kagaku-ken kyōju = Okayama," *Yomiuri Shimbun*, December 9, 1999.

48. Yasunori Matogawa, Mashiro Noda, and Akashi Kanda, "21 seiki no uchū kaihatsu," *Kagaku gijutsu jaanaru* 3, no. 9 (September 1994): 10–15; "Puranetariumu to tenmon kōenkai no yūbe," *Asahi Shimbun*, September 2, 1991; "Kanagawa kagaku gijutsu akademii kōenkai 'yume to roman no uchū kagaku' (moyōshi)," *Asahi Shimbun*, January 27, 1992; "Uchū ni semaru kyō Morioka de kagaku kōen-kai/ Iwate," *Asahi Shimbun*, December 6, 1997; "Ibento marion," *Asahi Shimbun*, January 29, 1999; "Shichōsonda yori/ Okayama," *Asahi Shimbun*, December 12, 1999; "Uchū fōramu chikyū wa hitotsu 12 gatsu 11nichi, Nagoya de," *Asahi Shimbun*, November 13, 1992.

49. Iida, "Kaze o yomeru ka," 4.

50. Takenaka, "N-1 roketto kaihatsu no ayumi," 131.

51. Neil Maher, "Shooting the Moon," *Environmental History* 9, no. 3 (July 2004): 528–531.

52. Iida, "Kaze o yomeru ka," 4.

53. Hiroshi Sakamoto, "Eisei yowa kiku 6-gō ni tazusawatte," *Space Japan Review* 43 (October–November 2005): 1.

54. Susumu Kitazume, "H-IIA roketto 3-gōki DRTS oyobi USERS no uchiage seikō," *Space Japan Review* 25 (October–November 2005): 1–3.

55. Hideki Mizuno, "Eisei Yōwa (2)," *Space Japan Review* 28 (April–May 2003): 2; Hideki Mizuno, "Eisei Yōwa (3)," *Space Japan Review* 29 (June–July 2003): 3.

56. "Shin roketto o tsukuru—rai aki no Itokawa kyōjū ga happyō," *Asahi Shimbun*, Akita suppl., February 9, 1956; "Asu Michikawa kaigan de—Itokawa kyōjura—Byōsoku sen mētoru no seinō," *Asahi Shimbun*, Akita suppl., February 5, 1958. The plastic rocket, dubbed the Pi, was launched from a balloon, technically making it part of what is known as a "rockoon."

57. Matogawa, "Uchū kaihatsu no rekishi to Nihon no yakuwari," 1109.

58. Kimura, "Kogata eisei tōsai no minseiyōki-ki riyō shisutemu de gazō shutoku ni seikō," 13.

59. Matogawa, "Uchū kaihatsu no rekishi to Nihon no yakuwari," 1109.

60. Matogawa, "Uchū kaihatsu no rekishi to Nihon no yakuwari," 1111; Matogawa, "Issho ni uchū e," 53; Matogawa, "Nihon uchū kagaku no saikin no kekka," 38.

61. *Space Japan Review* 90 (October 2015–January 2016): 12–13.

62. Yasushi Horikawa, "To Protect the Environment for a Peaceful Life," http://global.

jaxa.jp/article/special/eco/horikawa_e.html; Matogawa, "Uchū kaihatsu no rekishi to Nihon no yakuwari"; Matogawa, "Issho ni uchū e," 51–53.

63. Saito, "Japan's Space Policy," 193.

64. Matogawa, "Uchū kaihatsu no rekishi to Nihon no yakuwari," 1109–1113.

65. Mizuno, "Eisei Yōwa (4)," 1–2.

66. Matogawa, "Jimae roketto kaihatsu wa kaku ketsudan sareta," 57.

67. "Chikyū tobitatsu 'wagaya no aji' 'Mukai-san no uchū ryōri' nyūshō-saku kettei," *Yomiuri Shimbun*, May 28, 1994.

68. "Chikyū tobitatsu wagaya no aji Mukai-san no uchū ryōri nyūshō-saku kettei = tokushū," *Yomiuri Shimbun*, May 28, 1994.

69. Ezra F. Vogel, *Japan's New Middle Class* (Berkeley: University of California Press, 1965), 71–72.

70. Carol Gluck, "The Past in the Present," and William W. Kelly, "Finding a Place in Metropolitan Japan: Ideologies, Institutions, and Everyday Life," both in Andrew Gordon, ed., *Postwar Japan as History* (Berkeley, CA: University of California Press, 1993), 75 and 195, respectively.

71. "Uchū chūdan de oyako taimen," *Asahi Shimbun*, August 16, 1980.

72. Hideki Mizuno, "Eisei Yōwa (4)," *Space Japan Review* 30 (August–September 2003): 1.

73. Matogawa, "Uchū kaihatsu no rekishi to Nihon no yakuwari," 1109–1110.

74. Matsui and Shimizu, "Uchū kaihatsu jigyōdan no settei," 24, 33.

75. National Institute of Information and Communications Technology (Uchū tsūshin kenkyū kikō), "Kashima uchū gijutsu sentā," http://ksrc.nict.go.jp/pdf/pamphlet2017-2.pdf.

76. Hideki Mizuno, "Eisei Yōwa (2)," *Space Japan Review* 28 (April–May 2003): 2; Hideki Mizuno, "Eisei Yōwa (3)," *Space Japan Review* 29 (June–July 2003): 3.

77. Shin'ichi Kimura, "Kogata eisei tōsai no minseiyōki riyō shisutemu de gazō shutoku ni seikō," *Space Japan Review* 27 (February–March 2003): 13. The class of "microsatellites" is an amorphous one but is most often used to designate satellites weighing between 22 lb (10 kg) and 220 lb (100 kg); Micro-OLIVe weighed in at 135 lb (60 kg). Kimura Laboratory, "Minsei-hin o katsuyō shita jinkō eisei no kō chinō-ka jiritsu seigyo gazō shori ni kansuru jikken," https://www.rs.noda.tus.ac.jp/skimura/kimura_lab/Micro-OLIVe.html; Yoshiaki Suzuki, "H–IIA roketto 4-gōki uchiage seikō!," *Space Japan Review* 26 (December 2002–January 2003): 1.

78. 14th Noshiro Space Event, "Setsuritsu chōshi," http://www.noshiro-space-event.org/about_sub01.html.

79. Matogawa, "Uchū kaihatsu no rekishi to Nihon no yakuwari," 1113.

80. Itokawa, *Uchū o sanpo suru*, 36.

81. "Uchū ni nobiru erebētā chijō to seishi eisei o musubu kumo no ito," *Asahi Shimbun*, March 3, 2003.

82. Matogawa, "Issho ni uchū e," 51 (emphasis added).

83. Mizuno, "Eisei Yōwa (4)," 1–2.

84. William D. Wray, "Japanese Space Enterprise: The Problem of Autonomous Development," *Pacific Affairs* 64 (Winter 1991): 467.

85. Matogawa, "Uchū kaihatsu no rekishi to Nihon no yakuwari," 1111–1113.

86. Matogawa, "Issho ni uchū e," 48–49.

87. Matsui and Shimizu, "Uchū kaihatsu jigyōdan no settei," 26, 33.

88. Uchū kaihatsu jigyōdan hō (NASDA foundation law), June 23, 1964, https://www.shugiin.go.jp/internet/itdb_housei.nsf/html/houritsu/06119690623050.htm; Matsui and Shimizu, "Uchū kaihatsu jigyōdan no settei," 26.

89. Matsui and Shimizu, "Uchū kaihatsu jigyōdan no settei," 26–27; Takenaka, "N-1 roketto kaihatsu no ayumi," 127; Saito, "Japan's Space Policy," 193.

90. Mizuno, "Eisei Yōwa (4)," 1.

91. Matogawa, "Uchū gijutsu no gunji riyō," 41–43.

92. Matogawa, "Japan Liquid Rockets in World War II," 123.

93. Yasunori Matogwa, "The Shūsui Japanese Rocket Fighter in the Second World War"; Matogawa, "Japanese Solid Rockets in the Second World War," 123–136; Matogawa, "Japan Liquid Rockets in the Second World War," 111–125.

94. Logsdon, "U.S.–Japanese Space Relations," 299.

95. Kallendar, "Explaining the Logics of Japanese Space Policy Evolution 1969–2016," 95.

96. Sherwood S. Cordier, "Japan: Present and Potential Military Power," *United States Naval Institute Proceedings* 93, no. 777 (November 1967), https://www.usni.org/magazines/proceedings/1967/november/japan-present-and-potential-military-power.

97. Sabine Frühstück, *Playing War: Children and the Paradoxes of Modern Militarism in Japan* (Berkeley: University of California Press, 2017), 10, 166.

98. David Hunter-Chester, *Creating Japan's Ground Self-Defense Force, 1945–2015: A Sword Well Made* (London: Lexington Books, 2016), 173–222.

99. Takeo Ueda, "Uchū riyō to anzen hoshō," *Space Japan Review* 37 (October–November 2004): 1.

100. Iida, "Kaze o yomeru ka," 8.

101. Iida, "Kaze o yomeru ka," 3–7.

102. Kitazume, "Nihon no anzen hoshō to uchū kaihatsu," 2–4.

103. Luk and Wijeyeratne, "Alternatives to GPS," 57–62.

104. Kallendar, "Explaining the Logics of Japanese Space Policy Evolution 1969–2016," 17–18. Paul Kallendar quite rightly points out that, as far back as the 1990s, satellites such as Orihime had military-applicable anti-satellite capacities, indicating that this military turn was not entirely abrupt. This may be indicative of deep links to militarization, but the fact that these abilities were not openly identified is indicative of the narrative taboo against openly discussing the process at the time.

105. Watanabe, "The Evolution of Japanese Space Policy," 271; JAXA, "A Passion for Rocketry: Ryojiro Akiba."

106. Andrew J. Oros, "Explaining Japan's Tortured Course to Surveillance Satellites," *Review of Policy Research* 24, no. 1 (2007): 30–48.

107. Oros, "Explaining Japan's Tortured Course to Surveillance Satellites," 41.

108. Rieko Hayakawa, "Eisei tsūshin to watashi," *Space Japan Review* 16 (April–May 2001): 17.

109. Kalendar, "Explaining the Logics of Japanese Space Policy Evolution 1969–2016," 13.

Chapter 7

1. David E. Sanger, "A Japanese Innovation: The Space Antihero," *The New York Times*, December 8, 1990. Akiyama continues to be one of the most entertaining interviews in Japan's civilian space programs—opining, for example, that "it's normal to have a trace or two of a stomach ulcer by the time you are in your forties if you have a proper job. People who don't get stomach ulcers don't do a very good job." See Tomoko Otake, "Toyohiro Akiyama: Cautionary Tales from One Not Afraid to Risk All," *The Japan Times*, August 3, 2013. Akiyama is now a professor of agriculture at Kyoto University of Art and Design.

2. "Kakkidzuku jitsuyō eisei no kaihatsu—'gijutsu kakusa' no kabe o itomu," *Asahi Shimbun*, September 28, 1973.

3. Hideo Sudō and Masaki Katō, "Uchū kaihatsu jigyōdan Tsukuba uchū sentā no shōkai (toku ni uchū seibutsu kagaku ni kanren shite)," *Uchū seibutsu kagaku* 7, no. 1 (1993): 31; ISAS, "Sagamihara kyanpasu," https://www.isas.jaxa.jp/about/facilities/sagamihara.html.

4. Kenshi Funakawa, "Uchū kaihatsu jigyōdan no tuiseki kansei shisutemu," *Nihon kōkū uchū gakkaishi* 32, no. 364 (May 1984): 254–255; ISAS, "Usuda uchū kūkan kansokujō," https://www.isas.jaxa.jp/about/facilities/usuda.html.

5. Oda Mitsuhige. "Uchū robotto ni okeru ningen kikakei no kadai," *Nihonkōkū uchū gakkaishi* 40, no. 464 (September 1992): 486–490.

6. "Kōkū uchū gijutsu-ken-sho ga sōritsu 30 shūnen sekai ni hokoru dokuji no eisei seigyo sōchi," *Asahi Shimbun*, July 20, 1985.

7. Yoshida Tetsuya, "Daikikyū—uchū e no iriguchi," presentation at Kyoto University Cosmology Unit, Cosmology Seminar, November 28, 1999, http://www.usss.kyoto-u.ac.jp/wp-content/uploads/2021/03/20141128_yoshida.pdf; Masami Uchida, Michio Nakagawa, Hirohisa Sakurai, and Makoto Yamauchi, "Cyg X-1 no kata X-sen ryōiki ni okeru tanjikan hendō ni tsuite (II)," *Uchūkagakukenkyūjo hōkoku, tokushū: dai kikyū kenkyū hōkoku* 24 (December 1989): 103–112.

8. Masami Uchida, Michio Nakagawa, Hirohisa Sakurai, and Makoto Yamauchi, "Hokkyokuken ozon-sō kokusai kyōdō kikyū kyanpēn," *Uchūkagakukenkyūjo hōkoku, tokushū: dai kikyū kenkyū hōkoku* 24 (December 1989): 103–112; Akemi Izumi-Kurotani, Mayumi

Ooya, Yoshihiro Mogami, Masamichi Yamashita, Makoto Okuno, and Shoji A. Baba, "Genseidōbutsu zōrimushi no yūei kōdō to jūryoku," *Uchūkagakukenkyūjo hōkoku, tokushū: dai kikyū kenkyū hōkoku* 24, no. 24 (December 1989): 36.

9. Tsuguo Tadakawa, "Japan's Launch Vehicle Program Update," *SAE Transactions* 96, Section 6: Aerospace (1987): 219–228.

10. Yoshitaka Kurihara, "Wagakuni no seishi eisei," *Jōhō kanri* 21, no. 2 (May 1978): 81–91.

11. Katsuyuki Shimodaira and Takashi Hamasaki, "Eisei gijutsu no genjō to shōrai," *Terebishon gakkaishi* 41, no. 4 (1987): 306.

12. Saito Norio, "Uchū kaihatsu jigyōdan no gijitsu shiken eisei," *Nihon kikai gakkai ronbun shū* (*C shū*) 60, no. 580 (December 1994): 10.

13. Science and Technology Agency of Japan, *Chikyū kansoku eisei shisutemu no kaihatsu ni tsuite* (Tokyo: STA, July 12, 1977), 30–33.

14. Kenji Funakawa, "Uchū kaihatsu jigyōdan no kidō kanri shistemu," *Nihonkōkū uchū gakkaishi* 31, no. 364 (May 1984): 253.

15. M. Ogata, H. Mizusawa, and K. Irie, "Eisei tsūshin no ayumi to kongo no kadai," *Mitsubishi gunki gihō* 59, no. 6 (1985): 409.

16. Society of Japanese Aerospace Companies, *Report on the Present Status of the Japanese Space Industry* (Tokyo: Society of Japanese Aerospace Companies, October 1990).

17. "Nagoya yūdō suishin shisutemu seisakusho," Mitsubishi Heavy Industries, https://www.mhi.com/jp/company/location/nagoyaguidew.

18. Chikayoshi Higuchi, "Uchū sutēshon keikaku to Nihon jikken mojūru no kōsō," *Yōsetsu gakkaishi* 59, no. 6 (1990): 438.

19. *Asahi Shimbun*, September 28, 1973; MHI, "Nagoya yūdō suishin shisutemu seisakusho."

20. Mitsubishi Heavy Industries, *Umi ni riku ni soshite uchū e: zoku Mitsubishi Jūkōgyō shashi 1964–1989* (Tokyo: Mitsubishi jūkōgyō kabushiki kaisha, 1990), 736.

21. Sogame Eiji, "H-I roketto," *Nihonkōkū uchū gakkaishi* 36, no. 418 (November 1988): 488.

22. Peter G. Smith. "Non–US Approaches to Space Commercialisation," in *Second Symposium on Space Industrialisation* (Washington, DC: NASA, October 1984), 24. One reason for this is that the first stage has to be the most powerful part of a rocket, requiring more capable, sophisticated, and expensive engines.

23. MHI, *Umi ni riku ni soshite uchū e*, 736.

24. Sogame, "H-1 roketto," 488, 490–493, 495; Koji Shibato and Makoto Miwada, "H-II roketto ato no uchū yusō shisutemu 'roketto purēn,'" *Nihon kōkū uchū gakkaishi* 37, 427 (August 1989): 359–360.

25. ISAS, "The Age of Space Science—Mu Rockets," https://www.isas.jaxa.jp/e/japan_s_history/detail/mu.shtml.

26. Sogame Eiji, "H-I roketto dai ni dan suishin-kei no kaihatsu," *Nihonkōkū uchū gakkaishi* 34, no. 387 (April 1986): 175–185.

27. Asif Siddiqi, *The Challenge to Apollo: The Soviet Union and the Space Race, 1945–1974* (Washington, DC: NASA, 2000), 552–553.

28. Keidanren, *Space in Japan 1986–1987*, 5.

29. Logsdon, "U.S.–Japanese Space Relations," 294; Iida, "Kaze o yomeru ka," 3.

30. M. A. Cusumano, "Manufacturing Innovation: Lessons from the Japanese Auto Industry," *Sloan Management Review* 30, no. 1 (1988): 29–40.

31. Kuninori Uesugi, "Hiten kara dzubi teiryū e—Nihon no uchū kōgaku ni nokoshitamono," *Tokushū: Geotail kara no saki e—40 nenkan no kansha o komete,* https://www.isas.jaxa.jp/feature/special_issues/geotail/geotail_01.html; Kuninori Uesugi, "Kimo o hiyashita uchiage toki no antena no tsuibi," *Nihon uchū kaihatsu no rekishi (uchū kenbutsu monogatari),* https://www.isas.jaxa.jp/j/japan_s_history/chapter06/03/02.shtml; Kuninori Uesugi, "Space Odyssey of an Angel—Summary of HITEN's Three Year Mission," in *Advances in the Astronautical Sciences—Spaceflight Dynamics 1993: Proceedings of an AAS/NASA International Symposium Held April 26–30, 1993, Greenbelt, Maryland* (1993), 611.

32. Yasunori Matogawa, email to author, December 4, 2024 (emphasis added).

33. Yoshioka, "Uchū kaihatsu kyūsei no kakuritsu," 173.

34. ISAS, "Past: Hiten," https://www.isas.jaxa.jp/en/missions/spacecraft/past/hiten.html; NASA, *Galileo Jupiter Arrival Press Kit December 1995*, https://www.jpl.nasa.gov/news/press_kits/gllarpk.pdf; Asif Siddiqi, *Beyond Earth: A Chronicle of Deep Space Exploration, 1958–2016* (Washington, DC: NASA, 2018), 169.

35. Yasunori Matogawa, email to author, December 4, 2024.

36. C. L. Merkle, J. R. Brown, J. P. McCarty, G. B. Northam, A. Povinelli, M. L. Stangeland, and E. E. Zukoski, *JTEC Panel Report on Space and Transatmospheric Propulsion Technology: Final Report* (Baltimore: Loyola College, August 1990), xii.

37. Space Activities Commission, *Report of the Space Activities Commission* (Tokyo: Space Activities Commission, March 17, 1978).

38. Saito, "Japan's Space Policy, 194.

39. Saito, "Japan's Space Policy," 194–195.

40. Shōji Matsubara and Tetsuichi Itō, "The Conceptual Study of H-II Orbiting Plane (HOPE)," *SAE Transactions* 96, Section 6: Aerospace (1987): 1719–1728.

41. Keiji Nitta, "Shōrai no uchū riyō gijutsu," *Terebishon gakkaishi* 41, no. 4 (1987): 357; "Tsuki ni koronii keisei made 5 dankai kōsō happyō 'getsumen kichi to shigen kaihatsu' no kai," *Asahi Shimbun*, February 27, 1990.

42. "Yūjin uchū hikō shinpo (nyusu rain)," *Asahi Shimbun*, November 2, 1985.

43. Takao Doi, Shigeki Kamigauchi, and Yoshiyuki Hasegawa, "Uchū hikō-shi no kunren ni kansuru ichi kōsatsu," *Nihonkōkū uchū gakkaishi* 43, no. 496 (May 1995): 267. Interestingly enough, the single longest lasting module of this early training was English-

language practice, a subject on which the astronauts were given more than six years' worth of tutoring.

44. "Sayōnara 20 seiki: naka jiken sesō kagaku gijutsu no Nihon 10 dai nyūzu," *Asahi Shimbun*, December 27, 1999. Akiyama eventually became an academic focusing on horticulture.

45. Shibato and Miwada, "H-II roketto ato no uchū yusō shisutemu," 358, 360; "Dokushi kaihatsu no yūjin purattohōmu, Gainen sekkei wa hobo shūryō," *Asahi Shimbun*, June 2, 1990.

46. Vladimir V. Balepin, Makoto Yoshida, and Kenjiro Kamijo, "Rocket Based Combined Cycles for Single Stage Rocket," *SAE Transactions* 103, Section 1: Journal of Aerospace (1994): 174–188.

47. Matsumoto Shinji, "Uchū kaihatsu to kensetsu," *Doboku gakkai ronbun-shū* 422, no. 14 (October 1990): 1–10; Saito Takao, Kobayashi Hideo, and Takagi Kenji, "Sora chū kaihatsu no mirai kōsō nitsuite tsuki Wakusei no riyō o chūshin ni," *Nihonkōkū uchū gakkaishi* 42, 488 (September 1994): 511–522.

48. 120th NDHC S&T, No. 4, April 9, 1991, https://kokkai.ndl.go.jp/simple/detail?minId=112013928X00419910409&spkNum=45&current=13#s45.

49. "Junkokusan de taikei gijutsu eisei (nyūsu rain)," *Asahi Shimbun*, August 31, 1985.

50. Lynn, "Japanese Research and Technology Policy," 296.

51. Wray, "Japanese Space Enterprise," 464.

52. Matsubara and Itō, "Conceptual Study of H-II Orbiting Plane," 1719–1728.

53. *Asahi Shimbun*, August 31, 1985.

54. Akio Moro, "Kō shutsuryoku roketto nenryō," *Nenryō kyōkaishi* 68, no. 11 (1989): 993–994.

55. "'Eisei o yasuku uchiagemasu' Arian kaishachō ga rainichi sēru," *Asahi Shimbun*, April 27, 1985.

56. Ecosystems International, "Commercial Potential of European and Japanese Civilian Space Programs" (September 1987), 3, 60.

57. Shimodaira and Hamazaki, "Eisei gijutsu no genjō to shōrai," 299; Matsumoto, "Uchū kaihatsu to kensetsu," 1–10.

58. Shimodaira and Hamazaki, "Eisei gijutsu no genjō to shōrai," 299.

59. André Lebeau and Karl Reuter, "La place des techniques spatiales dans le développement économique. Réflexion sur des aspects actuels et futurs" (Paris: European Space Agency, 1980), 15.

60. Shibato and Miwada, "H-II roketto ato no uchū yusō shisutemu," 358, 360; Tetsuichi Itō and Hideki Nomoto, "Hōpu no kūryoku sekkei to gokuchō onsoku no kadai," *Nihonkōkū uchū gakkaishi* 38, no. 435-gō (April 1990): 194–198.

61. Sogame, "H-1 roketto," 496–497; Albert D. Wheelan. "A 'Born-Again' Space Program," *International Security* 11, no. 4 (Spring 1987): 148.

62. Michael Crichton. *Rising Sun: A Novel* (New York: Knopf, 1992).

63. Andrew C. McKevitt, *Consuming Japan* (Chapel Hill: University of North Carolina Press, 2017), 22–23.

64. "Uchū no kodō o kanchi," *Asahi Shimbun,* January 4, 1981; Logsdon, "U.S.–Japanese Space Relations," 294; Iida, "Kaze o yomeru ka," 3.

65. Nakabe et al., "Uchū kaihatsu 60 nen shi," 15.

66. Sato, "Contested Gift," 182–185.

67. Yoshioka, "Uchū kaihatsu kyūsei no kakuritsu," 179–180. The United States extracted similar guarantees from other recipients of American technology in East Asia at the same time, such as South Korea. See Peter B. Kwon, "The Anatomy of Chaju Kukpang: Military-Civilian Convergence in the Development of the South Korean Defense Industry under Park Chung Hee, 1968–1979," PhD diss., Harvard University, 2016.

68. Sato, "Managing the Interface," 177.

69. Oros, "Explaining Japan's Tortured Course to Surveillance Satellites," 33.

70. "Kyōsō to kyōryoku to uchū bijinesu josō," *Asahi Shimbun* January 26, 1991.

71. Takenaka, "N-I roketto kaihatsu no ayumi," 132–137. See Sato, "Contested Gift," 195; Ueda, "Uchū riyō to anzen hoshō," 4.

72. Dusko Doder, "Japan Buying U.S. Rocket System; Unions Assail Technology Export,' *Los Angeles Times,* March 7, 1973.

73. Wray, "Japanese Space Enterprise,"468.

74. Logsdon, "U.S.–Japanese Space Relations," 294–300 (emphasis added).

75. Nitta, "Shōrai no uchū riyō gijutsu," 355.

76. Sogame, "H-1 roketto," 487.

77. "Michinori kewashii junkokusan roketto toraburu tsudzuku H 2," *Asahi Shimbun,* June 29, 1992.

78. *Space Japan Review* 83 (June–September 2013): 1–4.

79. Kitazume, "Nihon no anzen hoshō to uchū kaihatsu," 3–4.

80. Ueda, "Uchū riyō to anzen hoshō," 3–5.

81. Sogame, "H-1 roketto," 487.

82. Sogame, "H-1 roketto," 487.

83. Guidance systems are particularly important to military usage and the conversion of rockets into missiles.

84. Sogame, "H-1 roketto," 487.

85. Watanabe "The Evolution of Japanese Space Policy," 285–286.

86. 63rd NDHR SST., No. 6, April 2, 1970, https://kokkai.ndl.go.jp/simple/detail?minId=106303913X00619700402&spkNum=5&current=1#s5; 63rd NDHR SST, Promotion Measures Subcommittee on Fundamental Issues in Space Development, No. 1, May 15, 1970, https://kokkai.ndl.go.jp/simple/detail?minId=106303917X00119700515&spkNum=2&current=2#s2.

87. Logsdon, "U.S.-Japanese Space Relations," 279, 298; *Keidanren Geppō* 19, no. 9 (September 1971): 60–61; "Nihon kigyō mo uchū jikken o Keidanren ga setsumeikai—uchūrenrakusen," *Asahi Shimbun*, September 14, 1978.

88. Itsurō Kimura, Hiroshi Nakakuchi, Yasuhiko Aihara, Kan'ichirō Katō, Kobayashi Shigeo, Tōru Tanabe, "Monbushō Kagaku kenkyū hiho jokin (ippan kenkyū A) shonendo hōkoku, sono 1: Supēsushatoru ni kansuru kenkyū" (Tokyo: Kimura Itsurō hoka, 1973); Kagakugijutsuchō kenkyū chōseikyoku, "Zairyō kagaku no bun'ya ni okeru supēsu shatoru no riyō ni kansuru chōsa no gaiyō" (Tokyo: Kagakugijutsuchō kenkyū chōsei-kyoku, 1973).

89. Watanabe "The Evolution of Japanese Space Policy," 281.

90. 84th NDHR SST, No. 6, March 29, 1978, https://kokkai.ndl.go.jp/simple/detail?minId=108403913X00619780329&spkNum=182&current=5#s182.

91. 84th NDHR Cab., No. 11, April 11, 1978, https://kokkai.ndl.go.jp/simple/detail?minId=108404889X01119780411&spkNum=46&current=7#s46.

92. Minoru Oda, "Uchū keikaku no mirai—supēsu shatoru nit suite (Danwa shitsu)," *Nihon butsurigakkaishi* 31, no. 2 (February 1976): 93–95.

93. Nitta, "Shōrai no uchū riyō gijutsu," 353; 84th NDHC, Settlement of Accounts Committee, No. 4 after adjournment, September 1, 1978, https://kokkai.ndl.go.jp/simple/detail?minId=108414103X00419780901&spkNum=113&current=8#s113; 63rd NDHR SST, Promotion Measures Subcommittee on Fundamental Issues in Space Development, No. 1, May 15, 1970, https://kokkai.ndl.go.jp/simple/detail?minId=106303917X00119700515&spkNum=2&current=2#s2; Teru Ariga, "Uchū jikken/Nihon no shatoru riyou," *Kesoku to seigyo* 23, no. 1 (January 1983): 43–48.

94. Wheelan. "A 'Born-Again' Space Program," 142–150.

95. Ogata et al., "Eisei tsūshin no ayumi to kongo no kadai," 408.

96. Wray, "Japanese Space Enterprise," 470, 474.

97. 104th Diet NDHC S&T, No. 4, April 11, 1986, https://kokkai.ndl.go.jp/simple/detail?minId=110413928X00419860411&spkNum=28&current=1504#s28.

98. "The Application of the Super 301 Provisions to Japan—Statement of Ambassador Carla A. Hills, May 25, 1989," in *Nichibei kankei shiryo shu: 1945–97* (Tokyo: Tōkyō Daigaku Shuppankai, 1999), 1163–1164.

99. "Jishu kaihatsu rosen ni tenkanki kanmin ainori fukanō ni Nichibei jinkō eisei mondai ga ketchaku," *Asahi Shimbun*, April 7, 1990.

100. Steven Brull, "Japan Satellite Market's Opening to U.S. Is a Dud for Both Sides," *The New York Times*, March 12, 1994.

101. 118th NDHR Bud., First Subcommittee, No. 2, April 27, 1990, https://kokkai.ndl.go.jp/simple/detail?minId=111805266X00219900427&spkNum=142&current=7#s142.

102. Masahiko Sato, Toshio Kosuge, and Peter van Fenema, *Legal Implications on Satellite Procurement and Trade Issues between Japan and the United States* (Paris: International Institute of Space Law, 1999), 235, 239.

103. Lebeau and Reuter, "La place des techniques spatiales dans le développement économique," 16.

104. "Supēsu shatoru gijutsu, nichiō de kyōdō kaihatsu ōshū uchūkikan to kagichō ga gōi," *Asahi Shimbun,* February 25, 1992.

105. Merkle et al., *JTEC Panel Report,* xii.

106. "H-2 roketto zunō mo shinzō mo kokusan de—Godai Tomifumi shi (90 nendai o niramu)," *Asahi Shimbun,* February 21, 1985

107. Nitta, "Shōrai no uchū riyō gijutsu," 353–355; Floyd A. Wyczalek, "Assessment of Aerospace Technology in Japan Viewed from an American Perspective," *SAE Transactions* 100, Section 1: Journal of Aerospace, Part 2 (1991): 1947–1954.

108. Higuchi, "Uchū sutēshon keikaku to Nihon jikken mojūru no kōsō," 438.

109. Wyczalek, "Assessment of Aerospace Technology in Japan," 1947–1954.

110. Ookami Yoshiaki, "Uchū robotto purojekkuto," *Nihon Robotto gakkaishi,* 7 no. 1 (February 1989): 82–83.

111. Sudō and Katō, "Tsukuba uchū sentā no shōkai," 31.

112. 96th NDHR Bud., No.19, March 9, 1982, https://kokkai.ndl.go.jp/simple/detail?minId=109605261X01919820309&spkNum=85&current=8#s85; 101st NDHR Bud., No. 7, February 18, 1982, https://kokkai.ndl.go.jp/simple/detail?minId=110105261X00719840218&spkNum=287&current=5#s287.

113. "Uchū geigeki heiki shisutemu kaihatsu-hi o morikomu—daitōryō yosan kyōsho," *Asahi Shimbun,* January 27, 1984; 101st NDHR SST, No. 6, March 27, 1980, https://kokkai.ndl.go.jp/simple/detail?minId=110103911X00619840327&spkNum=161&current=13#s161.

114. 101st NDHR Bud., No.7, February 18, 1984.

115. 101st NDHR Bud., No 16, March 3, 1984, https://kokkai.ndl.go.jp/simple/detail?minId=110105261X01619840303&spkNum=285&current=16#s285.

116. 101st NDHR Set., No. 3, March 26, 1984, https://kokkai.ndl.go.jp/simple/detail?minId=110104103X00319840326&spkNum=243&current=12#s243.

117. 118th NDHC S&T, No. 4, May 31, 1990, https://kokkai.ndl.go.jp/simple/detail?minId=111813928X00419900531&spkNum=14&current=10#s14.

118. H. Toyota, S. Tanaka, T. Sugimura, and Y. Nakayama, "Shōwa 61 nen Izu Ōshima funka ni kakari waru rimōto senshingu," *Nihon rimōto senshingugaku kaishi* 6, no. 4 (1986): 365–400.

119. Joanne Simpson, ed., *Report of the Science Steering Group for a Tropical Rainfall Measuring Mission* (Greenbelt, MD: Goddard Space Flight Center, August 1988), iv; NASA, "The Tropical Rainfall Measuring Mission," https://gpm.nasa.gov/missions/trmm.

120. "Uchū kaihatsu kikaku, minkan no eisei tōnyū ni wa furezu," *Asahi Shimbun,* March 14, 1985.

121. James L. Green, Robert E Maguire, and Brian Lopez-Swafford, *A Communications Model for an ISAS to NASA Span Link* (Washington, DC: NASA, January 1987).

122. 101st NDHR Bud., No.7, February 18, 1984.

123. 101st NDHR Set., No. 3, March 26, 1984.

124. "1996 nen uchū jikkenshitsu—kokusai kyōryoku no uchū kichi—Nihon no mojūru yosan keikaku ga katamaru," *Asahi Shimbun*, April 8, 1988.

125. "Shūin-i, uchū kichi kyōtei de 'heiwa riyō' o ketsugi konkokkaichū ni shōnin mo, 1989-nen 06 tsuki 14-nichi," *Asahi Shimbun*, April 6, 1989.

126. Tariq Malik, "Japan Prepares Space Station's Largest Laboratory for Flight," Space .com, May 2, 2007, https://www.space.com/3750-japan-prepares-space-station-largest-laboratory-flight.html.

127. Mark A. Garcia, ed., "Japanese Experimental Module Kibo," NASA, https://www .nasa.gov/international-space-station/japanese-experiment-module-kibo/.

128. 132nd NDHR For., No. 17, May 12, 1995, https://kokkai.ndl.go.jp/simple/detail?minId=113203968X01719950512&spkNum=18&current=26#s18.

129. Nitta, "Shōrai no uchū riyō gijutsu," 355.

130. Stephen Clark, "In Exchange for a Lunar Rover, Japan Will Get Seats on Moon-landing Mission," *Ars Technica*, April 11, 2024, https://arstechnica.com/space/2024/04/japan-will-be-first-among-nasas-partners-to-have-an-astronaut-on-the-moon/

Chapter 8

1. "H-2 roketto mata ashibumi kishōsei uchiage ni eikyō," *Asahi Shimbun*, June 30, 1992.

2. There is a convincing, but generally disregarded (in Japanese financial circles, at least), argument that the Bank of Japan's activities in this period were primarily intended to establish its own independence from political influence and control over key element of Japan's economy, best articulated in Richard A. Werner's *Princes of the Yen* (London: Routledge, 2003).

3. Reiko Aoki, "A Demographic Perspective on Japan's 'Lost Decades,'" *Population and Development Review* 38 (2012): 104.

4. Chiaki Moriguchi and Emmanuel Saez, "The Evolution of Income Concentration in Japan, 1885–2002: Evidence from Income Tax Statistics" (Econometrics Laboratory, University of California, Berkeley, August 25, 2005), table 1, https://eml.berkeley.edu/~saez/moriguchi-saez05japan.pdf.

5. 104th Diet NDHC S&T, No. 4, April 11, 1986; 120th NDHC Com., No. 12, April 25, 1991, https://kokkai.ndl.go.jp/simple/detail?minId=112014816X01219910425&spkNum=134&current=14#s134.

6. 112th NDHR S&T, No. 1, March 1, 1988, https://kokkai.ndl.go.jp/simple/detail?minId=111203911X00119880301&spkNum=6&current=1#s6; Hiroyuki Ueda, "Uchiage o

motsu H-II roketto—uchiage hi 1994 nen 8 gatsu 17 nichi," *Suiso enerugiishisutemu tokushū H-II roketto—Tanegashima no uchiage setsubi no shōkai* 19, no.1 (1994).

7. Shibato and Miwada, "H-II roketto ato no uchū yusō shisutemu," 355; "Kyōsō to kyōryoku to uchū bujinesu josō," *Asahi Shimbun*, January 26, 1991.

8. 104th Diet NDHC S&T, No. 4, April 11, 1986.

9. CIA Directorate of Intelligence, *Japan: Participation in the Civilian Space Programs*, 1–5.

10. Shibato and Miwada, ""H-II roketto ato no uchū yusō shisutemu 'roketto purēn,' " 357.

11. Justin Ray, "Atlas 5 Thunders to Milestone in U.S. Rocket History," *Spaceflight Now*, June 20, 2012, https://spaceflightnow.com/atlas/av023/#:~:text=%22The%20U.S.%20Air%20Force%20has,total%20%246%20billion%20development%20cost; G. Carra and J. de Dalmau, "Europe Ready for Ariane-5 Production," https://www.esa.int/esapub/bulletin/bullet93/b93carr.htm; "Bunkyō kagaku chūgaku 40 nin gakkyū dōnyū (6 1 nen yosando seifuan no naiyō)," *Asahi Shimbun*, December 29, 1985; "Senkyo to shakkin taisaku o yūsen kokusai jōsei no ninshiki futōmei 90nendo yosa," *Asahi Shimbun*, December 25, 1989; *Asahi Shimbun*, June 30, 1992.

12. ISAS, *Uchū kūkan kansoku 30 nen shi*, 62.

13. *Asahi Shimbun*, November 2, 1985; *Asahi Shimbun*, February 21, 1985. Ariane-IV was capable of lifting around 4,795 lb (2,175 kg) into GEO at a cost of around $85 million.

14. Shibato and Miwada, "H-II roketto ato no uchū yusō shisutemu 'roketto purēn,' " 354–355; Matsumoto Shinji, "Uchū kaihatsu to kensetsu," 2.

15. 118th Diet NDHC S&T, No. 3, May 30, 1990, https://kokkai.ndl.go.jp/simple/detail?minId=111813928X00319900530&spkNum=27&current=9#s27; 120th NDHC Com., No. 12, April 25, 1991.

16. NASDA, *H-II roketto kaihatsu ni okeru uchū kaihatsu jigyōdan to keiyaku kigyō no yakuwari* (December 22, 1999), https://warp.da.ndl.go.jp/info:ndljp/pid/233892/www.nasda.go.jp/press/1999/12/rocket_991222_2_j.html; "H 2 chōtatsu de shingaisha setsuritsu 77-sha no shusshi de, 7 gatsu ni," *Asahi Shimbun*, April 29, 1990.

17. "Japanese Firm to Launch Hughes Satellites," *Los Angeles Times*, November 27, 1996.

18. Kakuda City, "Kakudashi supēsutawā kosumohausu," https://www.city.kakuda.lg.jp/soshiki/14/624.html.

19. "Ochimasen yō ni Chikushino no JR Futsukaichi eki ni janbo ema Fukuoka," *Asahi Shimbun*, December 25, 1999.

20. NASDA, "Tanegashima 'uchū kagaku gijutsu-kan' no shinsō kaikan ni tsuite" (March 11, 1997), https://warp.da.ndl.go.jp/info:ndljp/pid/233892/www.nasda.go.jp/press/1997/03/tane970311-01.html.

21. MHI, *Umi ni riku ni soshite uchū e*, 1055.

22. Abe Shigeo, Tsuruta Michio, and Nakatani Tsuneshi, "Tanegashima uchū sentā H-II roketto sha za hontai kiso kōji," *Tsuchi to kiso,* 39, no. 5 (May 1991): 59–65.

23. Takuya Ara, *Uchū kaihatsu to ekitai suisō* (Tokyo: Uchū kaihatsu jigyōdan enjin kaihatsu gurūpu, July 15, 1977).

24. MHI, *Umi ni riku ni soshite uchū e,* 737. Iwatani has been producing commercial quantities of hydrogen since 1979, and remains one of the world's premier producers of hydrogen fuels.

25. *Asahi Shimbun,* June 30, 1992.

26. *Asahi Shimbun,* February 21, 1985.

27. Godai Tomifumi "Uchū ken to roketto no koto," *ISAS News* 100 (July 1989): 8.

28. Wyczalek, "Assessment of Aerospace Technology in Japan," 1949; "Rēzājairo (haitekku no kao)," *Asahi Shimbun,* November 14, 1985.

29. *Asahi Shimbun,* February 21, 1985.

30. Tadakawa Tsuguo, "Japan's Launch Vehicle Program Update," 219–220; Merkle et al., *JTEC Panel Report,* xii, 41.

31. *Asahi Shimbun,* June 29, 1992; Keiichi Hasegawa, Kiyoshi Ando, Shouji Kitade, Mitsumasa Sakamoto, Yukio Fukushima, and Koichi Okita, "INCONEL 718 yōsetsu tsugite kōzō moderu shiken oyobi kaiseki ni yoru LE—7 rokettoenjin-nushi funsha-ki no kurīpu hirō sonshō hyōka," *Nihon kikai gakkai ronbunshū (A)* 61, no. 588 (August 1995): 1695–1700.

32. Y. Fukushima and T. lmoto, "Lessons Learned in the Development of the LE-5 and LE-7," 30th AIA/ASME/ISAE/ASEE Joint Propulsion Conference (Indianapolis, June 27–29, 1994), 2.

33. *Asahi Shimbun,* June 29, 1992; "Enjin mata kaji H2 roketto," *Asahi Shimbun,* September 27, 1990; Kenjiro Kamijo, Hitoshi Yamada, Norio Sakazune, and Shogo Warashina, "Developmental History of Liquid Oxygen Turbopumps for the LE-7 Engine," *Transactions of the Japan Society of Aeronautical and Space Science* 44, no. 145 (2001): 155–163.

34. *Asahi Shimbun,* September 27, 1990; "H 2 roketto no nenshō shiken chūdan," *Asahi Shimbun,* April 14, 1992; Kamijo et al., "Developmental History of Liquid Oxygen Turbopumps for the LE-7 Engine," 155–163.

35. "Yōsetsu-bu kiretsu, hokano enjin kara mo H2 roketto uchiage enki," *Asahi Shimbun,* July 10, 1992; 126th NDHR S&T, No. 3, February 23, 1993, https://kokkai.ndl.go.jp/simple/detail?minId=112603911X00319930223&spkNum=145&current=46#s145; Fukushima and Imoto, "Lessons Learned in the Development of the LE-5 and LE-7," 1–3.

36. 123rd NDHR Science and Technology Committee, No. 6, May 26, 1992, https://kokkai.ndl.go.jp/simple/detail?minId=112303911X00619920526&spkNum=269&current=31#s269.

37. "H 2 roketto, shiken-chū ni kasai 93nen haru no uchiage kon'nan ni," *Asahi Shimbun,* June 19, 1992; "Bakuhatsu ga gen'in ka? Enjin ga shū-ka kokusan etchitsūroketto jikken jikō,"

Asahi Shimbun, June 22, 1992; *Asahi Shimbun,* June 30, 1992; "Roketto H 2 no 'yowai shinzō' bakuhatsu enjin no shasshin kōkai," *Asahi Shimbun,* July 1, 1992.

38. Ueda, "Uchiageomotsu H-II roketto," 8, 10.

39. *Asahi Shimbun,* June 29, 1992.

40. "Tanegashima roketto uchiage kikan, nen 130 nichi ni enchō," *Asahi Shimbun,* June 11, 1997; "H 2 roketto, nankan no shiken toppa tashi ga—Teimei tsudzukeru eisei ichiba," *Asahi Shimbun,* July 9, 1993.

41. 112th NDHC Com., No. 7, April 19, 1988, https://kokkai.ndl.go.jp/simple/detail?minId=111214816X00719880419&spkNum=164&current=4#s164.

42. "Yane, kanzen ni fukitobu Genba kenshō Hajimaru Kakuda shi no roketto jikenjō bakuhatsu," *Asahi Shimbun,* May 17, 1991.

43. "Roketto kaihatsu kankeisha ni shōgeki—Kokachi no Mitsubishi jūkō jikō," *Asahi Shimbun,* August 10, 1991; "Funsha-ki no haikan no yōsetsu ni kekkan," *Asahi Shimbun,* October 19, 1991; "Jiko wa yoken funō Aichi kenkei ketsugo H 2 roketto haretsu-shi jikobō," *Asahi Shimbun,* August 7, 1992.

44. Ueda, "Uchiageomotsu H-II roketto," 3–4.

45. "H 2 uchiage seikō chōnan tanjō, nijū no yorokobi," *Asahi Shimbun,* February 5, 1994.

46. "Roketto keiki kidō shūsei e Kagoshima Minamitanecho," *Asahi Shimbun,* February 5, 1994. Okuda politely turned down the suggestion.

47. "Wagamono ni suru juku shite ita shinkansen no gijutsu (Shima Hideo no sekai: 1 kataru)," *Asahi Shimbun,* July 18, 1994.

48. Smith, "Non-US Approaches to Space Commercialisation," 23; MHI, *Umi ni riku ni soshite uchū e,* 345.

49. *Asahi Shimbun,* July 9, 1993.

50. "Yume no kitai H 2 roketto—100 ten manten hikō—nenshō kanryō ni hakushu," *Asahi Shimbun,* February 2, 1994.

51. 136th NDHR S&T, No. 2, February 22, 1996, https://kokkai.ndl.go.jp/simple/detail?minId=113603911X00219960222&spkNum=9&current=8#s9.

52. 123rd NDHC S&T, No. 2, February 26, 1992, https://kokkai.ndl.go.jp/simple/detail?minId=112313928X00219920226&spkNum=4&current=5#s4; "Go etsu dōyō shinogi o kezuru 2 shōchō no uchū roketto," *Asahi Shimbun,* January 21, 1993; "Jiki roketto wa anka de kogata ni kairyō o keikaku Kagakugijutsuchō," *Asahi Shimbun,* May 26, 1997; NASDA Evaluation Committee, *Space Transport Subcommittee Evaluation Report* (June 1999), https://warp.da.ndl.go.jp/info:ndljp/pid/233892/www.nasda.go.jp/press/1999/06/hyouka_990608_b_01_j.html; JAXA, "About J-I Launch Vehicle," https://global.jaxa.jp/projects/rockets/j1/index.html. Interestingly enough, in its brief life, the J-1 did serve to attract the ire of some left-wing lawmakers who worried about its potential use as an ICBM.

53. 132nd Diet NDHC S&T, No. 4, March 20, 1995, https://kokkai.ndl.go.jp/simple/detail?minId=113213928X00419950320&spkNum=80&current=78#s80.

54. "'H 2' sekai mo chūmoku uchiage ni kaigai 33-sha no shuzai (media)," *Asahi Shimbun*, February 8, 1994.

55. "Machi kōjō (zoku kauntodaun H2:3) (Nagoya)," *Asahi Shimbun*, January 14, 1994; "Tōkai no gijutsu-ryoku o kesshō H 2 shissoku, machikōjō ni mo hamo," *Asahi Shimbun*, December 26, 1999. Isomura is still around, manufacturing airplane parts; *Asahi Shimbun*, February 8, 1994.

56. NASDA, "H-IIA roketto LE-7A enjin no nenshō shiken ni tsuite" (March 24, 1997), https://warp.da.ndl.go.jp/info:ndljp/pid/233892/www.nasda.go.jp/press/1997/03/le-7a_970324_j.html; "Yamada Takaaki-san roketto shisutemu shachō (henshū-chō intabiyū)," *Asahi Shimbun*, March 1, 1997.

57. "Uchū kaihatsu ootsun o kaitei eisei no kosuto-gen nado hakaru," *Asahi Shimbun*, January 24, 1996; 140th NDHR S&T, No. 2, February 20, 1997, https://kokkai.ndl.go.jp/simple/detail?minId=114003911X00219970220&spkNum=2&current=35#s2.

58. Jun Onoda, "Launch Clients Sought for H-2A: Rocket Faces Strong International Competition, Dearth of Non-Government Business," *Yomiuri Shimbun*, May 18, 2012.

59. *Asahi Shimbun*, July 9, 1993.

60. "Uchū gijutsu ōuridashi Roshia, gaika kakutoku sakusen," *Asahi Shimbun*, July 19, 1992.

61. "Eisei uchiage kigyō no Shyaruru bigo kaichō rainichi," *Asahi Shimbun*, March 29, 1993.

62. 132nd NDHR S&T, No. 3, February 21, 1995, https://kokkai.ndl.go.jp/simple/detail?minId=113203911X00319950221&spkNum=7&current=71#s7.

63. NASDA, "H-II roketto 5-gōki ni yoru tsūshin hōsō gijutsu eisei (COMETS) no uchiage kekka no hyōka ni tsuite (hōkoku)" (Tokyo: NASDA, May 19, 1999), https://www.mext.go.jp/component/b_menu/shingi/giji/__icsFiles/afieldfile/2013/06/10/1335987_009_1.pdf. NASDA engineers were in the end able to salvage the satellite somewhat, but it never attained the correct orbit, and its functions were terminated just over a year later.

64. "Eisei bijinesu ni an'un—H2 roketto uchiage shikkei," *Asahi Shimbun*, February 22, 1998; NASDA, *H-II roketto 5-gōki ni yoru tsūshin hōsō gijutsu eisei aishō: Kakehashi no uchiage ni tsuite* (February 21, 1998), https://warp.da.ndl.go.jp/info:ndljp/pid/233892/www.nasda.go.jp/press/1998/02/h25_980221_riji_j.html.

65. NASDA, "Tanegashima uchū sentā ni okeru jikō hassei ni tsuite" (March 10, 1999), https://warp.da.ndl.go.jp/info:ndljp/pid/233892/www.nasda.go.jp/press/1999/03/accident_990310_j.html.

66. NASDA, "H-II roketto 8 banki no fuguai ni kakaru chōken jōkyō ni tsuite" (September 20, 1999), https://warp.da.ndl.go.jp/info:ndljp/pid/233892/www.nasda.go.jp/press/1999/09/h28_990920_j.html; NASDA, "H-II roketto 8 ban ki ni kakaru gokuteion tenken no kekka ni tsuite" (October 8, 1999), https://warp.da.ndl.go.jp/info:ndljp/pid/233892/www.nasda.go.jp/press/1999/10/h28_991008_j.html. Such sensors are required to monitor levels of liquid hydrogen and ensure there are no leaks from the rocket's main tank.

67. NASDA, "H-II roketto 8banki no uchiage shippai ni tsuite (zokuhō)" (November 17, 1999), https://warp.da.ndl.go.jp/info:ndljp/pid/233892/www.nasda.go.jp/press/1999/11/h28_991117_j.html; H. Momma, M. Watanabe, K. Mitsuzawa, K. Danno, Masahiko Ida, M. Arita, and I. Ujino, "Search and Recovery of the H-II Rocket Flight No. 8 Engine," in *Proceedings of the 2000 International Symposium on Underwater Technology* (Cat. No.00EX418) (Tokyo, 2000), 19–23.

68. Uchū kaihatsu iinkai tokubetsu kaigō, *Uchū kaihatsu iinkai tokubetsu kaigō hōkokusho—shippai no saihatsu bōshi no tame no kaikaku hōsaku* (NASDA: Tokyo, May 1999), 5, 8.

69. 132nd NDHC For., No. 14, May 30, 1995, https://kokkai.ndl.go.jp/simple/detail?minId=113213968X01419950530&spkNum=30&current=29#s30. The issue of insuring government satellites became a topic of strangely heated debate for some time in the later 1990s.

70. "'Uchū no machi' yume bōshi o machi-okoshi kitai, jimoto rakutan H 2 shippai," *Asahi Shimbun*, November 16, 1998.

71. 146th NDHR S&T, No. 5, November 24, 1999, https://kokkai.ndl.go.jp/simple/detail?minId=114603911X00519991124&spkNum=193&current=17#s193.

72. *Asahi Shimbun*, March 1, 1997; "Japan Drops Cornerstone of Program for Rockets," *The New York Times*, December 10, 1999; Asako Saegusa, "Japan's NASDA under Fire on Launch Failure," *Nature* 392, 321 (1998).

73. Kiriko Nishiyama, "Hughes Deals Crippling Blow to Japan's Rocket Program," *Space Daily*, May 25, 2000.

74. Shimodaira and Hamasaki, "Eisei gijutsu no genjō to shōrai," 299.

75. *Asahi Shimbun*, June 29, 1992.

76. Yasunori Matogawa, "Hatsuyume: Henshūgoki," *ISAS News* 238 (January 2001): 24.

77. "H 2 hassha chūshi was 'shikkei de nai' Uchukaihatsujigyōdan ijō o hitei," *Asahi Shimbun*, August 19, 1994.

78. "Saigo wa kami nomizo shiru uchiage semaru H2 roketto—jisshi sekininsha ni kiku," *Asahi Shimbun*, January 24, 1994.

79. NASDA, "Enjin no kaihatsu oyobi jikki seisan ni okeru intāfēsu Chōsei kanri" (March 23, 2000), https://warp.da.ndl.go.jp/info:ndljp/pid/233892/www.nasda.go.jp/press/2000/03/engin_000323_j.html.

80. NASDA, "H-II roketto 8 banki ni kataru rokuteion tenken no jissei ni tsuite" (October 5, 1999), https://warp.da.ndl.go.jp/info:ndljp/pid/233892/www.nasda.go.jp/press/2000/03/engin_000323_j.html; NASDA, "H-II roketto 8 banki uchiage shippai ni tsuite (zokuhō)" (November 17, 1999), https://warp.da.ndl.go.jp/info:ndljp/pid/233892/www.nasda.go.jp/press/1999/11/h28_991117_j.html.

81. Uchū kaihatsu iinkai tokubetsu kaigō, *Uchū kaihatsu iinkai tokubetsu kaigō hōkokusho*, 3.

82. NASDA, "Beikoku no saikin no roketto shippai ni tsuite" (September 9, 1998),

https://warp.da.ndl.go.jp/info:ndljp/pid/233892/www.nasda.go.jp/press/1998/09/rocket_980909_j.html; Space Development Agency, *Beikoku roketto uchiage renzoku shikkei ni tsuite* (May 26, 1999), https://warp.da.ndl.go.jp/info:ndljp/pid/233892/www.nasda.go.jp/press/1999/05/rocket_990526_j.html.

83. Akira Konno, "H-II roketto shippai no genin to sono kyōkun," *Atsuryoku gijutsu* 41, no. 6 (2003): 45–54.

84. NASDA, "Kakō no jikō, toraburu genin ni tsuite" (January 17, 2000), https://warp.da.ndl.go.jp/info:ndljp/pid/233892/www.nasda.go.jp/press/2000/01/h28_000117_a_j.html.

85. NASDA, "H-IIA roketto no kaihatsu keikaku ni tsuite" (May 24, 2000), https://warp.da.ndl.go.jp/info:ndljp/pid/233892/www.nasda.go.jp/press/2000/05/h2a_000524_j.html; NASDA. LE-7A gōdō kaihatsu chiimu no settei ni tsuite (May 24, 2000), https://warp.da.ndl.go.jp/info:ndljp/pid/233892/www.nasda.go.jp/press/2000/05/le7a_000524_j.html.

86. *Asahi Shimbun*, November 16, 1998.

87. "Nagoyamaru," *Asahi Shimbun*, November 18, 1999.

88. 104th Diet NDHC S&T, No. 4, April 11, 1986; 147th NDHC TIC, No. 11, April 18, 2000, https://kokkai.ndl.go.jp/simple/detail?minId=114714197X01120000418&spkNum=37&current=34#s37; *Asahi Shimbun*, June 29, 1992; "Kokusan roketto kaihatsu wa chakujitsu ni," *Asahi Shimbun*, August 12, 1992; 144th NHDC Set., First Issue after closing, December 17, 1998, https://kokkai.ndl.go.jp/simple/detail?minId=114414103X00119981217&spkNum=41&current=95#s41.

89. "Chūgoku uchū jikken, seikō de Nihon ni mizu," *Asahi Shimbun*, December 6, 1999.

90. "Yotō ridatsu hatsugen (ka ekubota)," *Asahi Shimbun*, November 28, 1999.

91. NASDA, *Uchū kaihatsu ni okeru saikin no jikō, toraburu ni tsuite (besshi)* (December 22, 1999), https://warp.da.ndl.go.jp/info:ndljp/pid/233892/www.nasda.go.jp/press/1999/12/trouble_991222_b1_j.html#1; "Seishi eisei Kiku 6-gō no shippai wa chjō shiken ga fujūbun chōsa kaname ga shiteki," *Asahi Shimbun*, December 13, 1994; 132nd NDHR Science and Technology Committee, No. 3, February 21, 1995.

92. NASDA, *Chikyū kansoku eisei 'Midori' kansoku kiki no ichi-bu dōsa fuchō ni tsuite* (December 18, 1996), https://warp.da.ndl.go.jp/info:ndljp/pid/233892/www.nasda.go.jp/press/1996/12/adeos_961218_j.html.

93. NASDA, *Chikyū kansoku purattofōmu gijutsu eisei 'Midori' no un'yō dan'nen ni tsuite* (June 30, 1997), https://warp.da.ndl.go.jp/info:ndljp/pid/233892/www.nasda.go.jp/press/1997/06/adeos_970630_2_j.html; "Adios, ADEOS: Japanese Satellite Lost," *Science* (July 1, 1997); Japan Space Exploration Agency, *About Advanced Earth Observing Satellite "Midori" (ADEOS)*, https://global.jaxa.jp/projects/sat/adeos/index.html.

94. "M-3S II-8 banki no hishō jōkyō," *ISAS News* 167 (February 1995), 7; JAXA, "Express," https://www.isas.jaxa.jp/en/missions/spacecraft/past/express.html. Interestingly

enough, the satellite's remains were eventually uncovered in Ghana—see 136th NDHR Science and Technology Committee, No. 2, February 22, 1996.

95. Tsuruta Koichiro, "Dai 1-kai Planet-B no mezasu kasei no kagaku—Kasei tansa nyūmon," *ISAS News* 203 (February 1998): 11; Kawaguchi Jun'ichiro, "On the Lunar and Heliocentric Gravity Assist Experienced in the Planet-B ('Nozomi')" (ISAS: Sagamihara, 1999); Nakatani Ichiro, "'Nozomi' chikyū dasshutsu no tenmatsu," *ISAS News* 215 (February 1999): 6.

96. Yasunori Matogawa, "Hatsuyume: Henshūgoki," *ISAS News* 238 (January 2001): 24; CBS News, "Not Much 'Hope' for Mars Probe," December 9, 2003.

97. Tsuruta Koichiro, "'Nozomi' kasei tōnyū enki to gaikoku no han'nō," *ISAS News* 216 (March 1999): 4.

98. ISAS, *Launch Vehicles: M-V,* https://www.isas.jaxa.jp/en/missions/launch_vehicles/m-v.html.

99. See, for example, Naoyuki Yoshino and Farhad Taghizadeh-Hesary, eds., *Japan's Lost Decade: Lessons for Asian Economies* (Singapore: Springer Singapore, 2017); Naomi Griffin and Kazuhiko Odaki, *Reallocation and Productivity Growth in Japan: Revisiting the Lost Decade of the 1990s* (Washington, DC: Congressional Budget Office, 2006).

Chapter 9

1. 123rd NDHC S&T, No. 4, April 6, 1992, https://kokkai.ndl.go.jp/simple/detail?minId=112313928X00419920406&spkNum=11&current=22#s11.

2. 129th NDHR Bud., No. 3, February 22, 1994, https://kokkai.ndl.jp/simple/detail?minId=112905261X00319940222&spkNum=364&current=81#s364; 114th NDHC Research Committee on Industry, Natural Resources and Energy, No. 2, April 5, 1989, https://kokkai.ndl.go.jp/simple/detail?minId=111414379X00219890405&spkNum=8&current=6#s8.

3. 114th NDHC S&T, No. 3, June 21, 1989, https://kokkai.ndl.go.jp/simple/detail?minId=111413928X00319890621&spkNum=24&current=13#s24; 132nd NDHR S&T, No. 2, February 15, 1995, https://kokkai.ndl.go.jp/simple/detail?minId=113203911X00219950215&spkNum=180&current=5#s180; 132nd Diet Foreign Affairs Committee, May 12, 1995; 141st NDHR S&T, No. 2, November 6, 1997, https://kokkai.ndl.go.jp/simple/detail?minId=114103911X00219971106&spkNum=29&current=63#s29.

4. 131st NDHR S&T, No. 3, November 1, 1994, https://kokkai.ndl.go.jp/simple/detail?minId=113103911X00319941101&spkNum=81&current=94#s81.

5. 147th NDHR S&T, No. 2, March 14, 2000, https://kokkai.ndl.go.jp/simple/detail?minId=114703911X00220000314&spkNum=168&current=18#s175.

6. 51st Diet NDHC S&T, No. 7, June 3, 1966, https://kokkai.ndl.go.jp/simple/detail?minId=105113913X00719660603&spkNum=38&current=5#s38.

7. 55th NDHR Committee on Accounts, No. 4, March 28, 1967.

8. 48th NDHR Cab., No. 17, March 16, 1965, https://kokkai.ndl.go.jp/simple/detail?minId=104804889X01719650316&spkNum=75&current=45#s75; 151st NDHR Bud., Fourth Subcommittee, No. 2, March 2, 2001, https://kokkai.ndl.go.jp/simple/detail?minId=115105270X00220010302&spkNum=71&current=41#s75.

9. 126th Diet NDHC S&T, No. 3, February 26, 1993, https://kokkai.ndl.go.jp/simple/detail?minId=112613928X00319930226&spkNum=59&current=48#s59. While these were the two major sets of concerns, others existed as well—for example, the impact that the unreliability of Japan's rockets would have on the country's ability to launch spy satellites; see 151st NDHC Committee on Accounts, No. 2, April 2, 2001, https://kokkai.ndl.go.jp/simple/detail?minId=115114103X00220010402&spkNum=52&current=55#s52.

10. 61st NDHR Plenary Session, No. 14, March 14, 1969, https://kokkai.ndl.go.jp/simple/detail?minId=106105254X01419690314&spkNum=14&current=21#s14.

11. "Jinzai, haba hiroku kesshū Uchūkaihatsujigyōdan no kōsō Kagakugijutsuchō—uchū kaihatsu," *Asahi Shimbun*, September 1, 1967; "Zōshō, shōnin no ikō 'Uchūkaihatsujigyōdan' no shinsetsu eisei nado gyōmu o ipponka," *Asahi Shimbun*, January 13, 1969.

12. Sato, "Managing the Interface," 200.

13. Yasunori Matogawa, "Nihon uchū kagaku no saikin no kekka," *Gakujutsu kenkyū no saisen hata tokushū: jirei shōkai* 1411 (July 1994): 38.

14. Matsubara and Ito, "Conceptual Study of H-II Orbiting Plane," 1719–1728; Wyczalek, "Assessment of Aerospace Technology in Japan," 1947–1954.

15. 46th NDHR Cab., No. 46, June 18, 1964, https://kokkai.ndl.go.jp/simple/detail?minId=104604889X04619640618&spkNum=19&current=40#s19.

16. Merkle et al., *JTEC Panel Report*, xii, 8.

17. 104th NDHC S&T, No. 4, April 11, 1986.

18. 131st NDHR Science and Technology Committee, No. 3, November 1, 1994.

19. 112th NDHR Science and Technology Committee, No. 1, March 1, 1988.

20. 142nd NDHC S&T, No. 17, April 23, 1998, https://kokkai.ndl.go.jp/simple/detail?minId=114215074X01719980423&spkNum=32&current=52#s32.

21. 146th NDHC TIC, No. 2, November 16, 1999, https://kokkai.ndl.go.jp/simple/detail?minId=114614197X00219991116&spkNum=178&current=44#s178.

22. 142nd NDHC S&T, No. 12, April 7, 1998, https://kokkai.ndl.go.jp/simple/detail?minId=114215074X01219980407&spkNum=38&current=45#s38. ¥685 billion is $10 billion in 2024.

23. "Jigyō-dan no hihan, hatsukaigō de funshutsu uchū kaihatsu kihonmondai kondankai," *Asahi Shimbun*, July 16, 1998.

24. NASDA, "Uchū kaihatsu ni okeru saikin no jikō, toraburu ni tsuite," December 22, 1999, https://warp.da.ndl.go.jp/info:ndljp/pid/233892/www.nasda.go.jp/press/1999/12/trouble_991222_j.html.

25. NASDA, "Uchū kaihatsu ni okeru saikin no jikō."

26. 144th NDHC Set., First issue after closing, December 17, 1998.

27. 132nd NDHR Science and Technology Committee, No. 3, February 21, 1995; Onoda Junichiro, 'M-V-4 jikken shunin,' *ISAS News* 228 (March 2000): 4.

28. Central Intelligence Agency, Directorate of Intelligence, *Japan: Participation in the Civilian Space Programs* (Washington, DC: CIA, March 18, 1985), 1–2; Lynn, "Japanese Research and Technology Policy," 298.

29. "Chijō sōchi no toraburu ka H 2 roketto 2-gōki no hassha-chū tome," *Asahi Shimbun*, August 19, 1994.

30. "H 2 uchiage hiyō shiharae uchū kaihatsu jigyōdan ga 35 oku man matome minji chōtei," *Asahi Shimbun*, October 16, 2000; 150th NDHR Soc., No. 1, November 9, 2000, https://kokkai.ndl.go.jp/simple/detail?minId=115004127X00120001109&spkNum=0&current=13#s72.

31. "Aibanri eisei fuguai-ki ni kanryō no arasoi," *Asahi Shimbun*, February 20, 1995.

32. Wray, "Japanese Space Enterprise," 467.

33. Wray, "Japanese Space Enterprise," 466–467.

34. 146th NDHR Plenary Session, No. 4, November 16, 1999, https://kokkai.ndl.go.jp/simple/detail?minId=114605254X00419991116&spkNum=51&current=42#s51.

35. Diane Vaughn, *The Challenger Launch Decision: Risky Technology, Culture, and Deviance at NASA* (Chicago: University of Chicago Press, 1996).

36. NASDA, "Uchū kaihatsu jigyōdan ni okeru hinshitsu kakunin no genjō ni tsuite" (January 17, 2000), https://warp.da.ndl.go.jp/info:ndljp/pid/233892/www.nasda.go.jp/press/2000/01/h28_000117_b_j.html.

37. Uchū kaihatsu iinkai tokubetsu kaigō, *Uchū kaihatsu iinkai tokubetsu kaigō hōkokusho*, 1.

38. 146th NDHR S&T, No. 3, November 17, 1999, https://kokkai.ndl.go.jp/simple/detail?minId=114603911X00319991117&spkNum=5&current=47#s54.

39. "Musekinin jigyō-dan (Mado, Ronsetsu wa yōin ka mamora)," *Asahi Shimbun*, August 3, 2000.

40. NASDA, "Kakō no jikō, toraburu genin ni tsuite."

41. MEXT, "Shippai chishiki katsuyō kenkyū kaihōkokusho—shippai keiken no sekigyokuteki katsuyō no tame ni," August 10, 2001, https://www.mext.go.jp/b_menu/shingi/chousa/gijyutu/001/toushin/010801.html.

42. NASDA, "Kakō no jikō, toraburu genin ni tsuite."

43. 142nd NDHR S&T, No. 2, March 11, 1998, https://kokkai.ndl.go.jp/simple/detail?minId=114203911X00219980311&spkNum=15&current=36#s58.

44. NASDA, *NEC ta 2-sha ni yoru kadai seikyū ni kakaru kore made no chōsa kekka oyobi kore o fumaeta henkan seikyū ni tsuite* (April 16, 1999), https://warp.da.ndl.go.jp/info:ndljp/pid/233892/www.nasda.go.jp/press/1999/04/tyousa_990416_j.htm; 144th NDHR Science and

Technology Committee, No. 2, December 11, 1998, https://kokkai.ndl.go.jp/simple/detail?minId=114403911X00219981211&spkNum=148&current=94#s148.

45. *Asahi Shimbun* January 21, 1993; https://www.mhi.com/jp/company/location/nagoyaguidew.

46. Uchū kaihatsu iinkai tokubetsu kaigō, *Uchū kaihatsu iinkai tokubetsu kaigō hōkokusho—shippai no saihatsu bōshi no tame no kaikaku hōsaku* (NASDA: Tokyo, May 1999), 17.

47. 149th NDHR S&T, No. 1, August 4, 2000, https://kokkai.ndl.go.jp/simple/detail?minId=114903911X00120000804&spkNum=8&current=6#s69.

48. "Mitsubishi jōkō to Ishiha no shitsujō H2 roketto 8 ki shippai no genin," *Asahi Shimbun*, June 19, 2000.

49. Uchū kaihatsu iinkai tokubetsu kaigō, *Uchū kaihatsu iinkai tokubetsu kaigō hōkokusho*, 13–14, 17. The notion of having a prime contractor responsible for the sourcing and integration of parts also received some support in political circles, but NASDA reports later cast doubts on the efficacy of this system. See NASDA, *Enjin no kaihatsu oyobi jikki seisan ni okeru intāfēsu Chōsei kanri* (23 March 2000), https://warp.da.ndl.go.jp/info:ndljp/pid/233892/www.nasda.go.jp/press/2000/03/engin_000323_j.html.

50. 146th NDHR Science and Technology Committee, No. 5, November 24, 1999.

51. Uchū kaihatsu iinkai tokubetsu kaigō, *Uchū kaihatsu iinkai tokubetsu kaigō hōkokusho*, 17–18, 20, 24, 31.

52. "Tenkanki ni tatsu uchū kaihatsu (1) Amerika no jijō' ni yureru Nihon (rensai)," *Yomiuri Shimbun*, July 1, 1991.

53. "Uchū 3 kikan tōgō e no kadai wa mokuhyō, igi o meikaku ni gijutsu no minkan iten mo hitsuyō (kaisetsu)," *Yomiuri Shimbun*, August 8, 2001.

54. Merkle et al., *JTEC Panel Report*, xii, 8; NASDA Evaluation Committee, *Space Transportation Subcommittee Evaluation Report;* "Nipponban mujin shatoru no jikki seisaku o tōketsu kagi-ten hōshin," *Asahi Shimbun*, August 2, 2000.

55. Uchū kaihatsu iinkai tokubetsu kaigō, *Uchū kaihatsu iinkai tokubetsu kaigō hōkokusho*, 1, 3–4.

56. "Uchū kaihatsu no kiban gijutsu kyōka mezasu—seifu bukai hōkoku," *Asahi Shimbun*, November 9, 2000.

57. NASDA, *Uchū kaihatsu benchā haitekku kaihatsu seido no hassoku ni tsuite* (March 30, 2000), https://warp.da.ndl.go.jp/info:ndljp/pid/233892/www.nasda.go.jp/press/2000/03/venture_000330_j.html.

58. 148th NDHR Science and Technology Committee, No. 1 August 4, 2000; Budget Committee, Fourth Subcommittee, No. 2, March 2, 2001.

59. NASDA, "Renkei kyōryoku no suishin ni kansuru kyōtei' no teiketsu ni tsuite" (April 6, 2001), https://www.jaxa.jp/press/nasda/2001/renkei_010406_j.html; 153rd NDHC Set., First Issue after closing, December 11, 2001, https://kokkai.ndl.go.jp/simple/detail?minId=115314103X00120011211&spkNum=163&current=27#s163.

60. National Diet of Japan, Japan Aerospace Exploration Agency Act, 2002, Article 4, https://elaws.e-gov.go.jp/document?lawid=414AC0000000161; Toshio Matsumoto, "Uchū 3 kikan no tōgō ni tsuite," *ISAS News* 246 (September 2001): 5.

61. Yasunori Matogawa, *Nippon uchū kaihatsu hishi—ganso tori ningen kara minkan roketto e* (Tokyo: NHK Shuppan, 2017), 190–191.

62. 154th NDHR LIT, No. 5, March 29, 2002, https://kokkai.ndl.go.jp/simple/detail?minId=115404319X00520020329&spkNum=23&current=45#s23.

63. University Space Engineering Consortium (UNISEC), *Report on Japanese University Rocket Projects* (Tokyo: UNISEC, October 2011), 7–25.

64. Saadia M. Pekkanen, "Japan in Asia's Space Race: Directions and Implications," in *Rich Region, Strong States: The Political Economy of Security in Asia: Policy Brief July 2013* (San Diego: SITC Research Briefs, Institute on Global Conflict and Cooperation, University of California, April 2013), 4.

65. "Nissan Motors to Sell Aerospace Unit to IHI," *The Japan Times*, February 14, 2000.

66. Pekkanen and Kallender-Umezu, *In Defense of Japan*, 79–80.

67. Hitoshi Yoshioka, "Uchū kaihatsu kyūsei no kakuritsu," in *Nihon no kagaku gijutsushō dai san ki*, ed. Nakayama Shigeru (Tokyo: Gakuyō Shobō, 1995), 175.

68. Bingen and Young, *From Earth to Uchū*, 14.

69. Pekkanen and Kallender-Umezu, *In Defense of Japan*, 79–92.

70. 134th NDHR S&T, No. 3, November 7, 1995, https://kokkai.ndl.go.jp/simple/detail?minId=113403911X00319951107&spkNum=28&current=35#s28; Young Astronauts Club, *Gaiyō* (2012), https://www.yac-j.or.jp/outline.html; *Asahi Shimbun*, December 1, 1999.

71. "Kōkyō jigyō kaikaku, michi nakaba jūten haibun han'ei sezu 2000nen seifu yosan-an," *Asahi Shimbun*, December 24, 1999.

72. 151st NDHR Cab., No. 10, May 16, 2001, https://kokkai.ndl.go.jp/simple/txt/115104889X01020010516/8.

Conclusion

1. Carl Sagan, *Murmurs of Earth: The Voyager Interstellar Record* (New York: Ballantine Books, 1979).

2. Siddiqi, *Beyond Earth*, 175–177.

3. Semiconductor Museum of Japan, "32-bit RISC Microprocessors/Controllers with Original Architecture (Hitachi)—Integrated Circuit," 1992, https://www.shmj.or.jp/english/pdf/ic/exhibi735E.pdf; Makoto Yoshikawa, Junichiro Kawaguchi, Akira Fujiwara, and Akira Tsuchiyama, "Hayabusa Sample Return Mission," in *Asteroids IV*, ed. P. Michel, F. E. DeMeo, W. F. Bottke, and Renée Dotson (Tucson: University of Arizona Press, 2015), 408.

4. Hitoshi Kuninaka, "Maikuroha hōden-shiki ion enjin ni yoru Hayabusa shōwakusei tansa-ki no shin'uchū dōryoku kōkō," *Science and industry* 87, no. 2 (February 2013): 50–54.

5. ISAS, "Shōwakusei tansaki 'Hayabusa,'" http://www.isas.jaxa.jp/missions/spacecraft/past/hayabusa.html.

6. Takanori Maema, "Zenkan ga iku! Kokusanki o tsukuru otokotachi—Hideo Itokawa (Nihon uchū kaihatsu, roketto no oto) (ue) roketto hakase no kōkū gijutsusha jidai," *Kōkū Jōhō* 61, 4, no. 811 (April 2011): 83–84.

7. Mika McKinnon, "Everything That Could Go Wrong for Hayabusa Did, and Yet It Still Succeeded," October 15, 2010, https://gizmodo.com/everything-that-could-go-wrong-for-hayabusa-did-and-ye-1730940605.

8. ISAS, "Hayabusa tansaki no jōkyō ni tsuite," December 7, 2012, http://www.isas.ac.jp/j/snews/2005/1207.shtml.

9. JAXA, "Restoration of Asteroid Explorer, HAYABUSA's Return Cruise," November 19, 2009, https://global.jaxa.jp/press/2009/11/20091119_hayabusa_e.html.

10. ISAS, "'Hayabusa' ni yoru kikan shiryō ni mizu o hakken," May 21, 2019, http://www.isas.jaxa.jp/topics/002156.html.

11. "The Moment HAYABUSA Became a Meteor," Junichi Watanabe, https://global.jaxa.jp/article/special/hayabusareturn/watanabe01_e.html.

12. NEC, "JAXA Launches Hayabusa 2 Asteroid Probe—NEC Conducts Manufacturing and Testing as Probe System Coordinator" (December 3, 2014), https://www.nec.com/en/press/201412/global_20141203_01.html.

13. ISAS, "Asteroid Explorer Hayabusa2," http://www.isas.jaxa.jp/en/missions/spacecraft/current/hayabusa2.html.

14. Bruno Victorino Sarli and Yuichi Tsuda, "Hayabusa 2 Extension Plan: Asteroid Selection and Trajectory Design," *Acta Astronautica* 138 (September 2017): 225–232.

15. Beyond Gravity, "Launcher Structures," https://www.beyondgravity.com/en/launchers/launcher-structures.

16. "Ipushiron, H 3 shippai tsudzuite mo . . . 'nebāgibuappu' 'kunan no nochi ni wa kanarazu hana ga' roketto hasshaba no jimoto, jakusa shokuin ni ēru 800-ri messēji no ōdanmaku okuru Kagoshima Kimotsuki machi," *Mainichi Shimbun*, March 28, 2023.

BIBLIOGRAPHY

Japanese Sources

14th Noshiro Space Event. "Setsuritsu chōshi." http://www.noshiro-space-event.org/about_sub01.html.

Abe, Shigeo, Tsuruta Michio, Nakatani Tsuneshi. "Tanegashima uchū sentā H-II roketto sha za hontai kiso kōji." *Tsuchi to kisō* 39, no. 5 (May 1991): 59–65.

Ariga, Teru. "Uchū jikken/Nihon no shatoru riyou." *Kesoku to seigyo* 23, no. 1 (January 1983): 43–48.

Asahi Shimbun Urawa shikyoku, ed. *Arakawa: Sono tsuchi to kokoro.* Tokyo: Asahi Sonorama, 1972.

Atachi, Sukezō. "Rō go yaku." In *Nihon kaigun nenryō sho (jō)—nenryō konwakai.* Tokyo: Hara Shobō, 1972.

Bōeichō gijutsu kenkyū honbu. "Uchū kūkō kenkyū kaihatu kikō (JAXA) no Niijima shishō tōmon o ukete." *Bōei-shō gijutsu kenkyū honbu kōhō* 503 (March 8, 1995): 2.

Bōeichō hōsei kenshūjō senshishitsu. *Rikugun kōkū no gunbi to unyō—Shōwa 31-nen shūki made.* Tokyo: Choūn shimbunsha, 1971.

———. *Sōsho senshi rikugun kōkū heiki: seisan, hokyū no kaihatsu.* Tokyo: Tokyo Publishing, 1975.

———. *Kaigun kōkū gaishi.* Tokyo: Choūn shimbunsha, 1976.

Chichibu kankō kyōkai. "Ryūsei matsuri." http://www.chichibuji.gr.jp/event/ryūsei/.

Dai-ichi rikugun gijutsu kenkyūjō. "Ichi giken san kenhou dai kyū ban (hō 101) shisei yon shiki ni ju shiki senchi funshinhō kenkyū hōkoku." March 1945. https://www.digital.archives.go.jp/das/image/F0000000000000218330.

———. "Shisei yon shiki yonjū sencha mokusei funshinhō, shisei yon shiki yonjuu sencha funshin ryūdan setsumeisho." March 1945. https://www.digital.archives.go.jp/das/image/F0000000000000218310.

Doi, Takao, Shigeki Kamigauchi, and Yoshiyuki Hasegawa. "Uchū hikō-shi no kunren ni kansuru ichi kōsatsu." *Nihon kōkū uchū gakkaishi* 43, no. 496 (May 1995): 267–274.

Fujihira, Hitoshi, ed. *Kimitsu heiki no zenbō tosho shippitsu henshū*. Tokyo: Hara Shobō, 1976.

Funakawa, Kenji. "Uchū kaihatsu jigyōdan no tuiseki kansei shisutemu." *Nihon kōkū uchū gakkaishi* 32, no. 364 (May 1984): 253–263.

Godai, Tomifumi. "Uchū ken to roketto no koto." *ISAS News* 100 (July 1989): 8.

Harada, Kazuo. *Dare ni mo wakaru kagaku zenshū dai kyū ken: saikin hatsumei romansu*. Tokyo: Kokumin Chizu, 1929.

Harada, Mitsuo. *Kodomo rigaku e-banashi. Dai 1-kan (jidōsha no maki)*. Tokyo: Kodomo Rigaku kai, 1932.

Hasegawa, Keiichi, Kiyoshi Ando, Shouji Kitade, Mitsumasa Sakamoto, Yukio Fukushima, and Koichi Okita, "INCONEL 718 yōsetsu tsugite kōzō moderu shiken oyobi kaiseki ni yoru LE—7 rokettoenjin-nushi funsha-ki no kuriipu hirō sonshō hyōka." *Nihon kikai gakkai ronbunshū (A)* 61, no. 588 (August 1995): 1695–1700.

Hashimoto, Masai. "Tenbun kyūta no seimitsu koi wo seizō no kyūmu." *Tenbun Geppō* 2, no. 12 (March 1910): 135–136.

Hashimoto, Takehiko. *Hikōki no tanjō to kūkiryoku gaku no keisei: okkateki kenkyū kaihatsu no kigen o motomete*. Tokyo: Tokyo Daigaku Shuppankai, 2012.

Higuchi, Chikayoshi. "Uchū sutēshon keikaku to Nihon jikken mojūru no kōsō." *Yōsetsu gakkaishi* 59, no.6 (1990): 438–445.

Hino, Kumao. "Kayakurui no bakusoku ni kansuru kenkyū." PhD diss., Tokyo Imperial University [University of Tokyo], 1945.

———. "Kayōsei enrui no kyōshū sokudōshiki." *Kōgyō kayaku kyōkaishi* 10, no. 4 (March 1950): 142–145.

Hino, Kumao, and Mitarashi Isamu. "Shōsan anmon shokuen kangōbutsu no kyūshō sokudo." *Kōgyō kayaku kyōkaishi* 10, no. 4 (March 1950): 145–148.

———. "Kōgyō yōdō kasen no nenshō sokudo riron." *Kōgyō kayaku kyōkaishi* 11, no. 1 (June 1950): 36–41.

———. "Tankō yō bakuyaku no keitai." *Kōgyō kayaku kyōkaishi* 11, no. 2 (September 1950): 71–75.

Hino, Kumao, Ita Ichio, and Yamane Kōichi. "Kōgyō yōdō kasen no nenshō tokusei." *Kōgyō kayaku kyōkaishi* 11, no. 2 (September 1950): 99–101.

———. "Shūyū sō happa no kisoshiki ni kansuru kenkyū." *Kōgyō kayaku kyōkaishi* 11, no. 3 (December 1950): 162–170.

———. "Kōgyō yōdenki raikan no hakka riron." *Kōgyō kayaku kyōkaishi* 11, no. 3 (December 1950): 178–184.

Hirosige, Tetu, ed. *Nihon shihonshugi to kagaku gijutsu*. Tokyo: Sanichi Shobō, 1962.

———. *Kagaku no shakaishi—kindai nihon no kagaku taisei*. Tokyo: Chūō Kōron, 1973.

———. *Kagaku no shakaishi (ka)*. Tokyo: Iwanami Shōten, 2003.

Hori, Shinzaburō. *Saki dzu detarame nise bunka o hōmure: Katei keizai jukyū keizai kokka keizai no hatan o sukū ni wa seikō-ai katei-ai kokka ai no fusoku o oginau ni wa*. Tokyo: Nihon sōgō kagaku kenkyūjo, 1933.

Horikoshi, Jiro, and Eiichi Iwaya. "96 shiki kanjō sentōki." *Kōkū jōhō Rinji zōkan: Nihon Suguru-ki monogatari* (April 1959): 104–115.

———. "96 shiki kanjō sentōki (jō)." *Kōkū jōhō 64*, no. 2 (February 2014): 60–65.

———. "96 shiki kanjō sentōki (ge)." *Kōkū jōhō 64*, no. 3 (March 2014): 60–65.

Hosokawa, Takashi, ed. *Rekishi gunzō shiriizu [Ketteihan]—Taiheyō sensō 3: 'Nanpō shigen' to Ranin shisen*. Tokyo: Gakken, 2009.

Hosoya, Chihiro, ed. *Nichibei kankei shiryō hen 1945–1997*. Tokyo: Tokyo Daigaku Shuppatsukai, 1992.

Inō, Tadatoshi. "Nihon no tetsudō gijutsu wo rēdo shita hitotachi: tōkaidō shinkansen no gijutsu wo kumitateta Shima Hideo." *Shinsenrō* 61, no. 6 (2007): 46–49.

ISAS. "Ohsumi." https://www.isas.jaxa.jp/en/missions/spacecraft/past/ohsumi.html.

———. "Usuda uchū kūkan kansokujō." https://www.isas.jaxa.jp/about/facilities/usuda.html.

———. "Sagamihara kyanpasu." https://www.isas.jaxa.jp/about/facilities/sagamihara.html.

———. "Shōwakusei tansakki 'Hayabusa.'" http://www.isas.jaxa.jp/missions/spacecraft/past/hayabusa.html.

———. *Uchū kūkan kansoku 30-nenshi*. Tokyo: Uchū kūkan kansoku 30-nenshi henshū iinkai, 1987.

———. "Hayabusa tansakki no jōkyō ni tsuite." December 7, 2012. http://www.isas.ac.jp/j/snews/2005/1207.shtml.

———. "'Hayabusa' ni yoru kikan shiryō ni mizu o hakken." May 21, 2019. http://www.isas.jaxa.jp/topics/002156.html.

———. "Nihon no uchū Kaihatsu no rekishi dai 2 sho—Uchinoura no tōjō—chōsa no kaihatsu to ohagi." https://www.isas.jaxa.jp/j/japan_s_history/chapter02/01/05.shtml.

———. "Nihon no uchū kaihatsu no rekishi dai 10 sho—takumashiki nakamatachi—Oosumi." https://www.isas.jaxa.jp/j/japan_s_history/chapter10/01/01.shtml.

Ishihara, Shintaro, and Akio Morita. *"No" to ieru Nihon*. Tokyo: Kōbunsha, 1989.

Itani, Shigeo. "Ro-go yaku no omoidashi." In *Nihon kaigun nenryō shi (ge)*. Tokyo: Hara Shobō, 1972.

Itō, Tetsuichi, and Nomoto Hideki, "Hōpu no kūryoku sekkei to gokuchō onsoku no kadai." *Nihon kōkū uchū gakkaishi* 38, no. 435 (April 1990): 94–198.

Itokawa, Hideo, and Yu Taisaka. "Sō hatsuki no naseru no hanryū ga shōkōda kōka oyobi sōantei ni oyobi su eikyō—[suihei biyoku no jōge ichi kettei shiryō dai ni-pō]." *Tōkyō teikoku daigaku kōkū kenkyūjo ihō* 201 (May 1941): 103–134.

Itokawa, Hideo. "Tsubasa jōmen ni shōzuru henryū o riyō shita shi hayakei." *Tōkyō teikoku daigaku kōkū kenkyūjo ihō* 154 (June 1937): 348–351.

———. "Yoko oyobi hōkō no anteisei oyobi sōju sei." *Nakajima kenkyū jōhō* 4, no. 1 (March 1939): 19–32.

———. "Yoko oyobi hōkō no anteisei oyobi sōju sei dai-2 hō." *Nakajima kenkyū jōhō* 4, no. 3 (September 1939): 163–176.

———. *Kōkū rikugaku no kiso to ōyō*. Tokyo: Kyōritsu shuppan, 1943.

———. "Gakugei—saikin kōkūki no jūyō kadai (kōkū-bi tokushū)." *Asahi Shimbun*, September 19, 1943.

———. *Kōkūki no shomondai*. Tokyo: Meiji Shobō, 1944.

———. "Sudeni soku I de shiken-zumi Amerika no roketto hikōki." *Asahi Shimbun*, January 12, 1944.

———. "V 2-gō seisōken kara sakaotoshi tekichide wa kakkū suru kai 'denshinbashira'—Hideo Itokawa-shi-dan." *Asahi Shimbun*, November 13, 1944.

———. "Shineiki no miwake kata—senzoomotsu mono motanu mono." *Shūkan shokokumin* 3, no. 50 (December 1944): 15.

———. "B 24, B 29 no seinō nihon hondo kūshū-yō o kogō." *Yomiuri Shimbun*, June 18, 1944.

———. "Ni juppun de Taiheiyō kōdan." *Mainichi Shimbun*, January 3, 1955.

———. *Uchū o sampō suru: roketto no zuihitsu*. Tokyo: Ryūnan Shobō, 1957.

———. "Penshiru roketto Kara kappa 8 kei made," *Seisan kenkyū* 12, no. 12 (December 1960): 13–22.

———. *Roketto*. Tokyo: NHK Books, 1965.

———. *Nihon sōseiron*. Tokyo: Yoshimasado Insatsu, 1990.

Iwaya, Eiichi, "Taiheiyō sensō mitsuwa kaitei ichi-man go-sen ri no misshi." *Gekkan Yomiuri* (June 1951): 37.

———. "Jettoki 'Kitsuka' monogatari." *Sekai no kūkō* 1, no. 2 (December 1951): 10–15.

———. "Roketto ki Shūsui ni tsuite." *Sekai no kōkū* 22, no. 9 (January 1952): 82–88.

———. "Nihon sentōki no ayunda michi." *Kōkū jōhō* (October 1958): 39–49.

———. "Ichi—arishi nichi no waga kaigun kōkūki no zenbō to sono jittai." In *Kōkū gijutsu no zenbō (ge)*, ed. Jun Okamura, 231–246. Tokyo: Hara Shobō, 1976.

Izumi-Kurotani, Akemi, Mayumi Ooya, Yoshihiro Mogami, Masamichi Yamashita, Makoto Okuno, and Shoji A. Baba. "Genseidōbutsu zōrimushi no yūei kōdō to jūryoku." *Uchū kagaku kenkyūjo hōkoku, tokushū: dai kikyū kenkyū hōkoku* 24 (December 1989): 33–47.

JAXA. "Uchū kaihatsu jigyoudan hō." https://www.jaxa.jp/library/space_law/chapter_1/1-1-2-10/index_j.html.

———. "MMX—Martian Moons Explorer." https://www.mmx.jaxa.jp/en/#:~:text=Ap

proximately%20one%20year%20after%20leaving,formation%20in%20the%20Solar %20System.

Kagakugijutsuchō kenkyū chōseikyoku. "Zairyō kagaku no bun'ya ni okeru supēsu shatoru no riyō ni kansuru chōsa no gaiyō." Tokyo: Kagakugijutsuchō kenkyū chōsei-kyoku, 1973.

Kagoshima Prefectural Office. "Kakukai kokusei chōsa toki no shishō sonbetsu jinkō no suii." https://www.pref.kagoshima.jp/ac09/tokei/bunya/kokutyo/h27kokutyo/documents/55416_20161128120608-1.pdf.

Kagoshima Prefectural Office, Dai-4 no jinkō dōkō. http://www.pref.kagoshima.jp/ap01/chiiki/kumage/chiiki/documents/71700_20190409114440-1.pdf.

Kaigun gijutsu kenkyū shō. "Roketto hikōki ni tsuite." December 17, 1941. National Diet Library, Tokyo.

Kakuda City. "Kakudashi supēsutawā kosumohausu." https://www.city.kakuda.lg.jp/soshiki/14/624.html.

Kaneko, Shōichirō. "Rekuiem Itokawa, Hideo." *Shōron* 321 (May 1995): 386.

Kaneyama, Shuichi. *Saikin kodomo kagaku dokuhon.* Tokyo: Bunka shobō, 1931.

Kanezawa, Iwao, ed. *Ningen Hideo Itokawa hakase to wa.* Tokyo: Hirohide Kōgei, 2003.

Kaishima, Keitarō. *Saikin no hikōki to shōrai.* Tokyo: Kaishima Keitarō, 1930.

Kichinaga, Susumu. "Roketto no haitatsu kōshi." *Gunji to gijutsu* 1, no. 121 (January 1937): 95–101.

Kimotsuki jōhōkyoku. "Roketto kichi to chiiki no kizuna o musubu Hashimoto-san." July 31, 2015. https://kankou-kimotsuki.net/archives/10494.

Kimura, Hidemasa, Kazuo Ochiai, and Shinaki Tsukuda. "Shokyū guraidā no chijō shōgeki sokutei kekka." *Nihon kōkū gakkaishi* 5, no. 45 (1957): 29–30.

Kimura, Hidemasa. "N-58 Shigunetto keihikōki." *Nihon kōkū gakkaishi* 9, no. 84 (1961): 29–30.

Kimura, Itsurō, Hiroshi Nakakuchi, Yasuhiko Aihara, Kan'ichirō Katō, Kobayashi Shigeo, Tōru Tanabe. "Monbushō Kagaku kenkyū hiho jokin (ippan kenkyū A) shonendo hōkoku, sono 1: Supēsushatoru ni kansuru kenkyū." Tokyo: Kimura Itsurō hoka, 1973.

Kimura Laboratory. "Minsei-hin o katsuyō shita jinkō eisei no kō chinō-ka jiritsu seigyo gazō shori ni kansuru jikken." https://www.rs.noda.tus.ac.jp/skimura/kimura_lab/Micro-OLIVe.html.

Kozawa, Shichibei. "Senji chu no Itokawa sensei." In *Ningen Itokawa Hideo hakase to wa,* ed. Kanezawa Iwao. Tokyo: Hirohide Kōgei, 2003.

Kukimoto, Takashi. "Tsuisō." *ISAS News,* April 1999, 16.

Kuninaka, Hitoshi. "Maikuroha hōden-shiki ion enjin ni yoru Hayabusa shōwakusei tansa-ki no shin'uchū dōryoku kōkō." *Science and Industry* 87, no. 2 (February 2013): 50–54.

Kurihara, Yoshitaka. "Wagakuni no seishi eisei." *Jōhō kanri* 21, no.2 (May 1978): 81–91.

Kusayagi, Daizō. "'Suzō sangyō' no sutā Itokawa Hideo." *Bungei shunju* 47, no. 7 (July 1969): 260–273.

Konno, Akira. "H-II roketto shippai no genin to sono kyoukun." *Atsuryoku gijutsu* 41, no. 6 (2003): 45–54.

Maema, Takanori. *Dangan Ressha*. Tokyo: Jitsugyō no nihonsha, 1994.

———. "Maema ga iku! Kokusanki o tsukuru otokotachi dai 9 kai—Itokawa, Hideo (Nihon uchū kaihatsu, roketto no oto) (jō) roketto hakase no kōkū gijutsusha jidai." *Kōkū jōhō* 61, 4, no. 811 (April 2011): 83–88.

———. "Maema ga iku! Kokusanki o tsukuru otokotachi—Itokawa, Hideo (Nihon uchū kaihatsu, roketto no oto) (ge) roketto hakase no kōkū gijutsusha jidai." *Kōkū jōhō* 61, 5, no. 812 (May 2011): 82–87.

Matogawa, Yasunori. "Uchū gijutsu no gunji riyō." *Nihon kagakusha* 17, no. 7 (July 1982): 41–43.

———. *Harē suisei no kagaku hoshizora no pasuwāku*. Tokyo: Shinchō Bunko, 1984.

———. *Harē ni idomu—suisei ni kakeru roman*. Tokyo: Dōbun shōin, 1985.

———. *Harē suisei no nazo*. Tokyo: Popurā shagakushō bunko, 1986.

———. *Otona no tame no hontō no uchū no hanashi*. Toyosu: PHP Kenkyūjō, 1992.

———. *Uchū ni ichiban chikai machi—Uchinoura no roketto hasshaba*. Kagoshima: Shunendō Shuppan, 1994.

———. "Uchū kaihatsu no rekishi to Nihon no yakuwari." *Kikai no kenkyū* 46, no. 11 (1994): 1109–1113.

———. "Nihon uchū kagaku no saikin no kekka." *Gakujutsu kenkyū no saisen hata tokushū: jirei shōkai* 1411 (July 1994): 38–40.

———. "Issho ni uchū e—jinrui shinka ni kado ni okeru yūjin hikō." *Kikai no kenkyū* 48, no. 1 (1996): 47–58.

———. "Sengo Fukkoki no Roketto Gijutsu." In *Nihon Roketto monogatari: noroshi kara uchū kanko made*, ed. Hirokyuki Osawa, 76. Tokyo: Mita Shuppankai, 1996.

———. *Uchū de kurasu tame no 69 no kiso chishiki*. Tokyo: Yamato shohō, 1999.

———. "Hatsuyume: Henshūgoki." *ISAS News* 238 (January 2001): 24.

———. "Matogawa hakase no uchū yomoyama hanashi: jinrui no shōrai no tenbōshita Itokawa Hideo." *Sekai jōhō* 85, no. 17 (May 4, 2004): 62–63.

———. "Itokawa Hideo no baiorin dzukuri." *Sekai shūhō* 85, no. 15 (April 20, 2004): 64.

———. *Gyakuten no Tsubasa: Penshiru roketto monogatari*. Tokyo: Shin Nippon Shuppansha, 2005.

———"20 seiki no kōkūjin (sono 17) Hideo Itokawa—Nihon no uchū kaihatsu no oto," *Sora to Bunka* 93 (Spring 2006).

———. *Nippon uchū kaihatsu hishi—ganso tori ningen kara minkan roketto e*. Tokyo: NHK Shuppan, 2017.

Matogawa, Yasunori, Noda Mashiro, and Kanda Akashi. "21 seiki no uchū kaihatsu." *Kagaku gijutsu jānaru* 3, no. 9 (September 1994): 10–15.

Matsuda, Kōyaku. "Roketto no Shōrai." *Gunji to gijutsu* 5, no. 8 (August 1931): 43–48.

Matsui, Junji, and Masakane Shimizu. "Uchū kaihatsu jigyōdan no sesturitsu." *Toki no hōrei* 692 (October 1969): 24–33.

Matsukawa, Shōzō, Hideo Itokawa, Hiroshi Miyazaki. "Purantoru tiichensu ryūtai rikigaku." *Kikai gakkaishi* 252, no. 41 (April 1938): 178.

Matsumoto, Leiji, dir. *Space Battleship Yamato*. Toho, 1976.

———. *Space Battleship Yamato*. Yomiuri TV, 1974–1975.

Matsumoto Shinji. "Uchū kaihatsu to kensetsu." *Doboku gakkai ronbunshū* 422, no. 14 (October 1990): 1–10.

Matsumoto, Toshio. "Uchū 3 kikan no tōgō ni tsuite." *ISAS News* 246 (September 2001): 5.

Matsuoka, Hisamitsu. *Nihonsho no roketto sentōki Shūsui*. Tokyo: Miki Shobō, 2004.

Matsuura, Shinya. *Kokusan roketto hwas naze ochirunoka*. Tokyo: Nikkei BP Shuppansha, 2004.

MEXT [Monbukagakushō]. "Shippai chishiki katsuyō kenkyū kaihōkokusho—shippai keiken no sekigyokuteki katsuyō no tame ni." Tokyo: MEXT, August 10, 2001. https://www.mext.go.jp/b_menu/shingi/chousa/gijyutu/001/toushin/010801.html.

Mitsubishi Heavy Industries [Mitsubishi jukōgyō]. "Nagoya yūdō suishin shisutemu seisakusho." https://www.mhi.com/jp/company/location/nagoyaguidew.

———. *Mitsubishi jukōgyō kabushikigaisha shi*. Tokyo: Mitsubishi jūkōgyō kabushiki kaisha, 1953.

———. *Umi ni riku ni soshite uchū e: zoku Mitsubishi Jūkōgyō shashi 1964–1989*. Tokyo: Mitsubishi jūkōgyō kabushiki kaisha, 1990.

Miyazawa, Masafumi. "Waga kuni ni okeru jitsuyou roketto no Kaihatsu to gijutsu dōnyū." *Nihon kōkū uchū gakkaishi* 39, no. 445 (February 1992): 55–68.

Mizusawa, Akira. "Tenbō: Nihon no senji kagaku gijutsu taisei." *Kagakushi kenkyū* 52, no. 266 (2013): 65–69.

Mori, Masuo. "Kaiso dampen." *Nippon kaigun nenryōshi (ge)*. Tokyo: Hara Shobō, 1973.

Moro, Akio. "Kō shutsuryoku roketto nenryō." *Nenryō kyōkaishi* 68, no. 11 (1989): 994–1002.

Nabeshima, Shigeaki. "Kaigun nenryōshō chōkyū kaijukyoku chō jidai no omoidashi." *Nippon kaigun nenryōshi (ge)*. Tokyo: Hara Shobō, 1973.

Nakabe, Hakuo, Toshiaki Takesaki, Yuka Ōno. "Uchū Kaihatsu 60-nen shi. 1955-nen —2015-nen." In *Proceedings of Space Transportation Symposium FY2016*. Tokyo: Institute of Space and Astronautical Science, Japan Aerospace Exploration Agency, 2017.

Nakahara, Nobuhei. "Kaigun nenryō shō ni kanrensuru omoidashi." *Nippon kaigun nenryōshi (ge)*. Tokyo: Hara Shobō, 1973.

Nakanishi, Fujio. "On the Yield Point of Mild Steel." PhD diss., Tokyo Imperial University [University of Tokyo], 1931.

———. "Hikōki no roketto suishin ni tsuite." *Nihon kōkū gakkaishi* 5, no. 39 (July 1938): 669–678.

Nakanishi, Fujio, and Kazuo Sato. "Ōryoku jōtai to sōsei henkei no katachi." *Nihon kikaigaku hōron bunshū* 26, no. 170 (1960–1961): 1327–1332.

Nakatani, Ichiro. "'Nozomi' chikyū dasshutsu no tenmatsu." *ISAS News* 215 (February 1999): 6.

Nakayama, Shigeru, ed. "Haisen chokugo no kagaku gijutsukai no jittai." In *Nihon no kagaku gijutsu*, 179–188. Tokyo: Gakuyo Shobō, 1995.

NASDA [Uchū kaihatsu jigyōdan]. "Chikyū kansoku eisei 'Midori' kansoku kiki no ichi-bu dōsa fuchō ni tsuite." Tokyo: NASDA, December 18, 1996. https://warp.da.ndl.go.jp/info:ndljp/pid/233892/www.nasda.go.jp/press/1996/12/adeos_961218_j.html.

———. "Tanegashima 'uchū kagaku gijutsu-kan' no shinsō kaikan ni tsuite." Tokyo: NASDA, March 11, 1997. https://warp.da.ndl.go.jp/info:ndljp/pid/233892/www.nasda.go.jp/press/1997/03/tane970311-01.html.

———. "H-IIA roketto LE-7A enjin no nenshō shiken ni tsuite." Tokyo: NASDA, March 24, 1997. https://warp.da.ndl.go.jp/info:ndljp/pid/233892/www.nasda.go.jp/press/1997/03/le-7a_970324_j.html.

———. "Chikyū kansoku purattofōmu gijutsu eisei 'Midori' no un'yō dan'nen ni tsuite." Tokyo: NASDA, June 30, 1997. https://warp.da.ndl.go.jp/info:ndljp/pid/233892/www.nasda.go.jp/press/1997/06/adeos_970630_2_j.html.

———. "Beikoku no saikin no roketto shippai ni tsuite." Tokyo: NASDA, September 9, 1998. https://warp.da.ndl.go.jp/info:ndljp/pid/233892/www.nasda.go.jp/press/1998/09/rocket_980909_j.html.

———. "Tanegashima uchū sentā ni okeru jikō hassei ni tsuite." Tokyo: NASDA, March 10, 1999. https://warp.da.ndl.go.jp/info:ndljp/pid/233892/www.nasda.go.jp/press/1999/03/accident_990310_j.html.

———. "NEC ta 2-sha ni yoru kadai seikyū ni kakaru kore made no chōsa kekka oyobi kore o fumaeta henkan seikyū ni tsuite." Tokyo: NASDA, April 16, 1999. https://warp.da.ndl.go.jp/info:ndljp/pid/233892/www.nasda.go.jp/press/1999/04/tyousa_990416_j.html.

———. "H-II roketto 5-gōki ni yoru tsūshin hōsō gijutsu eisei (Komettsu) no uchiage kekka no hyōka ni tsuite (hōkoku)." Tokyo: NASDA, May 19, 1999. https://www.mext.go.jp/component/b_menu/shingi/giji/__icsFiles/afieldfile/2013/06/10/1335987_009_1.pdf.

———. "H-II roketto 8 banki no fuguai ni kakaru chōken jōkyō ni tsuite." Tokyo: NASDA, September 20, 1999. https://warp.da.ndl.go.jp/info:ndljp/pid/233892/www.nasda.go.jp/press/1999/09/h28_990920_j.html.

———. "H-II roketto 8-banki ni kataru rokuteion tenken no jissei ni tsuite." Tokyo: NASDA,

October 5, 1999. https://warp.da.ndl.go.jp/info:ndljp/pid/233892/www.nasda.go.jp/press/2000/03/engin_000323_j.html.

———. "H-II roketto 8 ban ki ni kakaru gokuteion tenken no kekka ni tsuite." Tokyo: NASDA, October 8, 1999. https://warp.da.ndl.go.jp/info:ndljp/pid/233892/www.nasda.go.jp/press/1999/10/h28_991008_j.html.

———. "H-II roketto 8 banki no uchiage shippai ni tsuite (zokuhō)." Tokyo: NASDA, November 17, 1999. https://warp.da.ndl.go.jp/info:ndljp/pid/233892/www.nasda.go.jp/press/1999/11/h28_991117_j.html.

———. "H-II roketto kaihatsu ni okeru uchū kaihatsu jigyōdan to keiyaku kigyō no yakuwari." Tokyo: NASDA, December 22, 1999. https://warp.da.ndl.go.jp/info:ndljp/pid/233892/www.nasda.go.jp/press/1999/12/rocket_991222_2_j.html.

———. "Uchū kaihatsu ni okeru saikin no jikō, toraburu ni tsuite." Tokyo: NASDA, December 22, 1999. https://warp.da.ndl.go.jp/info:ndljp/pid/233892/www.nasda.go.jp/press/1999/12/trouble_991222_j.html.

———. "Kakō no jikō, toraburu genin ni tsuite." Tokyo: NASDA, January 17, 2000. https://warp.da.ndl.go.jp/info:ndljp/pid/233892/www.nasda.go.jp/press/2000/01/h28_000117_a_j.html.

———. "Uchū kaihatsu jigyōdan ni okeru hinshitsu kakunin no genjō ni tsuite." Tokyo: NASDA, January 17, 2000. https://warp.da.ndl.go.jp/info:ndljp/pid/233892/www.nasda.go.jp/press/2000/01/h28_000117_b_j.html.

———. "Enjin no kaihatsu oyobi jikki seisan ni okeru intāfēsu Chōsei kanri." Tokyo: NASDA, March 23, 2000. https://warp.da.ndl.go.jp/info:ndljp/pid/233892/www.nasda.go.jp/press/2000/03/engin_000323_j.html.

———. "H-IIA roketto no kaihatsu keikaku ni tsuite." Tokyo: NASDA, May 24, 2000. https://warp.da.ndl.go.jp/info:ndljp/pid/233892/www.nasda.go.jp/press/2000/05/h2a_000524_j.html.

———. "Renkei kyōryoku no suishin ni kansuru kyōtei' no teiketsu ni tsuite." Tokyo: NASDA, April 6, 2001. https://www.jaxa.jp/press/nasda/2001/renkei_010406_j.html.

NASDA Evaluation Committee. *Uchū kaihatsu jigyōdan hyōka iinkai uchū yusō bukai hyōka hōkoku-sho*. Tokyo: NASDA, June 1999. https://warp.da.ndl.go.jp/info:ndljp/pid/233892/www.nasda.go.jp/press/1999/06/hyouka_990608_b_01_j.html.

National Diet of Japan. NASDA Foundation Law (Uchū kaihatsu jigyōdan hō). June 23, 1969. https://www.shugiin.go.jp/internet/itdb_housei.nsf/html/houritsu/06119690623050.htm.

———. Japan Aerospace Exploration Agency Act (Kokuritsu kenkyū kaihatsu hōjin uchū kōkū kenkyū kaihatsukikō-hō). 2002. https://elaws.e-gov.go.jp/document?lawid=414AC0000000161.

National Institute of Information and Communications Technology (Uchū tsūshin kenkyū kikō). "Kashima uchū gijutsu sentā." http://ksrc.nict.go.jp/pdf/pamphlet2017-2.pdf.

NHK purojekutoekkusu seisaku hand, ed. *Purojekutoekkusu dai 2-kan fukkatsu e no butaiura.* Tokyo: NHK Shuppan, 2003.

Nihon kagakushi gakkai, ed. *Nihon kagaku gijutsushi taikei dai-14 ken.* Tokyo: Dai ippō ki shuppan, 1964.

Nihon kōtsū kyōkai, ed. *Kōtsū kankei shiryō dai 6 shō—Kagaku no kyōi roketto.* Tokyo: Nihon kōtsū kyōkai, 1936.

Nishizawa, Yūshichi. *Hanabi no kenkyū.* Tokyo: Uchida rōkakuho, 1928.

———. "Uber die Strukturviskosimetrischen Untersuchungen." PhD diss., Tokyo Imperial University [University of Tokyo], 1932.

———. *Kōryō oyobi kashōhin.* Tokyo: Kyōritsusha, 1934.

———. *Kagaku sen heiki.* Tokyo: Sanshōdo, 1939.

Nitta, Keiji. "Shōrai no uchū riyō gijutsu." *Terebishon gakkaishi* 41, no. 4 (1987): 353–358.

Nimiya, Mayu. "Nihon no Da Winchi Itokawa Hideo hakase wo shinobu." In *Ningen Itokawa Hideo hakase to wa,* ed. Kanezawa Iwao, 17–18. Tokyo: Hirohide Kōgei, 2003.

No. 1 Naval Fuel Depot, Ōfuna. "Asetoarudihido kōsei ni kansuru kenkyū." In *Kenkyū jikken seiseki hōkoku* 15. Yokohama: No. 1 Naval Fuel Depot, Ōfuna, November 9, 1943.

———. "Asetoarudihaido yori kurtonaruderu no kōei ne kansuru kenkyū." *Kenkyū jikken seiseki hōkoku* 159. Yokohama: No. 1 Naval Fuel Depot, Ōfuna, November 11, 1943.

Oda, Minoru. "Uchuu keikaku no mirai—supēsu shatoru nit suite (Danwa shitsu)." *Nihon butsurigakkaishi* 31, no. 2 (February 1976): 93–95.

———. *Uchū kūkan kanshu 30-nenshi nenhyō.* Tokyo: Uchū kagaku kenkyūjō, 1987.

Oda, Mitsuhige. "Uchū robotto ni okeru ningen kikakei no kadai." *Nihon kōkū uchū gakkaishi* 40, no. 464 (September 1992): 486–490.

Ogata, M., H. Mizusawa, and K. Irie. "Eisei tsūshin no ayumi to kongo no kadai." *Mitsubishi gunki gihō* 59, no. 6 (1985): 408–412.

Okamura, Jun. *Kōkū hijutsu no zenbō (jō).* Tokyo: Shippitsusha shōkai, 1976.

———. *Kōkū gijutsu no zenbō (ge).* Tokyo: Hara Shobō, 1976.

Okamura, Jun, and Eiichi Iwaya. *Nihon no kōkūki kaigunki hen.* Tokyo: Shuppan kyōdō-sha, 1960.

Ōkawa, Ryūho. *Roketto Hakase Hideo Itokawa—dokuseiteki Mirai Kagaku hassōhō.* Tokyo: Kyubunsha, 2014.

Okuda, Hideo. *Nakahara Nobuhei den.* Tokyo: Tōa Nenryō Kōgyō, 1982.

Ōkami, Yoshiaki. "Uchū robotto purojekkuto." *Nihon Robotto gakkaishi* 7, no.1 (February 1989): 78–86.

Onda, Shigetaka. *Tokkō.* Tokyo: Kodansha, 1991.

Onoda, Junichiro. "M-V-4 jikken shunin." *ISAS News* 228 (March 2000): 4.

Onoda Junjiro. "Uchinoura to watashi—Uchinoura 50-shūnen ni yosete." In ISAS, *Uchinoura uchū kūkan kansokujo no 50 nen 1962–2012.* Uchinoura uchū kūkan kansokujo no 50-nen' kinen-shi henshū iinkai, November 2011.

Osawa, Hiroyuki, ed. *Shinban Nihon roketto monogatari.* Tokyo: Seibundo shinkosha, 2003.

Oyama, Kōichiro. "Omoide no Uchinoura." In ISAS, *Uchinoura uchū kūkan kansokujo no 50 nen 1962–2012.* Uchinoura uchū kūkan kansokujo no 50-nen' kinen-shi henshū iinkai, November 2011.

Saito, Akio. "Roketto wo kataru." *Gunji to gijutsu* 96 (December 1934): 38–44.

Saito, Norio. "Uchū kaihatsu jigyōdan no gijitsu shiken eisei." *Nihon kikai gakkai ronbun shū (C shū)* 60, no. 580 (December 1994): 10–16.

Saito, Shigebumi. *Nihon uchū kaihatsu monogatari—kokusan eisei ni kaketa senku-sha-tachi no yume.* Tokyo: Mita Shuppankai, 1991.

———. "Japan's Space Policy: Background and Outlook." *Space Policy* 5 (August 1989): 193–200.

Saito, Takao, Kobayashi Hideo, Takagi Kenji. "Sora chū kaihatsu no mirai kōsō nitsuite tsuki Wakusei no riyō o chūshin ni." *Nihon kōkū uchū gakkaishi* 42 (September 1994): 511–522.

Sakamoto, Seiichi. "Ginga Renpō 25-nen kinen fure-mu kitte no hakkō." *ISAS News* 378 (September 2012): 8.

Sawa, Takamitsu. *Bunka toshite no gijutsu.* Tokyo: Iwanami Shōten, 1991.

Sawai, Minoru. "Gijutsu-sha no gunmin tenkan to tetsudō gijutsu kenkyūjo." *Osaka daigaku keizaigaku* 59, no. 1 (2009): 1–19.

Science and Technology Agency of Japan. *Chikyū kansoku eisei shisutemu no kaihatsu ni tsuite.* Tokyo: Science and Technology Agency of Japan, July 12, 1977.

Shibato, Kōji and Makoto Miwada. "H-II roketto ato no uchū yusō shisutemu 'roketto purēn.'" *Nihon kōkū uchū gakkaishi* 37, no. 427 (August 1989): 354–363.

Shibuya, Atsushi. *Hino Kumazō den: Nihonkōkū shoki no shinsō.* Tokyo: Aoshio, 1977.

Shimodaira, Katsuyuki, and Hamasaki Takashi. "Eisei gijutsu no genjō to shōrai." *Terebishon gakkaishi* 41, no. 4 (1987): 299–308.

Shirasawa, Atsushi. "Denkan yōwa: Nihon ni okeru nōhakei no rekishi." *Denkan no sōgō gakkaishi* 14, no. 2 (November 2020): 130–133.

Shūkan shokokumin. *Roketto heiki no haitatsu* 4, no. 6 (February 1942): 2.

Sogame, Eiji. "H-I roketto daini dan suishin-kei no kaihatsu." *Nihon kōkū uchū gakkaishi* 34, no. 387 (April 1986): 175–185.

———. "H-I Roketto." *Nihon kōkū uchū gakkaishi* 36, no. 418 (November 1988): 487–497.

Space Development Agency. "Beikoku roketto uchiage renzoku shikkei ni tsuite." Tokyo: NASDA, May 26, 1999. https://warp.da.ndl.go.jp/info:ndljp/pid/233892/www.nasda.go.jp/press/1999/05/rocket_990526_j.html.

Sudō, Hideo, and Masaki Katō. "Uchū kaihatsu jigyōdan Tsukuba uchu sentā no shōkai (toku ni uchū seibutsu kagaku ni kanren shite)." *Uchū seibutsu kagaku* 7, no. 1 (1993): 31–34.

Sunō, Tarō. "Roketto." In *Tsūzoku kagaku: JOAK kaki tokubetsu kōza (dai ni hōsō tekisuto),* ed. Nihonhōsōkyōkai Kantō shibu, 280–287. Tokyo: Nippon hōsō shuppan kyōkai, 1930.

Suzuki, Gorō. "Jinbutsu fairu koksuan roketto no oto: Itokawa Hideo hakase to nihon no uchū kaihatsu." *Maru/Chō shobō* 65, no. 12 (December 2012): 151–157.

Suzuki, Masao. "Dai Tōa Sensō to Dai ichi kaigun nenryō shō." In *Nippon kaigun nenryōshi (jō)*, 680–688. Tokyo: Hara Shobō, 1973.

Takahashi, Dankichi. *Shinkansen o tsukutta otoko: densetsu no enjinia: Shima Hideo monogatari*. Tokyo: PHP Kenkyujō, 2012.

Takamoto, Akira. "Heiki toshite roketto." *Gunji to gijutsu* 111 (March 1936): 51–60.

Takata, Seiji. "Keisokugaku Genron Nōto: Keisokugaku no Ryōbun (6)." *Keisei* 12, no. 5 (May 1969): 71–73.

Takenaka, Yukihiko. "N-1 roketto kaihatsu no ayumi." *Nihon kōkū uchū gakkaishi* 23, no. 362 (1984): 127–141.

Takuya, Ara. *Uchū kaihatsu to ekitai suisō*. Tokyo: Uchū kaihatsu jigyōdan enjin kaihatsu gurūpu, July 15, 1977.

Tanaka, Kimi. "Omoidegusa (roketto obasan)." *ISAS News* (August 1985): 1.

Tanaka, Hiroshi. "Tokkyō kenkyū: roketto." *Hatsumei* 42, no. 1 (June 1946): 14–15.

Tani, Ichirō. "Kenkyū no kaiko." *Tōkyō daigaku uchū kōkū kenkyūjo hōkoku* 4, no. 2 (1968): 163–178.

———. "Kenkyū Kaihatsu to Gakkai—kaigun kōkū to no sōgū." In *Kaiwashi no kōseki—Nihon kaigun kenkyū gaishi*, ed. Nihon kaikūkai kaigun kōkūgaishi kankōkai, 43. Tokyo: Hara shobō, 1982.

Tawara, Hajime. "Itokawa sensei to Isuraeru." In *Ningen Hideo Itokawa hakase to wa*, ed. Iwao Kanezawa. Tokyo: Hirohide Kōgei, 2003.

Toda, Kazuya. "Sengo mo daikū wo yumemita oto tachi." In *Dai Nippon teikoku rikukaigun kōkū butai zenshi. Rekishi dokuhon*, 248–255. Tokyo: Henshūbu hen, 2015.

Tokyo University [Tōkyō daigaku]. "Daigaku sōchō kaigi ni okeru sankō shiryō." August 25, 1943. Uchida Yoshikazu Papers, Tokyo University archives (sorting no. 4–1).

Tokyo daigaku seisan gijutsu kenkyujō. *Tokyo daigaku daini kagakubushi*. Tokyo: Tokyo daigaku seisan gijutsu kenkyūjō, 1968.

Toyota, H., S. Tanaka, T. Sugimura, and Y. Nakayama. "Shōwa 61-nen Izu Ōshima funka ni kakari waru rimō to senshingu." *Nihon rimōto senshingugaku kaishi* 6, no. 4 (1986): 365–400.

Tsurihashi, Watarō. *Norakura shōi roketto tokkan-tai: Warai no manga bukku*. Tokyo: Taikodō, 1934.

Tsuruta, Koichiro. "Dai-1 kai Planet-B no mezasu kasei no kagaku—Kasei tansa nyūmon." *ISAS News* 203 (February 1998): 11.

Uchida, Masami, Michio Nakagawa, Hirohisa Sakurai, and Makoto Yamauchi. "Cyg X—1 no kata X-sen ryōiki ni okeru tanjikan hendō ni tsuite (II)." *Uchū kagaku kenkyūjo hōkoku, tokushū: dai kikyū kenkyū hōkoku* 24 (December 1989): 103–112.

Uchida, Toshikazu. "Kaigun nenryō shō to watashi." In *Nihon kaigun nenryō sho (jō): nenryō konwakai*, 812–814. Tokyo: Hara Shobō, 1972.

Uchū kaihatsu iinkai tokubetsu kaigō. *Uchū kaihatsu iinkai tokubetsu kaigō hōkoku-sho—shippai no saihatsu bōshi no tame no kaikaku hōsaku*. Tokyo: NASDA, May 1999.

Ueda, Hiroyuki. "Uchiageomotsu H-II roketto—uchiage hi 1994-nen 8-gatsu 17-nichi." *Suiso enerugii shisutemu tokushu H-II roketto—Tanegashima no uchiage setsubi no shōkai* 19, no.1 (1994): 1–11.

Uesugi, Kuninori. "Space Odyssey of an Angel—Summary of HITEN's Three Year Mission." In *Advances in the Astronautical Sciences—Spaceflight Dynamics 1993: Proceedings of an AAS/NASA International Symposium held April 26–30, 1993, Greenbelt, Maryland*, 607–621. AAS, 1993.

———. "Hiten kara Jioteiru—Nihon no uchū kōgaku ni nokoshitamono." *Tokushū: Geotail kara no saki e—40 nenkan no kansha o komete*. https://www.isas.jaxa.jp/feature/special_issues/geotail/geotail_01.html.

———. "Kimo o hiyashita uchiage toki no antena no tsuibi." *Nihon uchū kaihatsu no rekishi (uchū kenbutsu monogatari)*. https://www.isas.jaxa.jp/j/japan_s_history/chapter06/03/02.shtml

Whittle, Frank. *Jets* [*Jetto Furanku Hoittoru*]. Trans. Eiichi Iwaya. Tokyo: Hitotsuboshi, 1955.

Yamamoto, Chikao, ed. *Nihon kaigun kōkū shi 2—gunbi hen*. Tokyo: Jiji Press, 1969.

———. *Nihon kaigun kōkū shi 3—seido gijutsu hen*. Tokyo: Jiji Press, 1969.

Yamanaka, Tatsuo, and Yasunori Matogawa. "Uchū kaihatsu no ohanashi." *Yomiuri Shimbun*, November 4, 1991.

Yamazaki, Toshio. *Gijutsushi*. Tokyo: Tōyō keizai shimpōsha, 1961.

Yoshida Ryūsei hozonkai. "2023 matsuri jōhō." https://ryusei.biz/?page_id=21.

Yoshida, Tetsuya. "Daikikyū—uchū e no iriguchi." Presentation at Kyoto University Cosmology Unit Cosmology Seminar, November 28, 1999. http://www.usss.kyoto-u.ac.jp/wp-content/uploads/2021/03/20141128_yoshida.pdf.

Yoshidamachi kyōiku iinkai. *Yoshidamachi shi*. Yoshida: Yoshida City, 1982.

Yoshioka, Hitoshi. "Uchū kaihatsu kyūsei no kakuritsu." In *Nihon no kagaku gijutsushō dai-3 ki*, ed. Nakayama Shigeru, 172–183. Tokyo: Gakuyō Shobō, 1995.

Young Astronauts Club [Kōeki zaidanhōjin Nihon uchū shōnen-dan]. *Gaiyō* (2012). https://www.yac-j.or.jp/outline.html.

Non-Japanese Sources

Aldrich, Daniel. *Site Fights: Divisive Facilities and Civil Society in Japan and the West*. Ithaca: Cornell University Press, 2010.

An, Hyoung Joon An. "South Korea's Space Program: Activities and Ambitions." *Asia Policy* 15, no. 2 (2020): 34–42.

Aoki, Reiko. "A Demographic Perspective on Japan's 'Lost Decades.'" *Population and Development Review* 38 (2012): 103–112.

Balepin, Vladimir V., Makoto Yoshida, and Kenjiro Kamijo. "Rocket Based Combined Cycles for Single Stage Rocket." *SAE Transactions* 103, Section 1: Journal of Aerospace (1994): 174–188.

Bartholomew, James R. *The Formation of Science in Japan*. New Haven: Yale University Press, 1989.

Beckley, M., Y. Horiuchi, and J. M. Miller. "America's Role in the Making of Japan's Economic Miracle." *Journal of East Asian Studies* 18, no.1 (2018): 1–21.

Beggs, James M. "James M. Beggs to George P. Schultz. March 6, 1984." In *Exploring the Unknown: Selected Documents in the History of the U.S. Civilian Space Program Volume II: External Relationships*, ed. John M. Logsdon (Washington, DC: NASA History Office, 1996), 105–107.

Berner, Steven. *Japan's Space Program: A Fork in the Road?* (Santa Monica, CA: RAND Corp., 2005).

Bingen, Kari A., and Makena Young. *From Earth to Uchū: The Evolution of Japan's Space Security Policy and a Blueprint for Strengthening the U.S.–Japan Space Security Partnership.* (Washington, DC: Center for Strategic and International Studies, 2024).

Bin Umino, Kyo Kageura, and Shinichi Toda. "A Sixty Year History and Analysis of the Japanese Publishing Industry: A Statistical Analysis of Circulation." *Publishing Research Quarterly* 26, no. 4 (December 2010): 278–279.

Brown, Laurie M., and Yoichiro Nambu. "Physicists in Wartime Japan." *Scientific American* 279, no. 6 (December 1998): 96–103.

Burgess, W. W. "Solid Rocket Propellants Research." 1945. National Diet Library, Tokyo.

Caidin, Martin. "Japanese Guided Missiles in World War II." *Journal of Jet Propulsion* 26, no. 8 (1956): 691–694.

Callon, Michel. "Some Elements of a Sociology of Translation: Domestication of the Scallops and the Fisherman of St. Brieuc Bay." In *Power, Action, and Belief: A Sociology of a New Knowledge?* ed. J. Law, 196–223. London: Routledge, 1986.

Carra, G., and J. de Dalmau. "Europe Ready for Ariane-5 Production." https://www.esa.int/esapub/bulletin/bullet93/b93carr.htm.

Central Intelligence Agency, Directorate of Intelligence. *Japan: Participation in the Space Program*. Washington, DC: CIA, March 18, 1985.

Chang, Iris. *Thread of the Silkworm*. New York: Basic Books, 1995.

Clark, Stephen. "In Exchange for a Lunar Rover, Japan Will Get Seats on Moon-Landing Mission." *Ars Technica*, April 11, 2024. https://arstechnica.com/space/2024/04/japan-will-be-first-among-nasas-partners-to-have-an-astronaut-on-the-moon/.

Cohen, Theodore. *Remaking Japan*. New York: Free Press, 1987.

Compton, Karl T. "Mission to Tokyo." Office of the President Records: Jonathan Dickinson to Harold W. Dodds Subgroup, Box 205, Folder 11. Princeton University Archives, Department of Rare Books and Special Collections, Princeton University Library.

Cordier, Sherwood S. "Japan: Present and Potential Military Power." *United States Naval Institute Proceedings* 93, no. 777 (November 1967). https://www.usni.org/magazines/proceedings/1967/november/japan-present-and-potential-military-power.

Crail, Peter. "South Korea Attempts First Space Launch." *Arms Control Today* 39, no. 7 (September 2009): 37–39.

Crichton, Michael. *Rising Sun: A Novel*. New York: Knopf, 1992.

Cusumano, M. A. "Manufacturing Innovation: Lessons from the Japanese Auto Industry." *Sloan Management Review* 30, no. 1 (1988): 29–40.

Danquah, Francis K. "Japan's Food Farming Policies in Wartime Southeast Asia: The Philippine Example, 1942–1944." *Agricultural History* 64, no. 3 (Summer 1990): 60–80.

Dateki, Ryoko. "Space Industry Promotion in Japan—The Role of JAXA." JAXA, July 22, 2024. https://earth.jaxa.jp/conseo/news/20240618-1/02_JAXA_Business_Development%20and_Industrial_Relations_Dept.pdf.

Dick, Steven J., ed. *Remembering the Space Age: Proceedings of the 50th Anniversary Conference*. Washington, DC: NASA, 2008.

Dick, Steven J., and Roger D. Launius, eds. *Critical Issues in the History of Spaceflight*. Washington, DC: NASA, 2006.

Dower, John. *War without Mercy*. New York: Pantheon, 1986.

———. *Empire and Aftermath: Yoshida Shigeru and the Japanese Experience, 1878–1954*. Cambridge, MA: Harvard University Asia Center, 1988.

———. *Japan in War and Peace*. New York: New Press, 1993.

———. "Occupied Japan in the Cold War and Asia." In *Japan in War and Peace: Selected Essays*. New York: New Press, 1993.

———. *Embracing Defeat: Japan in the Wake of World War II*. New York: W. W. Norton, 2000.

Drier, Casey. "An Improved Cost Analysis of the Apollo Program." *Space Policy* 60 (2022). https://www.sciencedirect.com/science/article/pii/S0265964622000029.

Dusinberre, Martin. *Hard Times in the Hometown: A History of Community Survival in Modern Japan*. Honolulu: University of Hawaii Press, 2012.

Eberstadt, George A. "Japan's High Frontier: Viewing the Space Program as Industrial Policy." M.A. thesis, Harvard College, 1989.

Ecosystems International. *Commercial Potential of European and Japanese Space Programs*. September 1987.

Elleman, Michael. "Prelude to an ICBM? Putting North Korea's Unha-3 Launch Into Context." *Arms Control Today* 43, no. 2 (March 2013): 8–13.

European Space Agency. *Convention for the Establishment of a European Space Agency*. Noordwick: ESA Publications Division, 2007.

Federation of Economic Organizations [Keidanren]. "Ensuring a Viable and Productive Future for Japan's Space Programs." In *Report of the Space Activities Council of the Keidanren*. Tokyo: Keidanren, October 25, 1988.

Fritzsche, Peter. *A Nation of Flyers: German Aviation and the Popular Imagination*. Cambridge, MA: Harvard University Press, 1992.

Fukushima, Y., and T. Imoto. "Lessons Learned in the Development of the LE-5 and LE-7." In *30th AIAA/ASME/ISAE/ASEE Joint Propulsion Conference* (Indianapolis, June 27–29, 1994), 1–7.

Frühstück, Sabine. *Playing War: Children and the Paradoxes of Modern Militarism in Japan*. Berkeley: University of California Press, 2017.

Garcia, Mark A., ed. "Japanese Experimental Module Kibo." NASA. https://www.nasa.gov/international-space-station/japanese-experiment-module-kibo/.

General Headquarters, Supreme Commander for the Allied Powers. *History of the Nonmilitary Activities of the Occupation of Japan. Vol. I—Introduction*. National Diet Library, Tokyo.

———. *History of the Nonmilitary Activities of the Occupation of Japan. Vol. IX—Reparations and Property Administration—Part A: Reparations* (1945 through January 1951). National Diet Library, Tokyo.

General Headquarters, Supreme Commander for the Allied Powers. SCAPIN-301 (AG 360 [18 Nov 45])—Commercial and Civil Aviation. https://en.wikisource.org/wiki/SCAPIN-301.

General Headquarters, United States Army Forces, Pacific, Scientific and Advisory Section. *Report on Scientific Intelligence Survey in Japan, September and October 1945* (November 1, 1945). National Diet Library, Tokyo.

Geoghegan, John J. *Operation Storm: Japan's Top Secret Submarines and Its Plan to Change the Course of World War II*. New York: Crown, 2013.

George, Timothy S. *Minamata: Pollution and the Struggle for Democracy in Postwar Japan*. Cambridge, MA: Harvard University Asia Center, 2002.

Geppert, Alexander C. T. "Space *Personae:* Cosmopolitan Networks of Peripheral Knowledge, 1927–1957." *Journal of Modern European History* 6, no. 2 (September 2008): 262–286.

Gerovitch, Slava. *Soviet Space Mythologies: Public Images, Private Memories, and the Making of a Cultural Identity*. Pittsburgh: University of Pittsburgh Press, 2015.

Gluck, Carol. "The Past in the Present." In *Postwar Japan as History*, ed. Andrew Gordon, 64–98. Berkeley: University of California Press, 1993.

Goddard, Robert H. *Rockets—Two Classic Papers*. Mineola, NY: Dover, 2002.

Gordon, Andrew. *A Modern History of Japan: From Tokugawa Times to the Present*. Oxford: Oxford University Press, 2020.

Gorn, Michael. *Harnessing the Genie: Science and Technology Forecasting for the Air Force, 1944–1986*. Washington, DC: Office of Air Force History, United States Air Force, 1988.

Green, James L., Robert E. Maguire, and Brian Lopez-Swafford. *A Communications Model for an ISAS to NASA Span Link*. Washington, DC: NASA, January 1987.

Griffin, Naomi, and Kazuhiko Odaki. *Reallocation and Productivity Growth in Japan: Revisiting the Lost Decade of the 1990s.* Washington, DC: Congressional Budget Office, 2006.

Griffin, Stuart. "Japan Asks for Help." *Science News* 97, no. 21 (May 23, 1970): 516.

Grunden, Walter. *Secret Weapons and the Second World War.* Lawrence: University Press of Kansas, 2005.

Grunden, Walter E., Yutaka Kawamura, Eduard Kolchinsky, Helmut Maier, and Masakatsu Yamazaki. "Laying the Foundation for Wartime Research: A Comparative Overview of Science Mobilization in National Socialist Germany, Japan, and the Soviet Union." *Osiris* 20 (2005): 79–106.

Gruntman, Mike. *Blazing the Trail: The Early History of Spacecraft and Rocketry.* Reston, VA: American Institute of Aeronautics and Astronautics, 2004.

Harrison, Todd, Kaitlyn Johnson, Thomas G. Roberts, Madison Bergethon, and Alexandra Coultrup. "Space Threat Assessment 2019: Iran." Washington, DC: CSIS, 2019.

Harvey, Brian. *The Chinese Space Program from Conception to Manned Flight.* New York: Wiley, 1998.

———. *Japan in Space: Past, Present, and Future.* (Chichester, UK: Springer Praxis, 2023.

Havens, Thomas R. H. *Valley of Darkness: The Japanese People and World War Two.* New York: W. W. Norton, 1978.

Heppenheimer, T. A. *Countdown: A History of Space Flight.* New York: Wiley, 1997.

Hickman, John. "Opening Gambit: Red Moon Rising." *Foreign Policy* 194 (July–August 2010): 20–21.

Hills, Carla A. "The Application of the Super 301 Provisions to Japan—Statement of Ambassador Carla A. Hills, May 25, 1989." In *Nichibei kankei shiryō shu: 1945–97,* 1163–1164. Tokyo: Tōkyō Daigaku Shuppankai, 1999.

Hino, Kumao. "On the Combustion and Detonation Velocities of Solid Explosives (including the Applications of the Principles)." Records of General Headquarters Far East Command, Assistant Chief of Staff (November 22, 1945). National Diet Library, Tokyo.

———. Letter from Hino Kumao to Dr. F. Zwicky. Records of the US Strategic Bombing Survey, Extracts from Intelligence Bulletin pertaining to the Bakam Report (November 21, 1945). National Diet Library, Tokyo.

———. "Theory of the Burning Rate of Industrial Safety Fuse." *Kōgyō kayaku kyōkaishi* 15, no. 1 (March 30, 1954): 12–21.

———. "Theory of Blasting with Concentrated Charge." *Kōgyō kayaku kyōkaishi* 14, no. 4 (December 1954): 233–250.

———. "Concentrated Type of No-cut Round of Blasting." *Kōgyō kayaku kyōkaishi* 16, no. 3 (September 1955): 166–179.

———. "Fragmentation of Rock through Blasting." *Kōgyō kayaku kyōkaishi* 17, no. 1 (March 1956): 2–11.

———. "Shock Wave Theory of Blasting with Cylindrical Charge." *Kōgyō kayaku kyōkaishi* 18, no. 1 (March 1957): 2–21.

———. "Theory of Relations between the Detonation Velocity of Solid Explosives and the Thickness of Cases or the Diameter of the Charge." *Kōgyō kayaku kyōkaishi* 19, no. 3 (August 1958): 169–180.

———. *Theory and Practice of Blasting.* Asa: Nippon Kayaku Kabushiki-gaisha, 1959.

Hodogaya Chemical Company [Hodogaya kagaku kōgyō]. "History of Hodogaya Chemical Group." https://www.hodogaya.co.jp/english/company/history.

Home, R. W., and Morris F. Low. "Postwar Scientific Intelligence Missions to Japan." *Isis* 84 (September 1993): 527–537.

Horikawa, Yasushi. "To Protect the Environment for a Peaceful Life." http://global.jaxa.jp/article/special/eco/horikawa_e.html.

International Astronautical Federation. "Biographies—Kiyoshi Higuchi." https://www.iafastro.org/biographie/kiyoshi-higuchi.html.

Hunter-Chester, David. *Creating Japan's Ground Self-Defense Force, 1945–2015: A Sword Well Made.* London: Lexington Books, 2016.

ISAS. "The Age of Space Science—Mu Rockets." https://www.isas.jaxa.jp/e/japan_s_history/detail/mu.shtml.

———. "History of Japanese Space Research—Pencil at Michikawa." http://www.isas.jaxa.jp/e/japan_s_history/profito/pencil.shtml

———. "Asteroid Explorer Hayabusa2." http://www.isas.jaxa.jp/en/missions/spacecraft/current/hayabusa2.html.

———. "Hayabusa." https://www.isas.jaxa.jp/en/missions/spacecraft/past/hayabusa.html.

———. "History of Japanese Space Research." 2008. https://www.isas.jaxa.jp/e/japan_s_history/detail/backgr.shtml.

———. "Launch Vehicles: M-V." https://www.isas.jaxa.jp/en/missions/launch_vehicles/m-v.html.

———. "Past: Hiten." https://www.isas.jaxa.jp/en/missions/spacecraft/past/hiten.html.

Ito, Kenji. "Values of 'Pure Science': Nishina Yoshio's Wartime Discourse between Nationalism and Physics, 1940–1945." *Historical Studies in the Physical and Biological Sciences* 33, no. 1 (2002): 61–86.

Itokawa, Hideo. "The Japanese Sounding Rockets." *Journal of Jet Propulsion* 27, no. 3: 286.

Japan Ministry of Posts and Telecommunications. "Radio Research Laboratory." Pamphlet. Tokyo: Ministry of Posts and Telecommunications, 1985.

JAXA [Uchū kōkū kenkyū kaihatsu kikō]. "About Advanced Earth Observing Satellite 'Midori' (ADEOS)." https://global.jaxa.jp/projects/sat/adeos/index.html.

———. "About J-I Launch Vehicle." https://global.jaxa.jp/projects/rockets/j1/index.html.

———. "Express." https://www.isas.jaxa.jp/en/missions/spacecraft/past/express.html.

———. "Hayabusa2 Project" (2013). http://www.hayabusa2.jaxa.jp.

———. "Humans in Space—Mukai Chiaki." https://humans-in-space.jaxa.jp/en/astronaut/mukai-chiaki/.

———. "Humans in Space—Yamazaki Naoko." https://humans-in-space.jaxa.jp/en/astronaut/yamazaki-naoko/.

———. "JAXA President Regular Monthly Press Conference" (July 18, 2013). http://global.jaxa.jp/about/president/presslec/201307_e.html.

———. "JAXA President Regular Monthly Press Conference" (January 13, 2017). http://global.jaxa.jp/about/president/presslec/201701.html/.

———. "A Passion for Rocketry: Ryōjirō Akiba." http://global.jaxa.jp/article/interview/2013/vol77/index_e.html.

———. "Restoration of Asteroid Explorer, HAYABUSA's Return Cruise" (November 19, 2009). https://global.jaxa.jp/press/2009/11/20091119_hayabusa_e.html.

———. "Space Transportation Systems: H-IIB Launch Vehicle in Operation" (October 17, 2016). http://global.jaxa.jp/projects/rockets/h2b/.

———. "Transition of Number of Staff and Budget." http://global.jaxa.jp/about/transition/index.html/.

———. "Treaty on Principles Governing the Activities of States in the Exploration and Use of Outer Space, including the Moon and Other Celestial Bodies." https://www.jaxa.jp/library/space_law/chapter_1/1-2-2-5_e.html#:~:text=The%20Moon%20and%20other%20celestial%20bodies%20shall%20be%20used%20by,celestial%20bodies%20shall%20be%20forbidden.

———. "Uchū kaihatsu jigyōdan hō, dai 1-shō—sōsoku." https://www.jaxa.jp/library/space_law/chapter_1/1-1-2-10/1-1-2-101_j.html.

Jenkins, Dennis R. *Space Shuttle: Developing an Icon: 1972–2013*. Vol. 1. Forest Lake, MN: Specialty Press, 2016.

Johnson, Chalmers. *MITI and the Japanese Miracle*. Stanford: Stanford University Press, 1982.

Johnson-Freese, Joan. *Space As a Strategic Asset*. New York: Columbia University Press, 1998.

Kallendar-Umezu, Paul. "Explaining the Logics of Japanese Space Policy Evolution 1969–2016, Combining Macro- and Microtheories, Notably the Strategic Action Field Framework." PhD diss., Keio University, 2017.

Kamijo, Kenjiro, Hitoshi Yamada, Norio Sakazune, and Shogo Warashina. "Developmental History of Liquid Oxygen Turbopumps for the LE-7 Engine." *Transactions of the Japan Society of Aeronautical and Space Science* 44, no. 145 (2001): 155–163.

Kawaguchi Jun'ichiro. "On the Lunar and Heliocentric Gravity Assist Experienced in the Planet-B ('Nozomi')." Sagamihara: ISAS, 1999.

Keidanren. *Space in Japan 1986–1987*. Tokyo: Keidanren, 1987.

Kelly, William W. "Finding a Place in Metropolitan Japan: Ideologies, Institutions, and Everyday Life." In *Postwar Japan As History*, ed. Andrew Gordon, 189–238. Berkeley: University of California Press, 1993.

Kim, Dong-Won. "Nishina Yoshio and the Two Cyclotrons." *Historical Studies in the Physical and Biological Sciences* 36, no. 2 (March 2006): 243–273.

Kochhar, Rajesh. "Zinc, Steel and Rocketry: Western Mainstreaming of Empirical Indian Technologies." *Current Science* 113, no. 9 (November 10, 2017): 1791–1796.

Koizumi, Kenkichiro. "In Search of 'Wakon': The Cultural Dynamics of the Rise of Manufacturing Technology in Postwar Japan." *Technology and Culture* 43, no. 1 (January 2002): 29–49.

Koschmann, J. Victor, Ryūichi Narita, and Yasushi Yamanouchi, eds. *Total War and "Modernization."* Ithaca: Cornell University Press, 1998.

Krammer, Arnold. "Transfer As War Booty: The US Technical Oil Mission to Europe, 1945." *Technology and Culture* 22, no. 1 (January 1981): 68–103.

Kwon, Peter B. "The Anatomy of Chaju Kukpang: Military-Civilian Convergence in the Development of the South Korean Defense Industry under Park Chung Hee, 1968–1979." PhD diss., Harvard University, 2016.

Launius, Roger D. *Historical Analogs for the Stimulation of Space Commerce* (Washington, DC: NASA, 2014.

Lebeau, André, and Karl Reuter. "La place des techniques spatiales dans le développement économique. Réflexion sur des aspects actuels et futurs." Paris: European Space Agency, 1980.

Logsdon, John M. "U.S.–Japanese Space Relations at a Crossroads." *Science* 255, no. 5042 (January 17, 1992): 294–300.

Low, Morris, Shigeru Nakayama, and Hitoshi Yoshioka, eds. *Science, Technology, and Society in Contemporary Japan*. Cambridge: Cambridge University Press, 1999.

Luk, Christine Y. L., and Subodhana Wijeyeratne. "Alternatives to GPS: Space Infrastructure in China and Japan." *Roadsides* 3 (August 2020): 57–62.

Lynn, Leonard. "Japanese Research and Technology Policy." *Science* 233, no. 4761 (July 18, 1986): 296–301.

Mahajan, Shobjit. "ISRO's Mars Mission." *Economic and Political Weekly* 49, no. 40 (October 4, 2014): 10–12.

Maher, Neil. "Shooting the Moon." *Environmental History* 9, no. 3 (July 2004): 526–531.

Malik, Tariq. "Japan Prepares Space Station's Largest Laboratory for Flight." Space.com. May 2, 2007. https://www.space.com/3750-japan-prepares-space-station-largest-laboratory-flight.html).

Matogawa, Yasunori. "A Survey of Rocketry for Space Science in Japan." In *History of Rocketry and Astronautics: Proceedings of the History Symposia of the International Academy of Astronautics* 22, 208–214. San Diego: American Astronautical Society, 1993.

———. "'Pencil' Rocket and Itokawa Hideo: A Pioneering Work of Japanese Rocketry." In *History of Rocketry and Astronautics: Proceedings of the History Symposia of the International Academy of Astronautics*, 121–132. San Diego: AAS, 1994–1995).

———. "Another Destiny of Rocketry in Japan—Festival Rockets in Japanese Shrines." In *History of Rocketry and Astronautics: Proceedings of the History Symposia of the International Academy of Astronautics*, 315–326. San Diego: AAS, 1994–1995.

———. "Yasunori Matogawa hakase no uchū yomoyama hanashi: jimae roketto kaihatsu wa kaku ketsudan sareta." *Sekai shūhō* 79, no. 38 (October 20, 1998): 56.

———. "The Shūsui Japanese Rocket Fighter in the Second World War." *History of Rocketry and Astronautics* 28 (1999): 177–187.

———. "Japanese Solid Rockets in the Second World War." In *History of Rocketry and Astronautics—AAS History Series 25*, ed. Herve Moulin and Donald C. Elder, 123–136. San Diego: AAS, 2003.

———. "Japan Liquid Rockets in the Second World War." In *History of Rocketry and Astronautics—AAS History Series 26*, ed. Donald C. Elder and George C. Jams, 111–125. San Diego: AAS, 2005.

———. "Halley's Comet Exploration in Japan—Japan's First Interplanetary Flight." *49th International Astronautical Congress*, 223–240. International Astronautical Federation, 2007.

———. "Lessons from Half a Century's Experience of Japanese Solid Rocketry Since Pencil Rocket." *Acta Astronautica* 61 (2007): 1107–1115.

Matsubara, Shōji, and Tetsuichi Itō. "The Conceptual Study of H-II Orbiting Plane (HOPE)." *SAE Transactions* 96, Section 6: Aerospace (1987): 1719–1728.

McCurdy, Howard. *Space and the American Imagination*. Washington, DC: Smithsonian Institution Press, 1997.

McDougall, Walter A. *The Heavens and the Earth: A Political History of the Space Age*. New York: Basic Books, 1985.

McKevitt, Andrew C. *Consuming Japan*. Chapel Hill: University of North Carolina Press, 2017.

McKinnon, Mika. "Everything That Could Go Wrong for Hayabusa Did, and Yet It Still Succeeded." October 15, 2010. https://gizmodo.com/everything-that-could-go-wrong-for-hayabusa-did-and-ye-1730940605.

Melzer, Jürgen P. *Wings for the Rising Sun: A Transnational History of Japanese Aviation*. Cambridge, MA: Harvard University Press, 2020.

Merkle, C. L., J. R. Brown, J. P. McCarty, G. B. Northam, A. Povinelli, M. L. Stangeland, and E. E. Zukoski. *JTEC Panel Report on Space and Transatmospheric Propulsion Technology: Final Report*. Baltimore: Loyola College, August 1990.

Middleton, Sallie. "Space Rush: Local Impact of Federal Aerospace Programs on Brevard and Surrounding Counties." *The Florida Historical Quarterly* 87, No. 2 (Fall 2008): 258–289.

Milani, Livia Peres. "Brazil's Space Program: Finally Taking Off?" The Wilson Center. October 24, 2019. https://www.wilsoncenter.org/blog-post/brazils-space-program-finally-taking.

Mimura, Janis. *Planning for Empire: Reform Bureaucrats and the Japanese Wartime State.* Ithaca: Cornell University Press, 2011.

Ministry of Finance of Japan [Zaimushō]. "(1) General Account." In *Budget Reference Documents according to Article 28 of the Finance Law, Records on National Treasury Debt Obligation Acts since 1971.* Tokyo: Ministry of Finance, 1972.

———. *1972 General Account Budget Reference Book Ministry of Posts and Telecommunications.* Tokyo: Ministry of Finance, 1973.

Mizuno, Hiromi. *Science for the Empire: Scientific Nationalism in Modern Japan.* Stanford: Stanford University Press, 2009.

Momma, H., M. Watanabe, K. Mitsuzawa, K. Danno, Masahiko Ida, M. Arita, and I. Ujino. "Search and Recovery of the H-II Rocket Flight No. 8 Engine." In *Proceedings of the 2000 International Symposium on Underwater Technology,* 19–23. Tokyo, 2000.

Moore, Aaron. *Constructing East Asia Technology, Ideology, and Empire in Japan's Wartime Era, 1931–1945.* Stanford: Stanford University Press, 2013.

Moriguchi, Chiaki, and Emmanuel Saez. "The Evolution of Income Concentration in Japan, 1885–2002: Evidence from Income Tax Statistics." Econometrics Laboratory, University of California, Berkeley. August 25, 2005. https://eml.berkeley.edu/~saez/moriguchi-saez05japan.pdf.

Nakayama, Brian. *Emerging Space Powers.* Berlin: Springer, 2010.

Nakayama, Shigeru, ed. *A Social History of Science and Technology in Contemporary Japan, Vol. 1: The Occupation Period, 1945–1952.* Melbourne: Trans Pacific Press, 2006.

———. "The Central Laboratory Boom and the Rise of Corporate R&D." In *A Social History of Science and Technology in Contemporary Japan, Vol. 3: High Economic Growth Period,* ed. Shigeru Nakayama and Kunio Gotō. Melbourne: Trans Pacific Press, 2006.

NASA. Galileo Jupiter Arrival Press Kit. December 1995. https://www.jpl.nasa.gov/news/press_kits/gllarpk.pdf.

———. "The Tropical Rainfall Measuring Mission." https://gpm.nasa.gov/missions/trmm.

———. "Nasa Spinoff." https://spinoff.nasa.gov/.

NEC [Nippon Denki Kabushiki-gaisha]. "JAXA Launches Hayabusa 2 Asteroid Probe—NEC Conducts Manufacturing and Testing As Probe System Coordinator." December 3, 2014. https://www.nec.com/en/press/201412/global_20141203_01.html.

Needham, Joseph. "The Fire-lance, Ancestor of All Gun-barrels." In *Chinese Ideas About Nature and Society: Studies in Honour of Derk Bodde,* ed. C. Le Blanc and Susan Blader. Hong Kong: Hong Kong University Press, 1987.

Neufeld, Michael. *Von Braun: Dreamer of Space, Engineer of War.* New York: Knopf, 2007.

———. "Three Heroes of Spaceflight: The Rise of the Tsiolkovskii-Goddard-Oberth Interpretation and Its Current Validity." *Quest: The History of Spaceflight Quarterly* 19, no. 4 (2012): 4–14.

Nippon Kayaku Kabushiki-gaisha. "A Cap Sensitive Ammonium Nitrate-Fuel Oil Explosive and a Method of Manufacturing the Same." US patent no. 3,111,437. Filed September 6, 1960, issued November 19, 1963.

———. Email to author. October 3, 2016.

Nishiyama, Kiriko. "Hughes Deals Crippling Blow to Japan's Rocket Program." *Space Daily*, May 25, 2000.

Nishiyama, Takashi. "Swords into Plowshares: Civilian Application of Wartime Military Technology in Modern Japan, 1945–1964." PhD diss., Ohio State University, 2005.

Oganesoff, Igor. "Japan Builds Rocket Industry, Emphasises Peaceful Use, Exports." *Wall Street Journal*, October 22, 1962.

Okazaki, Tetsuji. "The Supplier Network and Aircraft Production in Wartime Japan." *Economic History Review* 54, no. 3 (August 2011): 973–994.

Onoda, Jun. "Launch Clients Sought for H-2A: Rocket Faces Strong International Competition, Dearth of Non-Govt Business." *Yomiuri Shimbun*, May 18, 2012.

Oros, Andrew J. "Explaining Japan's Tortured Course to Surveillance Satellites." *Review of Policy Research* 24, no. 1 (2007): 30–48.

O'Sullivan, John. *Japanese Missions to the International Space Station: Hope from the East.* Cham, Switzerland: Springer Praxis, 2019.

Otake, Tomoko. "Toyohiro Akiyama: Cautionary Tales from One Not Afraid to Risk All." *The Japan Times*, August 3, 2013.

Pan, Jixing. "On the Origin of Rockets." *T'oung Pao*, col. 73, livr. 1/3 (1987): 2–15.

Pape, Robert A. "Why Japan Surrendered." *International Security* 18, no. 2 (Fall 1993): 154–201.

Pekkanen, Saadia. *Picking Winners? From Technology Catch-Up to the Space Race in Japan.* Stanford: Stanford University Press, 2003.

———. "Japan in Asia's Space Race: Directions and Implications." In *Rich Region, Strong States: The Political Economy of Security in Asia: Policy Brief July 2013.* San Diego: SITC Research Briefs, Institute on Global Conflict and Cooperation, University of California, April 2013.

Pekkanen, Saadia, and Paul Kallender-Umezu. *In Defense of Japan: From the Market to the Military in Space Policy.* Stanford: Stanford University Press, 2010.

The Planetary Society. "The Japan Aerospace Exploration Agency (JAXA)." https://www.planetary.org/the-japan-aerospace-exploration-agency-jaxa.

Pomarański, Marcin. "Science and Technology in Russian Cosmic Utopias from the Beginning of the Twentieth Century: Konstantin Tsiolkovsky and Alexander Bogdanov." *Utopian Studies* 33, no. 1 (2022): 36–53.

Prime Minister's Office of Japan. "The Constitution of Japan." https://japan.kantei.go.jp/constitution_and_government_of_japan/constitution_e.html.

Ray, Justin. "Atlas 5 Thunders to Milestone in U.S. Rocket History." *Spaceflight Now*, June 20, 2012. https://spaceflightnow.com/atlas/av023.

Records of the US Strategic Bombing Survey. "Principal Army Ground Force Laboratories." *Report on Scientific Intelligence Survey in Japan II* (1946). National Diet Library, Tokyo.

———. 2 USSBS YD-208 (1947). National Diet Library, Tokyo.

Redfield, Peter. "Beneath a Modern Sky: Space Technology and Its Place on the Ground." *Science, Technology, & Human Values* 21, no. 3 (Summer 1996): 251–274.

Reilly, Robin. *Kamikaze Attacks of World War II: A Complete History of Japanese Suicide Strikes on American Ships, by Aircraft and Other Means.* London: McFarland, 2010.

Robertson, Andrew John. "Mobilizing for War, Engineering the Peace: The State, the Shop Floor, and the Engineer, 1965–1960." Phd diss., Harvard University, 2000.

Saegusa, Asako. "Japan's NASDA under Fire on Launch Failure." *Nature* 392 (1998): 321.

Sagan, Carl. *Murmurs of Earth: The Voyager Interstellar Record.* New York: Ballantine Books, 1979.

Saito, Shigefumi. "Japan's Space Policy: Background and Outlooks." *Space News* 5, no. 3 (August 1989): 193–200.

Samuel, Richard J. *Rich Nation, Strong Army: National Security and the Technological Transformation of Japan.* Ithaca: Cornell University Press, 1996.

Sarli, Bruno Victorino, and Yuichi Tsuda. "Hayabusa 2 Extension Plan: Asteroid Selection and Trajectory Design." *Acta Astronautica* 138 (September 2017): 225–232.

Sato, Masahiko, Toshio Kosuge, and Peter van Fenema. *Legal Implications on Satellite Procurement and Trade Issues between Japan and the United States.* Paris: International Institute of Space Law, 1999.

Sato, Yasushi. "A Contested Gift of Power: American Assistance to Japan's Space Launch Vehicle Technology, 1965–1975." *Historia Scientiarum* 11, no. 2 (2001): 176–204.

———. "Managing the Interface between Politics and Technology: Itokawa, Hideo, Shima Hideo, and the Early Japanese Space Programs." *Historia Scientiarum* 21, no. 3 (2012): 193–210.

Semiconductor Museum of Japan. "32-bit RISC Microprocessors/Controllers with Original Architecture (Hitachi)—Integrated Circuit." 1992. https://www.shmj.or.jp/english/pdf/ic/exhibi735E.pdf.

Shillony, Ben-Ami. "Universities and Students in Wartime Japan." *The Journal of Asian Studies* 45, no. 4 (August 1986): 769–787.

Siddiqi, Asif. A. *The Challenge to Apollo: The Soviet Union and the Space Race, 1945–1974.* Washington, DC: NASA, 2000.

———. "Spaceflight in the National Imagination." In *Remembering the Space Age*, ed. Steven J. Dick, 17–35. Washington, DC: NASA Office of External Relations History Division, 2008.

———. *The Red Rockets' Glare: Spaceflight and the Soviet Imagination, 1857–1957.* Cambridge: Cambridge University Press, 2010.

———. *Sputnik and the Soviet Space Challenge.* Gainesville: University of Florida Press, 2010.

———. "Fighting Each Other: The N-1, Soviet Big Science, and the Cold War at Home." In *Science and Technology in the Global Cold War*, ed. Naomi Oreskes and John Krige, 189–225. Cambridge, MA: MIT Press, 2014.

———. *Beyond Earth: A Chronicle of Deep Space Exploration, 1958–2016*. Washington, DC: NASA, 2018.

———. "Soviet Secrecy: Toward a Social Map of Knowledge." *American Historical Review*, 126, no. 3 (September 2021): 1046–1071.

Simpson, Joanne, ed. *Report of the Science Steering Group for a Tropical Rainfall Measuring Mission*. Greenbelt, MD: Goddard Space Flight Center, August 1988.

Smith, Josh. "North Korea's First Spy Satellite Is 'Alive', Can Manoeuvre, Expert Says." *Reuters*, February 28, 2024. https://www.reuters.com/technology/space/north-koreas-first-spy-satellite-is-alive-can-manoeuvre-expert-says-2024-02-28/.

Smith, Peter G. "Non–US Approaches to Space Commercialisation." In *Second Symposium on Space Industrialisation*, 2–24. (Washington, DC: NASA, October 1984.

Society of Japanese Aerospace Companies. *Report on the Present Status of the Japanese Space Industry*. Tokyo: Society of Japanese Aerospace Companies, October 1990.

Sogame, Eiji. "A History of the Tanegashima Space." In *History of Rocketry and Aeronautics: Proceedings of the Thirty-Ninth History Symposium of the International Academy of Astronautics: Fukuoka, Japan (2005)*, 255–270. San Diego: AAS, 2011.

Space Activities Commission. *Report of the Space Activities Commission*. Tokyo: Space Activities Commission, March 17, 1978.

Stanley, Amy. *Stranger in the Shogun's City: A Japanese Woman and Her World*. New York: Scribner, 2020.

Steinhoff, Ernst. "The Development of the German A-4 Guidance and Control System 1939–1945: A Memoir." In *Essays on the History of Rocketry and Astronautics : Proceedings of the Third through the Sixth History Symposia of the International Academy of Aeronautics*, 206–214. Washington, DC, 1977.

Swope, Kenneth M. "Crouching Tigers, Secret Weapons: Military Technology Employed during the Sino-Japanese-Korean War, 1592–1598." *The Journal of Military History* 69, no. 1 (January 2005): 11–41.

Tadakawa, Tsuguo. "Japan's Launch Vehicle Program Update." *SAE Transactions* 96, Section 6: Aerospace (1987): 219–228.

Takahashi, Yuzo. "A Network of Tinkerers: The Advent of the Radio and Television Receiver Industry in Japan." *Technology and Culture* 41, no. 3 (July 2000): 460–484.

Tipton, Elise K. *Modern Japan: A Social and Political History*. New York: Routledge, 2008.

Tsutsui, William M. "Landscapes in the Dark Valley: Toward an Environmental History of Wartime Japan." *Environmental History* 8, no. 2 (April 2003): 294–311.

Ulmer, S. Sidney. "Local Authority in Japan since the Occupation." *The Journal of Politics* 19, no. 1 (February 1957): 46–65.

Umino, Bin, Kyo Kageura, and Shinichi Toda. "A Sixty Year History and Analysis of the Japanese Publishing Industry: A Statistical Analysis of Circulation." *Publishing Research Quarterly* 26, no. 4 (December 2010): 272–286.

United Nations Office for Outer Space Affairs. "Law concerning the National Space Development Agency of Japan." June 23, 1969. https://www.unoosa.org/oosa/en/ourwork/spacelaw/nationalspacelaw/japan/nasda_1969E.html.

US Air Force—Biographies. "'General Henry. H. Arnold." http://www.af.mil/AboutUs/Biographies/Display/tabid/225/Article/107811/general-henry-h-arnold.aspx.

US Naval Technical Mission to Japan. "Japanese Bombs." *Report of the US Naval Technical Mission to Japan*. San Francisco: US Naval Technical Mission to Japan, September 4, 1945. National Diet Library, Tokyo.

———. "Japanese Propellants—Article 1: Use and Manufacture of Ortho Tolyl Urethane for Stabilizing Rocket and Gun Propellants." *Report of the US Naval Technical Mission to Japan, Intelligence Targets Japan*. San Francisco: US Naval Technical Mission to Japan, November 1945. National Diet Library, Tokyo.

———. "Japanese Propellants—Article 2: Rocket and Gun Propellants—General." *Report of the US Naval Technical Mission to Japan, Intelligence Targets Japan*. San Francisco: US Naval Technical Mission to Japan, November 1945. National Diet Library, Tokyo.

———. "Japanese Guided Missiles." *Report of the US Naval Technical Mission to Japan*. San Francisco: US Naval Technical Mission to Japan, December 1945. National Diet Library, Tokyo.

———. "Enclosure (A): Historical Narrative of the US Naval Technical Mission to Japan." *Report of the US Naval Technical Mission to Japan—History of Mission*. San Francisco: US Naval Technical Mission to Japan, January 1946. National Diet Library, Tokyo.

———. "Shipboard Rocket Launchers." *Report of the US Naval Technical Mission to Japan*. San Francisco: US Naval Technical Mission to Japan, February 1946. National Diet Library, Tokyo.

———. "Japanese Fuels and Lubricants—Article 5. Research on Rocket Fuels of the Hydrogen Peroxide-Hydrazine Hydrate Type." *Report of the US Naval Technical Mission to Japan*. San Francisco: US Naval Technical Mission to Japan, February 1946. National Diet Library, Tokyo, Japan.

———. "Pharmacology and Malariology in Japan: Civilian and Naval." *Report of the US Naval Technical Mission to Japan*. San Francisco: US Naval Technical Mission to Japan, February 21, 1946. National Diet Library, Tokyo.

———. "Japanese Fuels and Lubricants, Article 1- Fuel and Lubricant Technology." *Report of the US Naval Technical Mission to Japan, Ship and Related Targets—Japanese Navy Diesel Engines*. San Francisco: US Naval Technical Mission to Japan, February 21, 1946. National Diet Library, Tokyo.

———. "Japanese Torpedoes and Tubes, Article 2: Aircraft Torpedoes." *Report of the US*

Naval Technical Mission to Japan. San Francisco: US Naval Technical Mission to Japan, March 1946. National Diet Library, Tokyo.

United States Pacific Fleet and Pacific Ocean Areas. "Japanese Artillery Weapons." CINCPAC-CINCPOC Bulletin 152, vol. 45 (July 1, 1945). https://www.bulletpicker.com/pdf/CINCPAC-Bulletin-152-45-Japanese-Artillery-Weapons.pdf.

University of Tokyo. "Ohsumi, Japan's First Satellite." http://www.u-tokyo.ac.jp/en/whyutokyo/hongo_hi_009.html.

University Space Engineering Consortium (UNISEC). *Report on Japanese University Rocket Projects*. Tokyo: UNISEC, October 2011.

Vaughn, Diane. *The Challenger Launch Decision: Risky Technology, Culture, and Deviance at NASA*. (Chicago: University of Chicago Press, 1997).

Vogel, Ezra F. *Japan's New Middle Class*. Berkeley: University of California Press, 1965.

Walthall, Anne. *The Weak Body of a Useless Woman*. Chicago: University of Chicago Press, 1998.

Watanabe, Hirotaka. "Japan–US Space Relations during the 1960s: Dependence or Autonomy?" In *54th International Astronautical Congress: 29 September—2 October, Bremen, Germany*, 1–16. Paris: International Astronautical Federation, 2003.

———. "The Evolution of Japanese Space Policy: Autonomy and International Cooperation." In *History of Rocketry and Astronautics: Proceedings of the Thirty-Ninth History Symposium of the International Academy of Astronautics: Fukuoka, Japan (2005)*, 271–295. San Diego: AAS, 2011.

Watanabe, Junichi. "The Moment HAYABUSA Became a Meteor." https://global.jaxa.jp/article/special/hayabusareturn/watanabe01_e.html.

Webb, James E. *Space Age Management: A Large Scale Approach*. New York: McGraw Hill, 1969.

Werner, Richard A. *Princes of the Yen*. London: Routledge, 2003.

Wheelan, Albert D. "A 'Born-Again' Space Program." *International Security* 11, no. 4 (Spring 1987): 142–150.

Wijeyeratne, Subodhana. "Between the Rocket and the Deep Blue Sea: Space Facilities and the 'Fishing Problem' in Southern Japan, 1950–1990." *Historia Scientiarum* 31, no. 22 (January 2022): 106–127.

Willard-Foster, Melissa. "Planning the Peace and Enforcing the Surrender: Deterrence in the Allied Occupation of Germany and Japan." *The Journal of Interdisciplinary History* 40, no. 1 (Summer 2009): 43, 50–51.

Williams, Frank. "Japanese Aeronautical Research Program and Achievements." In *Towards New Horizons: Technical Intelligence Supplement: A Report Prepared for the AAF Scientific Advisory Group*. Dayton, OH: Army Air Force Headquarters Air Materiel Command, May 1946. National Diet Library, Tokyo.

Wilson Center Digital Archive. "Report on Meetings between Chinese and Soviet Representatives on Rocket Production." September 23, 1957. Wilson Center Digital Archive, RGAE f. 8157, op. 1, 1957, d. 1991, l. 77–80. Obtained and translated for CWIHP by Austin Jersild. https://digitalarchive.wilsoncenter.org/document/116821.

Wohl, Robert. *A Passion for Wings: Aviation and the Western Imagination, 1908–1918.* New Haven: Yale University Press, 1994.

World Bank. "Life Expectancy at birth, Female (Years)—Japan." https://data.worldbank.org/indicator/SP.DYN.LE00.FE.IN?locations=JP&locale=ja.

World Economic Forum. "Which Countries Spend the Most on Space Exploration?" January 11, 2016. https://www.weforum.org/agenda/2016/01/which-countries-spend-the-most-on-space-exploration.

Wray, William D. "Japanese Space Enterprise: The Problem of Autonomous Development." *Pacific Affairs* 64 (Winter 1991): 463–488.

Wyczalek, Floyd A. "Assessment of Aerospace Technology in Japan Viewed from an American Perspective." *SAE Transactions* 100, Section 1: Journal of Aerospace, Part 2 (1991): 1947–1954.

Yoshikawa, Makoto, Junichiro Kawaguchi, Akira Fujiwara, and Akira Tsuchiyama. "Hayabusa Sample Return Mission." In *Asteroids IV*, ed. P. Michel, F. E. DeMeo, W. F. Bottke, and Renée Dotson, 408. Tucson: University of Arizona Press, 2015.

Yoshino, Naoyuki, and Farhad Taghizadeh-Hesary, eds. *Japan's Lost Decade: Lessons for Asian Economies.* Singapore: Springer Singapore, 2017.

Yoshioka, Hitoshi. "The Establishment of Space Development Organisations." In *A Social History of Science and Technology in Contemporary Japan, Vol. 3: High Economic Growth Period*, ed. Shigeru Nakayama, 243–256. Melbourne: Pan-Pacific Press, 2006.

Xin, Ling. "China's Moon Atlas Is the Most Detailed Ever Made." *Nature*, April 25, 2024. https://www.nature.com/articles/d41586-024-01223-0.

Zorn, E. L. "Israel's Quest for Satellite Intelligence." *Studies in Intelligence* 10 (Winter–Spring 2001): 33–38.

Zwicky, Fritz. "Remarks on the Japanese War Technical Report." December 1945. Archives of the Secretariat of the US Air Force Scientific Advisory Board, Washington DC.

Magazines and Newspapers

Aera

"Chūgoku uchū jikken, seikō de Nihon ni mizu." December 6, 1999.

"Mitsubishi jōkō to Ishiha no shitsujō H2 roketto 8 ki shippai no genin." June 19, 2000.

"Uchū ni nobiru erebētā chijō to seishi eisei o musubu kumo no ito." March 3, 2003.

Arms Control Today. "Factfile: The Proliferation of Ballistic Missiles." *Arms Control Today* 22, no. 3 (April 1992): 28–29.

Asahi Shimbun

"Hikōki kenbutsu-ki Yoyogi Hara no shiyō Hino taii Tokugawa taii (Yoyogi renpeiba de shiken hikō)." December 15, 1910.

"Tokugawa taii sōjū no fuaruman-shiki fukuyō hikōki." December 16, 1910.

"Abura nure no hikō-fuku no naka ni hana no Hyōdō-san jū ni-ki tsubasa o nabete Shimoshidzu no kyōgi taikai kowaretaru do kei hikō-ka ni fuhei ga ōi Negishi-kun wa yarinaoshi / hikō kyōgi taikai." June 3, 1922.

"Kyōdan kara sora e Ferisu-kō jokyōyu Mabuchi Chōko jō." July 10, 1933.

"Kaigun kōkūgijutsushō-chō ninmei Wada Misao shōshō / Wada shōshō ryakureki." May 24, 1940.

"'Heiki no senshi' ni hatsu no kōshō gijutsu no homare 100 yo-mei 'fuhai no sōbi' ni kagayaku rikugun yūkōshō." November 15, 1941.

Sen koto hikkei / bōgyoryoku ni nayami saikin no bakugeki-ki kaisetsu (1)." June 21, 1943.

"Sen koto hikkei / kogata de kōsoku-ryoku e saikin no bakugeki-ki kaisetsu (2)." June 22, 1943.

"Sen koto hikkei / sokuryoku to tōsai karyoku saikin no bakugeki-ki kaisetsu (3)." July 23, 1943.

"Sen koto hikkei / kōzokuryoku no jūyō-sei saikin no bakugeki-ki kaisetsu (4)." July 24, 1943.

"Kōsō no kiryū—kinō Michikawa de Kappa uchiage." Akita supplement. December 24, 1953.

"Shin chōkyori roketto-hō butai Ōshū e rikugun tōkyoku happyō gunji." October 31, 1954.

"Iwaki chō Michikawa kaigun de—ken mo kyōryoku jissoku wa 32-nen kara." Akita supplement. August 2, 1955.

"Roketto—watashi no hanashi." August 19, 1955.

"'Bebii-s' kei mo jōjō—kokusan roketto tesuto." Akita supplement. August 24, 1955.

"Chōon roketto-ki Tōdai Itokawa kyōjura no kenkyū kagaku." December 12, 1955.

"Kotoshi no hanashi." Akita supplement. December 17, 1955.

"Ichigatsu ni 'Kappa' jikken." Akita supplement. December 28, 1955.

"Shin roketto o tsukuru—rai aki no Itokawa kyōjū ga happyō." Akita supplement. February 9, 1956.

"Jinkō kōu, kairyū chōsa ni." Akita supplement. February 10, 1956.

"24 nichi no Kappa jikken ni sonaete." Akita supplement. September 21, 1956.

"Roketto jikken ni yosete." Akita supplement. September 23, 1956.

"Sensei to seito no gassoku—Yokote ni-chū ni yon tsubō no seiza." Akita supplement. November 2, 1956.

"Isshun ni kumoma ni—'Kappa' roketto jikken." Akita supplement. December 9, 1956.
"Kappa-3 ki tetsu kan hikō jikken 'Chōonsoku no hashiri ya'—Michikawa kaigan kenbutsujin mo odoroku." Akita supplement. June 24, 1957.
"Roketto kansoku dōryōkukai tansei." Akita supplement. July 25, 1957.
"Shigoto shiyasui yo—roketto kansoku dōryōku kai hassoku." Akita supplement. July 30, 1957.
"Kokusai chikyū kansoku nen—sho no roketto kansoku." Akita supplement. September 3, 1957.
"Kansoku jin zokuzoku Michikawa e—roketto 16 nichi, honban no ichiban ki." Akita supplement. September 12, 1957.
"Mujōken de kyōryukō—gyogyōsha ga mōshiawasu." Akita supplement. September 12, 1957.
"Tōbu kaishū testo—17 nichi Kappa go-kei ni sonaete." Akita supplement. September 13, 1957.
"Kōkei roketto bakuhatsu jikken—chūgakusei hitori shinu." Akita supplement, September 25, 1957.
"Jinkō eisei no X-masu kēki—Akita, onedan wa 20 man en." Akita supplement. December 10, 1957.
"Kokusai kyōsō ni osore? Roketto yosan, sanbai heru." Akita supplement. January 22, 1958.
"Asu Michikawa kaigan de—Itokawa kyōjura—Byōsoku sen mētoru no seinō." Akita supplement. February 5, 1958.
"Roketto asobi de jūshō." Akita supplement. March 4, 1958.
"Kōha roketto to zengeru fusai Doitsu kagaku no monogatari." June 20, 1958.
"Chūgakusei roketto asobi de jūshō." Akita supplement. January 30, 1959.
"Sanin ga jūkeishō, Hanawa." Akita supplement. June 26, 1959.
"Kon go wa tsūshin eisei—uchū danwa-shitsu." April 18, 1960.
"Mitsubishi denki de jūchū—roketto tsuibi rēdā." November 29, 1960.
"Keidanren ni uchū heiwa riyō—keizai dantai." May 17, 1961.
"Tōdai uchū kūkan kansoku jo o kikō—Kagoshima, Uchinoura." February 3, 1962.
"Tsūshin eisei to kokusai chūkei—nyūsu no me." February 13, 1962.
"Tōkyō orinpikku ni eisei de terebi hōsō o Okazaki taishi enzetsu—taikiken-gai heiwa riyō-i; Tōkyō orinpikku to terebi eisei." March 23, 1962.
"Roketto jikken shippai de jimoto no fuan takamaru—michikawa kaigan." Akita supplement. May 26, 1962.
"Noshiro e iten ga ketteiteki—Roketto sentā Tōdai seiken ga setsumei kai." July 12, 1962.
"Kagodai ni kyōryoku tanomu—Hideo Itokawa roketto uchiage de." April 27, 1963.
"Uchinoura roketto kaihatsujō kaisho." December 9, 1963.
"Boku no yume, watashi no yume." Kagoshima supplement. January 6, 1964.

"Honnen do ni mo kōji susumu—roketto dōro." Kagoshima supplement. May 7, 1964.
"Roketto memo—jikkenjō." Kagoshima supplement. July 5, 1964.
"Roketto memo—tantō suru hitotachi." Kagoshima supplement. July 7, 1964.
"Kenbutsu mo yōryō yoku—roketto nare no machi no hito." Kagoshima supplement. October 30, 1964.
"Kondo wa Mu da—yume wa jinko eisei o." Kagoshima supplement. January 1, 1965.
"Uchinoura roketto yon nen." Kagoshima supplement. February 3, 1966.
"Kensetsu susumu Mu roketto daichi—Uchinoura." Kagoshima supplement. April 6, 1966.
"Tanegashima no roketto kichi—uchū kaihatsu." May 7, 1966.
"Mitsubishi denki mo gijutsu teikei e tsūshin eisei setsubi o buntan seisan—uchū kaihatsu." May 17, 1966.
"Tōmin, minna ga kangei." Kagoshima supplement. May 24, 1966.
"Taibei yushutsu ni noridasu—Mitsubishi jūkō, Purinsu ryōsha kei no roketto." July 22, 1965.
"Nihon nado sangoku ni denwa sābisu maitsūshin eisei kaisha—uchū kaihatsu." July 3, 1966.
"Nihondenki, Beisha to gōben tsūshin eisei chijō-kyoku." August 3, 1966.
"Futatsu no uchū genchi o miru." September 30, 1966.
"Tanegashima wa secchi būmu." Kagoshima supplement. October 1, 1966.
"Uchū sentā o shisatsu—secchi hantai no Miyazaki gyomin." Kagoshima supplement. November 13, 1966.
"Eikyō kensa nado yōbō—gyomin, uchū sentā de hanashiai." Kagoshima supplement. November 17, 1966.
"Kankō shiki ni fukikin atsume." November 28, 1966.
"Roketto no machi Uchinoura genkon—'Kotoshi koso' no kitai." Kagoshima supplement. January 2, 1967.
"Ranibādo 2-gō kinen kitte o hatsubai—Yūseishō." January 20, 1967.
"Gyogyō chōsa oeru kaiketsu o—Tanegashima roketto uchiage ni." Kagoshima supplement. February 8, 1967.
"Raigatsu chū ni wa ichibu kansei—'roketto dōro' settei kyū picchi." Kagoshima supplement. February 24, 1967.
"17 nichi kara uchiage Kagakugijutsuchō no roketto—Tanegashima no roketto jikken." March 3, 1967.
"Beichū kankei kaizen mo tōron Miki gaishō to Rasuku chōkan uchūchūkei o tsūjite." March 5, 1967.
"Gimon-darake 'hinomaru eisei' gakujutsu kaigi ga shiteki—Tōdai uchū kōkū-ken mondai." March 21, 1967.
"Kaku hōmen kara giwaku ni memo." March 21, 1967.

"Chigen ga gakkuri—Uchinoura, Tanegashima, 'Yota yota roketto.'" March 28, 1967.

"Tōdai uchū ken—giwaku wa fumon—Itokawa kyōjū no jinin ga moto kimari." March 28, 1967.

"Cha no ma ni chikyū no kao Sōren no jinkō eisei kara TV chūkei NHK—uchūchūkei." June 9, 1967.

"'Jūbun kenkyū o kasaneta,'—machi wa onegau." Kagoshima supplement. July 5, 1967.

"Nagare hōdai no uchū sentā Tanegashima." Kagoshima supplement. July 13, 1967.

"Mitooshi shita tanu uchiage." Kagoshima supplement. July 15, 1967.

"Jinzai, haba hiroku kesshū Uchū kaihatsu jigyōdan no kōsō kagaku gijutsuchō—uchū kaihatsu." September 1, 1967.

"Tsūshin eisei keiyu hayaku naru 'moshimoshi' Yamaguchi ni chijō-kyoku Nihon—Ōshū Chūkintō." November 3, 1967.

"Roketto mondai ni gatsu ni ugoku ka." Kagoshima supplement. January 25, 1968.

"Roketto jikken no gyogyō hoshō—seifu no taisoku ni manzoku." Kagoshima supplement. August 18, 1968.

""Hana hiraku uchū sangyō eisei tsūshin jidai hikaete—sangyō." September 1, 1968.

"Roketto taidan—uchiage matsu Uchinoura Tanegashima." Kagoshima supplement. September 1, 1968.

"Mekishiko Orinpikku Nihon ni mo karā de maitsūshin eisei kaisha keikaku." September 4, 1968.

"Yonki mei uchiage chūshi—Uchinoura Tōkyō daigaku roketto Miyazaki no gyosen ga bōgai." September 13, 1968.

"Ichibu gyomin no hyōjo kataku—uchiage ni yahari fukuzatsu." September 17, 1968.

"Isshun kieta shimpai—kurōhō irareta to yorokobi." Kagoshima supplement. September 18, 1968.

Chūshi ni gakkari—roketto uchiage ni kengaku jidō." Kagoshima supplement. September 19, 1968.

"Roketto 2-ki no tesshū." Kagoshima supplement. September 21, 1968.

"Hasō kōji isogu—roketto dōro." Kagoshima supplement. October 15, 1968.

"Rokkuheedo roketto urikomi—Mitsubishi jūkō nado mōshire." November 15, 1968.

"Chikadzuku roketto shiizon—Uchinoura, Tanegashima kichi." Kagoshima supplement. December 23, 1968.

"Ryokō wa roketto ni notte—ōbei wa wazuka ichi jikan." Kagoshima supplement. January 1, 1969.

"Zōshō, shōnin no ikō 'Uchū kaihatsu jigyōdan' no shinsetsu eisei nado gyōmu o ipponka." January 13, 1969.

"Yamakawa minato ni senin no shukusha." January 21, 1969.

"Tōdai no jinkō eisei uchiage chūshi shoho-teki misu tsudzuku." January 22, 1969.

"Jiko ni sonae Shobōsha mo—Roketto uchiage, hama no ue no kengakusha." Kagoshima supplement. February 7, 1969.
"Ayabumareru roketto uchiage—'jitsujō mushi' okoru gyomin." Kagoshima supplement. May 18, 1969.
"Rēzā de kansoku Hitachi seisakusho jinkō eisei tsuiseki ni seikō uchū kaihatsu." June 13, 1969.
"Uchinoura—roketto hassha jumbi wa OK." Kagoshima supplement. August 4, 1969.
"Jinkō eisei no shisei kenshutsu sōchi yūshūna kokusan-hin ga kansei—uchū kaihatsu." August 15, 1969.
"Roketto kichi no meian—Tanegashima." Kagoshima supplement. September 8, 1969.
"Chū ni uita shukuga hanabi—'yasumi mo muda' ukanu gyomin." Kagoshima supplement. September 23, 1969.
"Umetatechi no sōzei hajimaru." Kagoshima supplement. August 13, 1969.
"Roketto kichi chūshin ni kankō kaihatsu." Kagoshima supplement. January 1, 1970.
"Kore de jikken mo kidō ni—kakushu shikenshitsu ga totonou." January 14, 1970.
"Mitsubishi jū ga Amerika-sha to gōi roketto nenryō gijutsu dōnyū—uchū kaihatsu." February 15, 1970.
"Roketto yūdō seigyo no gijutsu dōnyū de gōben kaisha o setsuritsu Nihondenki to Haniueru." June 19, 1970.
"Eisei tsūshin nōritsu yoku kokusai denden ga shin hōshiki—tsūshin." July 5, 1970.
"Yūgo ni chijō eisei kichi kensetsu Nichiden to hishi-shō ga kyōdō de." August 19, 1972.
"Zanbia no tsūshin eisei shisetsu Nihondenki ga juchū—yushutsu." November 19, 1972.
"Eisei chūkei no ryōkin nesage—terebi." September 22, 1973.
"Kakkidzuku jitsuyō eisei no kaihatsu—'gijutsu kakusa' no kabe o itomu." September 28, 1973.
"Arabu no hōsō tsūshinmō seibi Nihon no kyōryoku gutai-ka eisei chijō-kyoku shuto-kan denwa terekkusu Nichiden teian, Ōbei-sha o shirizokeru." March 15, 1974.
"Shinkansen no kotsu ikashite kidou eisei ga uchigeta yo." March 7, 1977.
"Nihon kigyō mo uchū jikken o Keidanren ga setsumeikai—uchūrenrakusen." September 14, 1978.
"Amerika IBM kei no eisei tsūshin kaisha kara juchū Fujitsū ga henfukuchō sōchi—nichibeibōeki." July 12, 1979.
"Uchū chūdan de oyako taimen." August 16, 1980.
"Uchū no kodō o kanchi." January 4, 1981.
"Sangyō kai, jimae no eisei kōsō sōki uchiage hitsuyō Keidanren hōkoku—tsūshin eisei 1983-nen." January 5, 1983.
"Uchū geigeki heiki shisutemu kaihatsu-hi o morikomu—daitōryō yosan kyōsho." January 27, 1984.

"H 2 roketto zunō mo shinzō mo kokusan de—Godai Tomifumi shi (90 nendai o niramu)." February 21, 1985.
"Uchū kaihatsu kikaku, minkan no eisei tōnyū ni wa furezu." March 14, 1985.
"'Eisei o yasuku uchiagemasu' Arian kaishachō ga rainichi sēru." April 27, 1985.
"Kōkū uchū gijutsu-ken-sho ga sōritsu 30 shūnen sekai ni hokoru dokuji no eisei seigyo sōchi." July 20, 1985.
"Junkokusan de taikei gijutsu eisei (nyūsu rain)." August 31, 1985.
"Yūjin uchū hikō shinpo (nyūsu rain)." November 2, 1985.
"Rēzājairo (haitekku no kao)." November 14, 1985.
"Bunkyō kagaku chūgaku 40 nin gakkyū dōnyū (61-nen yosando seifuan no naiyō)." December 29, 1985.
"1996 nen uchū jikkenshitsu kokusai kyōryoku no uchū kichi Nihon no mojūru yosan keikaku ga katamaru." April 8, 1988.
"Senkyō to shakkin taisaku o yūsen kokusai jōsei no ninshiki futōmei 90 nendo yosa." December 25, 1989.
"Shūin-i, uchū kichi kyōtei de 'heiwa riyō' o ketsugi konkokkaichū ni shōnin mo, 1989-nen 06 tsuki 14-nichi." April 6, 1989.
"Wakusei tansa e shin roketto Uchūken ga 150 oku-en tōji 'M 5' kaihatsu e." January 20, 1990.
"Tsuki ni koronii keisei made 5 dankai kōsō happyō 'getsumen kichi to shigen kaihatsu' no kai." February 27, 1990.
"Jishu kaihatsu rosen ni tenkanki kanmin ainori fukanō ni Nichibei jinkō eisei mondai ga ketchaku." April 7, 1990.
"Dai-10 kai Asahi yasashii kagaku no kyōshitsu." April 21, 1990.
"H 2 chōtatsu de shingaisha setsuritsu 77-sha no shusshi de, 7 gatsu ni." April 29, 1990.
"Dokushi kaihatsu no yūjin purattohōmu, Gainen sekkei wa hobo shūryō." June 2, 1990.
"Kyōsō to kyōryoku to uchū bijinesu josō." January 26, 1991.
"Yane, kanzen ni fukitobu genba kenshō hajimaru Kakuda shi no roketto jikenjō bakuhatsu." May 17, 1991.
"Roketto jikkenjō de bakuhatsu—H2 keikaku jikō tsuzuki." May 17, 1991.
"Roketto kaihatsu kankeisha ni shōgeki—Kokachi no Mitsubishi jūkō jikō." August 10, 1991.
"Puranetariumu to tenmon kōen kai no yūbe." September 2, 1991.
"Funsha-ki no haikan no yōsetsu ni kekkan." October 19, 1991.
"Supēsu shatoru gijutsu, nichiō de kyōdō kaihatsu ōshū uchūkikan to kagichō ga gōi." February 25, 1992.
"Kanagawa kagaku gijutsu akademii kōen kai 'yume to roman no uchū kagaku' (moyōshi)." January 27, 1992.
"H 2 roketto no nenshō shiken chūdan." April 14, 1992.

"Dai-14 kai Asahi yasashii kagaku no kyōshitsu 'koko made kita uchū kaihatsu.'" April 28, 1992.

"H 2 roketto, shiken-chū ni kasai 93-nen haru no uchiage kon'nan ni." June 19, 1992.

"Roketto H 2 no 'yowai shinzō' bakuhatsu enjin no shasshin kōkai.' July 1, 1992.

"Yōsetsu-bu kiretsu, hokano enjin kara mo H2 roketto uchiage enki." July 10, 1992.

"Roshia no uchū gijutsu sokodjikara to kibishii genjitsu kanmin gōdō no Nihon chōsa-dan ga genchi shisatsu." July 20, 1992.

"Bakuhatsu ga gen'in ka? Enjin ga shū-ka kokusan etchitsūroketto jikken jikō." June 22, 1992.

"Michinori kewashii junkokusan roketto toraburu tsudzuku H 2." June 29, 1992.

"H-2 roketto mata ashibumi kishōsei uchiage ni eikyō." June 30, 1992.

"Uchū gijutsu ōuridashi Roshia, gaika kakutoku sakusen." July 19, 1992.

"Jiko wa yoken funō Aichi kenkei ketsugo H 2 roketto haretsu-shi jikobō." August 7, 1992.

"Kokusan roketto kaihatsu wa chakujitsu ni." August 12, 1992.

"Ōte denki mēkā, eiseibijinesu o kyū kakudai kaden fushin o oginau nerai mo." November 11, 1992.

"Uchū fōramu chikyū wa hitotsu 12 gatsu 11-nichi, Nagoya de." November 13, 1992.

"Go etsu dōyō shinogi o kezuru 2 shōchō no uchū roketto." January 21, 1993.

"Eisei uchiage kigyō no Shyaruru bigo kaichō rainichi." March 29, 1993.

"H 2 roketto, nankan no shiken toppa tashi ga—Teimei tsudzukeru eisei ichiba." July 9, 1993.

"Deruta kurippā retorona sugata ni shingijutsu mansai tan-danshiki 'jisedai shatoru' fujō." September 18, 1993.

"Machi kōjō (zoku kauntodaun H2:3) (Nagoya)." January 14, 1994.

"Saigo wa kami nomizo shiru uchiage semaru H2 roketto—jisshi sekininsha ni kiku." January 24, 1994.

"Yume no kitai H 2 roketto—100 ten manten hikō—nenshō kanryō ni hakushu." February 2, 1994.

"H 2 uchiage seikō chōnan tanjō, nijū no yorokobi." February 5, 1994.

"Roketto keiki kidō shūsei e Kagoshima Minamitanecho." February 5, 1994.

"'H2' seikō de hirogaru yume kokusai kyōryoku de kankyō kanshi eisei hōpu no kaihatsu mo." February 7, 1994.

"'H 2' sekai mo chūmoku uchiage ni kaigai 33-sha no shuzai (media)." February 8, 1994.

"Wagamono ni suru juku shite ita shinkansen no gijutsu (Shima Hideo no sekai: 1 kataru)." July 18, 1994.

"Chijō sōchi no tora buru ka H 2 roketto 2-gōki no hassha-chū tome." August 19, 1994.

"H 2 hassha chūshi was 'shikkei de nai' Uchū kaihatsu jigyōdan ijō o hitei." August 19, 1994.

"Seishi eisei Kiku 6-gō no shippai wa chjō shiken ga fujūbun chōsa kaname ga shiteki." December 13, 1994.

"Aibanri eisei fuguai-ki ni kanryō no arasoi." February 20, 1995.
"Dai-20 kai Asahi yasashii kagaku no kyōshitsu." May 16, 1995.
"Harē suisei' no kinen botoru hatsubai (nyūsurain)." September 7, 1995.
"Uchū kaihatsu ootsun o kaitei eisei no kosuto-gen nado hakaru." January 24, 1996.
"Yamada Takaaki-san roketto shisutemu shachō (henshū-chō intabiyū)." March 1, 1997.
"Jiki roketto wa anka de kogata ni kairyō o keikaku Kagakugijutsuchō." May 26, 1997.
"Tanegashima roketto uchiage kikan, nen 130 nichi ni enchō." June 11, 1997.
"Uchū ni semaru kyō Morioka de kagaku kōen kai / Iwate." December 6, 1997.
"Eisei bijinesu ni an'un—H2 roketto uchiage shikkei." February 22, 1998.
"Yasunori Matogawa-san 'anata no namae o kasei e' no kagaku-sha (hito)." June 20, 1998.
"Jigyō-dan no hihan, hatsukaigō de funshutsu uchū kaihatsu kihonmondai kondan kai." July 16, 1998.
"'Uchū no machi' yume boshi o machi-okoshi kitai, jimoto rakutan H 2 shippai." November 16, 1998.
"Ibento marion." January 29, 1999.
"Nagoyamaru." November 18, 1999.
"Yotō ridatsu hatsugen (ka ekubota)." November 28, 1999.
"Shichōsonda yori / Okayama." December 12, 1999.
"Kōkyō jigyō kaikaku, michi nakaba jūten haibun han'ei sezu 2000-nen seifu yosan-an." December 24, 1999.
"Ochimasen yō ni Chikushino no JR Futsukaichi eki ni janbo ema Fukuoka." December 25, 1999.
"Tōkai no gijutsu-ryoku o kesshō H 2 shissoku, machikōjō ni mo hamo." December 26, 1999.
"Sayōnara 20 seiki: naka jiken sesō kagaku gijutsu no Nihon 10-dai nyūzu." December 27, 1999.
"Chūgoku, uchū kara 'niihao' yūjin hikō madjika? Junbi wa chakuchaku." January 24, 2000.
"Nipponban mujin shatoru no jikki seisaku o tōketsu kagi-ten hōshin." August 2, 2000.
"Musekinin jigyō-dan (Mado, Ronsetsu wa yōin ka mamora)." August 3, 2000.
"H 2 uchiage hiyō shiharae uchū kaihatsu jigyōdan ga 35 oku man matome minji chōtei." October 16, 2000.
"Uchū kaihatsu no kiban gijutsu kyōka mezasu—seifu bukai hōkoku." November 9, 2000.
"Kokusan shin roketto 'H2A' no kumitate susumu jigyō-dan ga setsumei / kagoshima." July 18, 2001.
"Uchū kaihatsu ni daidageki teitai, Amerika ni todomarazu shatoru tsuiraku jikō." February 3, 2003.
CBS News. "Not Much 'Hope' for Mars Probe." December 9, 2003.

Chōsa to jōhō. "Uchū seisaku no shireitō kinō o meguru giron." No. 748 (April 5, 2012): 1–12.

Chūō Kōron. "Gurabia han: saishin kagakkai no kyōi." Vol. 46, no. 1 (January 1, 1931): 453.

Gunji to gijutsu. No. 82 (October 1933): frontispiece.

Hatsumei. "Kanten mo nan'nosono roketto riyō no jinkō kōu." Vol. 28, no. 8 (August 1931): 47.

Ichinen gakushū

"Sutekina roketto." June 1959, 36–39.

"Tsuki e iku roketto basu." February 1960, 2–3.

ISAS News. "M-3S II-8 banki no hishō jōkyō." No. 167 (February 1995): 7.

Japan Times. "Nissan Motors to Sell Aerospace Unit to IHI." February 14, 2000.

Kagaku Zasshi. "Roketto no kenkyū." Vol. 18, no. 4 (April 1933): 33.

Keidanren Geppō

"Uchū kaihatsu chōki keikaku chōsa-dan shoken oyobi reigen." Vol. 19, no. 9 (September 1971): 60–61.

"Uchū kaihatsu ni kansuru wareware no kenkai / keizai dantai rengō-kai uchū kaihatsu suishin kaigi." Vol. 25, no. 2 (February 1977): 43–47.

Los Angeles Times

Arthur J. Dommen. "Japanese Rocketeers Have Come a Long Way." October 3, 1965./

"Much Talk, Little New: U.S.–Japan Talk." July 18, 1966.

Dusko Doder. "Japan Buying U.S. Rocket System; Unions Assail Technology Export." March 7, 1973.

"Japanese Firm to Launch Hughes Satellites." November 27, 1996.

Mainichi Shimbun

"Chūkyō ki, roketto shiyō ka." October 4, 1958.

"Gaiden memo—roketto seisaku de Seidoku ga Chūgoku o kyōso." January 10, 1969.

"Kaigai bunka uchū kaihatsu: Chūgoku, 5-nen inai ni ningen eisei!? Amerika gakusha no yosoku." May 7, 1977.

"Kaigai bunka uchū kaihatsu: Chūgoku ni mo yūjin eisei uchiage nōryoku." June 7, 1985.

"Ipushiron, H 3 shippai tsudzuite mo . . . 'nebāgibuappu' 'kunan no nochi ni wa kanarazu hana ga' roketto hasshaba no jimoto, jakusa shokuin ni ēru 800-ri messēji no ōdan-maku okuru Kagoshima Kimotsuki machi." March 28, 2023.

New York Times

"Japan Tests Rocket." October 30, 1957.

"Honeywell Link to NEC." July 2, 1986.

David E. Sanger. "A Japanese Innovation: The Space Antihero." December 8, 1990.

Andrew Pollack. "Japan Set to Launch Rocket to Match West: Japan Set to Launch Its Own Rocket." January 25, 1994.

Steven Brull. "Japan Satellite Market's Opening to U.S. Is a Dud for Both Sides." March 12, 1994.

"Japan Drops Cornerstone of Program for Rockets." December 10, 1999.

Nihon Keizai Shimbun. "Uchū kaihatsu jishu rosen hashiru." October 27, 1969.

Science. "Adios, ADEOS: Japanese Satellite Lost." July 1, 1997.

Shōnen gonensei. "Roketto sono hatsumei ka." Vol. 17, no. 3 (June 1936): 215.

Shōnen kurabu. "Yūbin roketto." Vol. 23, no. 10 (September 1936): 9.

Shūkan shokokumin. "Shin eiki no hanashi—Itokawa sensei kakonde." Vol. 3, no. 1 (January 1944): 24.

Space Japan Review

Rieko Hayakawa. "Eisei tsūshin to watashi." No. 16 (April–May 2001): 15–19.

Susumu Kitazume. "Mitsubishi jūkō meiyū ni okeru H-IIA roketto enjin no kaihatsu." No. 19 (October–November 2001): 1–12.

Fumihiko Tomita. "Chūgoku no saikin no uchū kaihatsu kenmonroku." No. 20 (December 2001–January 2002).

"Sekai no eisei kigyō no CEO ni kiku." No. 22 (April–May 2002): 1–2.

"Sekai no eisei kigyō no CEO ni kiku." No. 23 (June–July 2002): 1–6.

"Sekai no eisei kigyō no CEO ni kiku." No. 26 (December 2002–January 2003): 1–13.

Yūko Yoshikawa. "Eisei tsūshin to watashi: Miryoku o saidaigen ni tsutaeru koto ga kōhō senden no shimei." No. 26 (December 2002–January 2003): 13–16.

Yoshiaki Suzuki. "H–IIA roketto 4-gōki uchiage seikō!" No. 26 (December 2002–January 2003): 1.

Hideki Mizuno. "Rensai tokushū. Eisei yōwa." No. 27 (February–March 2003): 1–2.

Shin'ichi Kimura. "Kogata eisei tōsai no minseiyōki riyō shisutemu de gazō shutoku ni seikō." No. 27 (February–March 2003): 13.

Hideki Mizuno. "Eisei Yōwa (2)." No. 28 (April–May 2003): 2.

Hideki Mizuno. "Eisei Yōwa (3)." No. 29 (June–July 2003): 3.

Hideki Mizuno. "Eisei Yōwa (4)." No. 30 (August–September 2003): 1.

Takashi Iida. "Kaze wo yomeru ka." No. 35 (June–July 2004): 3–6.

Susumu Kitazume. "Nihon no anzen hoshō ro uchū kaihatsu." No. 36 (August–September 2004): 1–4.

Takeo Ueda. "Uchū riyō to anzen hoshō." No. 37 (October–November 2004): 1–10.

Akira Meguro. "Eisei Yōwa—subarashiki eisei jinsei." No. 42 (August–September 2005): 1–3.

Susumu Kitazume. "H–IIA roketto 3-gōki DRTS oyobi USERS no uchiage seikō." No. 25 (October–November 2005): 1–3.

Sakamoto Hiroshi. "Eisei yowa kiku 6-gō ni tazusawatte." No. 43 (October–November 2005).

Kazuhiko Hashimoto. "Aru eisei tsūshin gijutsusha no omoide—dai-2 kai." No. 66 (February–March 2010): 1–2.

Kazuhiko Hashimoto. "Aru eisei tsūshin gijutsusha no omoide—dai-3 kai.' No. 67 (April–May 2010): 1–3.

Kazuhiko Hashimoto. "Aru eisei tsūshin gijutsusha no omoide—dai-4 kai." No. 68 (June–July 2010): 1–2.

Takashi Hirose, Yutaka Imaizumi, and Hideyoshi Yoshida. "Higashinihon daishinsai de riyōsareta NTT saigai taisaku eisei tsūshin shisutemu." No. 76 (October–November 2011): 1–4.

Masahiro Nakao, Naoya Tomii, Shinichiro Takayama, Tsuyoshi Horishi, Takashi Horiuchi, Hiroki Hakuchi, Takashi Ikemi, Tadashi Ikema, Takashi Ikekawa, and Kei Sunagawa. "Higashinihon daishinsai ni okeru 'kizuna' oyobi 'kiku 8-gō' ni yoru jakusa taiō ni tsuite." No. 77 (December 2011–January 2012): 1–18.

"Eisei yōwa. moto tsūshin sōgō kenkyūsho-chō Fugono Nobuyoshi 1." No. 82 (February–May 2013): 2–4.

"Eisei yōwa: moto tsūshin sōgō kenkyūsho-chō Fugono Nobuyoshi 2." No. 83 (June–September 2013).

Sora

"Shinsekkeika no yume." September 1936, 261–263.

"Shinsekkeika no yume." February 1938, 202–204.

"Shinsekkeika no yume." March 1938, 306–309.

"Shinsekkeika no yume." October 1938, 210–213.

"Shinsekkeika no yume." December 1938, 418–421.

"Shinsekkeika no yume." April 1939, 114–117.

"Shinsekkeika no yume." January 1941, 98–101.

Yomiuri Shimbun

"Jūgo no gunji hatsumei ni hatsu no kōshō senshō in no shukun kō 'rikugun gijutsu yūkōshō' kagayaku 28-ken." November 15, 1941.

"Bōei kiki no kaihatsu wa, minkan to kyōdō sagyō de Nakasone chōkan—kokubō." February 13, 1970.

"'Kasei no nazo ni semaru' (jō) eisei tansa-ki fobosu 7 tsuki uchiage (rensai)." July 4, 1988.

"Uchū e no kauntodaun fōramu to ronbun kaiga sakuhin boshū (shakoku)." May 27, 1991.

"Tenkanki ni tatsu uchū kaihatsu (1) Amerika no jijō' ni yureru Nihon (rensai)." July 1, 1991.

"Chikyū tobitatsu wagaya no aji Mukai-san no uchū ryōri nyūshō-saku kettei = tokushū." May 28, 1994.

"Kokusaiuchū kichi' tanan'na shuppatsu 'yume no nedan' kakkoku no omoni ni 29-nichi ni mo kyōtei shomei." January 23, 1998.

"Sentan gijutsu no kōkai kōza 26-nichi, Kawasaki de kagaku gijutsu akademii ga kaisai = Kanagawa." August 20, 1998.

"Nosutoradamusu no daiyogen' no kenkyū kyōfu no daiō no 'shōtai' o kenshō." February 2, 1999.

"Bisei de ten mondai kōza 12-nichi ni hiraku kōshi wa Matogawa Yasunori uchū kagaku-ken kyōju = Okayama." December 9, 1999.
"Uchūken no 'M 5' uchiage shippai 1-dan-me ni atsuryoku ijō tenmon eisei, kidō tōnyū dekizu." February 10, 2000.
"Uchū sutēshon 'Miiru' Nihon ni hahen rakka no osore mo saishū funsha shippai no baai." March 22, 2001.
"Uchū 3 kikan tōgō e no kadai wa mokuhyō, igi o meikaku ni gijutsu no minkan iten mo hitsuyō (kaisetsu)." August 8, 2001.
"'Kokoro no fudoki' Kureshi (Hiroshima ken) uchū kōgaku-sha Yasunori Matogawa-san." April 7, 2004.
Uchū. "Homi uchu kichi keika chosatan hokokusho." No. 23 (1985).

Minutes of the National Diet of Japan

26th NDHR Cab., No. 46, October 8, 1957. https://kokkai.ndl.go.jp/simple/detail?minId=102604889X04619571008&spkNum=240&current=49#s240.
26th NDHR Com., No. 34, October 11, 1957. https://kokkai.ndl.go.jp/simple/detail?minId=102604816X03419571011&spkNum=34&current=52#s34.
26th NDHR For., No. 28, October 15, 1957. https://kokkai.ndl.go.jp/simple/detail?minId=102603968X02819571015&spkNum=76&current=53#s76.
27th NDHC Com., No. 2, November 7, 1957. https://kokkai.ndl.go.jp/simple/detail?minId=102714816X00219571107&spkNum=12&current=68#s12.
27th NDHR SST, No. 3, November 9, 1957. https://kokkai.ndl.go.jp/simple/detail?minId=102703913X00319571109&spkNum=84&current=10#s84.
31st NDHR Bud., Subcommittee, No. 3, February 27, 1959. https://kokkai.ndl.go.jp/simple/detail?minId=103105266X00319590227&spkNum=55&current=11#s55.
32nd NDHR SST, No. 4, August 11, 1959. https://kokkai.ndl.go.jp/simple/detail?minId=103203913X00419590811&spkNum=91&current=4#s91.
34th NDHR Com., No. 3, February 10, 1960. https://kokkai.ndl.go.jp/simple/detail?minId=103404816X00319600210&spkNum=32&current=3#s32.
34th NDHR Cab., No. 3, February 12, 1960. https://kokkai.ndl.go.jp/simple/detail?minId=103404889X00319600212&spkNum=12&current=4#s12.
40th NDHR E&C, No. 8, February 28, 1962. https://kokkai.ndl.go.jp/simple/detail?minId=104005077X00819620228&spkNum=99&current=28#s99.
46th NDHR Cab., No. 46, June 18, 1963. https://kokkai.ndl.go.jp/simple/detail?minId=104604889X04619640618&spkNum=19&current=40#s19.
48th NDHR Com., No. 8, March 12, 1965. https://kokkai.ndl.go.jp/simple/detail?minId=104804816X00819650312&spkNum=64&current=13#s64.
48th NDHR Cab., No. 17, March 16, 1965. https://kokkai.ndl.go.jp/simple/detail?minId=104804889X01719650316&spkNum=75&current=45#s75.

48th NDHC Com., No. 12, March 30, 1966. https://kokkai.ndl.go.jp/simple/detail?minId=104814816X01219650330&spkNum=178&current=19#s178.

48th NDHR Cab., No. 17, March 16, 1965. https://kokkai.ndl.go.jp/simple/detail?minId=104804889X01719650316&spkNum=75&current=45#s75.

48th NDHR SST, No. 5, February 17, 1965. https://kokkai.ndl.go.jp/simple/detail?minId=104803913X00519650217&spkNum=3&current=4#s3.

51st NDHC SST, No. 7, June 3, 1966. https://kokkai.ndl.go.jp/simple/detail?minId=105113913X00719660603&spkNum=38&current=5#s38.

51st NDHR SST, No. 23, June 2, 1966. https://kokkai.ndl.go.jp/simple/detail?minId=105103913X02319660602&spkNum=71&current=2#s71.

51st NDHR E&C, No. 31, June 27, 1966. https://kokkai.ndl.go.jp/simple/detail?minId=105105077X03119660627&spkNum=0&current=53#s0.

52nd NDHR Bud. No. 3, October 20, 1966. https://kokkai.ndl.go.jp/simple/detail?minId=105205261X00319661020&spkNum=60&current=58#s60.

52nd NDHR S&T, No. 1 after closing, November 29, 1966. https://kokkai.ndl.go.jp/simple/detail?minId=105213913X00119661129&spkNum=4&current=60#s4.

55th NDHR Com., No. 3, March 24, 1967. https://kokkai.ndl.go.jp/simple/detail?minId=105504816X00319670324&spkNum=22&current=54#s22.

55th NDHR Set., No. 4, March 28, 1967. https://kokkai.ndl.go.jp/simple/detail?minId=105504103X00419670328&spkNum=132&current=17#s132.

55th NDHR Bud., No. 10, April 3, 1967. https://kokkai.ndl.go.jp/simple/detail?minId=105505261X01019670403&spkNum=181&current=63#s181.

55th NDHR, Bud., Subcommittee, No. 3, April 21, 1967. https://kokkai.ndl.go.jp/simple/detail?minId=105505266X00319670421&spkNum=182&current=64#s182.

55th NDHR For., No. 18, July 13, 1967. https://kokkai.ndl.go.jp/simple/detail?minId=105503968X01819670713&spkNum=67#s67.

58th NDHC Cab., No. 14, April 25, 1968. https://kokkai.ndl.go.jp/simple/detail?minId=105814889X01419680425&spkNum=182&current=2#s182.

58th NDHR Bud. No. 2, February 6, 1968. https://kokkai.ndl.go.jp/simple/detail?minId=105805261X00219680206&spkNum=158&current=324#s158.

61st NDHR, Bud., Subcommittee, No. 3, February 26, 1969. https://kokkai.ndl.go.jp/simple/detail?minId=106105266X00319690226&spkNum=244&current=13#s244.

61st NDHR, Com., No. 3, February 27, 1969. https://kokkai.ndl.go.jp/simple/detail?minId=106104816X00319690227&spkNum=237&current=15#s237.

61st NDHR Plenary Session, No. 14, March 14, 1969. https://kokkai.ndl.go.jp/simple/detail?minId=106105254X01419690314&spkNum=14&current=21#s14.

61st NDHC Com., No. 7, March 25, 1969. https://kokkai.ndl.go.jp/simple/detail?minId=106114816X00719690325&spkNum=251&current=29#s251.

63rd NDHR SST., No. 6, April 2, 1970. https://kokkai.ndl.go.jp/simple/detail?minId=

106303913X00619700402&spkNum=5&current=1#s5.

63rd NDHR SST., Promotion Measures Subcommittee on Fundamental Issues in Space Development, No. 1, May 15, 1970. https://kokkai.ndl.go.jp/simple/detail?minId=106303917X00119700515&spkNum=2&current=2#s2.

84th NDHR SST., No. 6, March 29, 1978. https://kokkai.ndl.go.jp/simple/detail?minId=108403913X00619780329&spkNum=182&current=5#s182.

84th NDHR Cab., No. 11, April 11, 1978. https://kokkai.ndl.go.jp/simple/detail?minId=108404889X01119780411&spkNum=46&current=7#s46.

84th NDHC, Settlement of Accounts Committee, No. 4 after adjournment, September 1, 1978. https://kokkai.ndl.go.jp/simple/detail?minId=108414103X00419780901&spkNum=113&current=8#s113.

84th NDHR Bud., 1st subcommittee, No. 3, March 1, 1978. https://kokkai.ndl.go.jp/simple/detail?minId=108405266X00319780301&spkNum=5&current=11#s5.

96th NDHR Bud., No. 19, March 9, 1982. https://kokkai.ndl.go.jp/simple/detail?minId=109605261X01919820309&spkNum=85&current=8#s85.

101st NDHR Bud., No. 7, February 18, 1984. (https://kokkai.ndl.go.jp/simple/detail?minId=110105261X00719840218&spkNum=287&current=5#s287).

101st NDHR Bud., No. 16, March 3, 1984. https://kokkai.ndl.go.jp/simple/detail?minId=110105261X01619840303&spkNum=285&current=16#s285.

101st NDHR Set., No. 3, March 26, 1984. https://kokkai.ndl.go.jp/simple/detail?minId=110104103X00319840326&spkNum=243&current=12#s243.

101st NDHR SST, No. 6, March 27, 1984. https://kokkai.ndl.go.jp/simple/detail?minId=110103911X00619840327&spkNum=161&current=13#s161.

104th NDHC SST, No. 4, April 11, 1986. https://kokkai.ndl.go.jp/simple/detail?minId=110413928X00419860411&spkNum=28&current=1504#s28.

112th NDHR S&T, No. 1, March 1, 1988. https://kokkai.ndl.go.jp/simple/detail?minId=111203911X00119880301&spkNum=6&current=1#s6.

112th NDHC Com., No. 7, April 19, 1988. https://kokkai.ndl.go.jp/simple/detail?minId=111214816X00719880419&spkNum=164&current=4#s164.

114th NDHC Investigation Committee on Industry, Resources and Energy, No. 2, April 5, 1989. https://kokkai.ndl.go.jp/simple/detail?minId=111414379X00219890405&spkNum=8&current=6#s8.

114th NDHC SST, No. 3 June 21, 1989. https://kokkai.ndl.go.jp/simple/detail?minId=111413928X00319890621&spkNum=24&current=13#s24.

118th NDHR Bud., First Subcommittee, No. 2, April 27, 1990. https://kokkai.ndl.go.jp/simple/detail?minId=111805266X00219900427&spkNum=142&current=7#s142.

118th NDHC SST, No. 3, May 30, 1990. https://kokkai.ndl.go.jp/simple/detail?minId=111813928X00319900530&spkNum=27&current=9#s27.

118th NDHC SST, No. 4, May 31, 1990. https://kokkai.ndl.go.jp/simple/detail?minId=111813928X00419900531&spkNum=14&current=10#s14.

120th NDHC SST, No. 4, April 9. https://kokkai.ndl.go.jp/simple/detail?minId=112013928X00419910409&spkNum=45&current=13#s45.

120th NDHC Com., No. 12, April 25, 1991. https://kokkai.ndl.go.jp/simple/detail?minId=112014816X01219910425&spkNum=134&current=14#s134.

123rd NDHC SST, No. 4, April 6, 1992. Accessed June 2, 2023. https://kokkai.ndl.go.jp/simple/detail?minId=112313928X00419920406&spkNum=11&current=22#s11.

123rd NDHC SST, No. 2, February 26, 1992. https://kokkai.ndl.go.jp/simple/detail?minId=112313928X00219920226&spkNum=4&current=5#s4.

123rd NDHR S&T, No. 6, May 26, 1992. https://kokkai.ndl.go.jp/simple/detail?minId=112303911X00619920526&spkNum=269&current=31#s269.

126th NDHR S&T, No. 3, February 23, 1993. https://kokkai.ndl.go.jp/simple/detail?minId=112603911X00319930223&spkNum=145&current=46#s145.

126th NDHCSST, No. 3, February 26, 1993. https://kokkai.ndl.go.jp/simple/detail?minId=112613928X00319930226&spkNum=59&current=48#s59.

129th NDHR Bud., No. 3, February 22, 1994. https://kokkai.ndl.go.jp/simple/detail?minId=112905261X00319940222&spkNum=364&current=81#s364.

131st NDHR S&T, No. 3, November 1, 1994. https://kokkai.ndl.go.jp/simple/detail?minId=113103911X00319941101&spkNum=81&current=94#s81.

132nd NDHR S&T, No. 2, February 15, 1995. https://kokkai.ndl.go.jp/simple/detail?minId=113203911X00219950215&spkNum=180&current=5#s180.

132nd NDHR S&T, No. 3, February 21, 1995. https://kokkai.ndl.go.jp/simple/detail?minId=113203911X00319950221&spkNum=7&current=71#s7.

132nd NDHR For., No. 17, May 12, 1995. https://kokkai.ndl.go.jp/simple/detail?minId=113203968X01719950512&spkNum=18&current=26#s18.

132nd NDHCSST, No. 4, March 20, 1995. https://kokkai.ndl.go.jp/simple/detail?minId=113213928X00419950320&spkNum=80&current=78#s80.

132nd NDHC For., No. 14, May 30, 1995. https://kokkai.ndl.go.jp/simple/detail?minId=113213968X01419950530&spkNum=30&current=29#s30.

134th NDHR S&T, No. 3, November 7, 1995. https://kokkai.ndl.go.jp/simple/detail?minId=113403911X00319951107&spkNum=28&current=35#s28.

136th NDHR S&T, No. 2, February 22, 1996. https://kokkai.ndl.go.jp/simple/detail?minId=113603911X00219960222&spkNum=9&current=8#s9.

140th NDHR S&T, No. 2, February 20, 1997. https://kokkai.ndl.go.jp/simple/detail?minId=114003911X00219970220&spkNum=2&current=35#s2.

141st NDHR S&T, No. 2, November 6. https://kokkai.ndl.go.jp/simple/detail?minId=114103911X00219971106&spkNum=29&current=63#s29.

142nd NDHR S&T, No. 2, March 11, 1998. https://kokkai.ndl.go.jp/simple/detail?minId=114203911X00219980311&spkNum=15&current=36#s58.

142nd NDHC Committee on Education and Science, No. 12, April 7, 1998. https://kokkai.ndl.go.jp/simple/detail?minId=114215074X01219980407&spkNum=38&current=45#s38.

142nd NDHC Committee on Education and Science, No. 17, April 23, 1998. https://kokkai.ndl.go.jp/simple/detail?minId=114215074X01719980423&spkNum=32&current=52#s32.

144th NDHR S&T, No. 2, December 11, 1998. https://kokkai.ndl.go.jp/simple/detail?minId=114403911X00219981211&spkNum=148&current=94#s148.

144th NDHC Set., No. 1 after closing, December 17, 1998. https://kokkai.ndl.go.jp/simple/detail?minId=114414103X00119981217&spkNum=41&current=95#s41.

145th NDHR S&T, No. 2, February 9, 1999. https://kokkai.ndl.go.jp/simple/detail?minId=114503911X00219990209&spkNum=147&current=3#s193.

146th NDHC TIC, No. 2, November 16, 1999. https://kokkai.ndl.go.jp/simple/detail?minId=114614197X00219991116&spkNum=178&current=44#s178.

146th NDHR S&T, No. 3, November 17, 1999. https://kokkai.ndl.go.jp/simple/detail?minId=114603911X00319991117&spkNum=5&current=47#s54.

146th NDHR S&T, No. 5, November 24, 1999. https://kokkai.ndl.go.jp/simple/detail?minId=114603911X00519991124&spkNum=193&current=17#s193.

147th NDHR S&T, No. 2, March 14, 2000. https://kokkai.ndl.go.jp/simple/detail?minId=114703911X00220000314&spkNum=168&current=18#s175.

147th NDHC TIC, No. 11, April 18, 2000. https://kokkai.ndl.go.jp/simple/detail?minId=114714197X01120000418&spkNum=37&current=34#s37.

149th NDHR S&T, No. 1, August 4, 2000. https://kokkai.ndl.go.jp/simple/detail?minId=114903911X00120000804&spkNum=8&current=6#s69.

150th NDHR SOC, No. 1, November 9, 2000. https://kokkai.ndl.go.jp/simple/detail?minId=115004127X00120001109&spkNum=0&current=13#s72.

151st National NDHR Bud., Fourth Subcommittee, No. 2, March 2, 2001. https://kokkai.ndl.go.jp/simple/detail?minId=115105270X00220010302&spkNum=71&current=41#s75.

151st NDHC Set., No. 2, April 2, 2001. https://kokkai.ndl.go.jp/simple/detail?minId=115114103X00220010402&spkNum=52&current=55#s52.

151st National NDHR Cab., No. 10, May 16, 2001. https://kokkai.ndl.go.jp/simple/txt/115104889X01020010516/8.

153rd NDHC Set. ,No. 1 after closing, December 11, 2001. https://kokkai.ndl.go.jp/simple/detail?minId=115314103X00120011211&spkNum=163&current=27#s163.

154th NDHR LIT Committee, No. 5, March 29, 2002. https://kokkai.ndl.go.jp/simple/detail?minId=115404319X00520020329&spkNum=23&current=45#s23.

Visual and Cinematic Sources

Frau im Mond (Woman in the Moon). Dir. Fritz Lang. UFA, 1929.

Hayabusa: Harukanaru kikan (Hayabusa: The Long Voyage Home). Dir. Tomoyuki Takimoto, Japan, 2012.

Momotarō no Umiwashi (Momotaro's Sea Eagles). Dir. Seo Mitsuyo, Geijutsu Eigasha, 1943.

Momotarō: Umi no Shimpei (Momotaro: Sacred Sailors). Dir. Seo Mitsuyo, Shochiku, 1945.

Okaeri, Hayabusa (Welcome Home, Hayabusa). Dir. Katsuhide Motoki, Japan, 2012.

Space Battleship Yamato. Dir. Leiji Matsumoto, Yomiuri TV, 1974–1975.

Space Battleship Yamato. Dir. Leiji Matsumoto, Tōhō, 1976.

Space Battleship Yamato. Dir. Takashi Yamazaki, Tōhō, 2010.

Things to Come. Dir. William Cameron Menzie, United Artists, 1936.

INDEX

A6M Zero, *see Mitsubishi A6M*
Acta Astronautica (journal), 130
accidents, 176, 188, 190,
ADEOS I/Midori (satellite), 180, 186
Aerojet (manufacturer), 104
Aeronautical Research Institute (Tokyo University), 86
Aeronautical Research Laboratory (Tokyo Imperial University), 48–49
aeronautics: ban on, 49, 80; development and funding for, 35, 37; public opinion of, 19; technology and, 80. *See also specific aspects*
Aerospace Technology Research Institute, 183
air raids, 27, 36-39, 75.
Akiba, Ryōjirō, 10, 16, 49, 50, 87, 127, 141
Akita prefecture, 109–12
Akiyama, Toyohiro, 143, 144, 151, 165
Aldrich, Daniel, 11, 112, 117
Allied aircraft, Japanese perceptions of 64, 73
ALOS-1 (satellite), 141
American Astronomical Society, 224n23
Andō, Sukejirō, 234n7
Ariane-4, 169, 176
Ariane-5, 169z
Arms Control and Disarmament Agency, 156
Armstrong, Neil, 126, 128
Army Air Force Scientific Advisory Group (AAF), 29
Army Institute of Scientific Research, 33
Arnold, H. H. "Hap," 29, 39
Artemis program, 166
ASI *see Italian sprogram*
Asahi Shimbun (newspaper), 16, 63, 75, 112, 125, 177, 178, 189, 199
Ashita no Hikōki (Tomorrow's Aircraft) (Kaishima book), 58, 59
asteroid 6526 Matogawa, 132
asteroid 25143 Itokawa, 10–11, 128–29, 198
asteroid 162173 Ryūgu, 200
ASTRO-E (satellite), 180
Asuka (satellite), 134, 203
Atlas V (rocket), 169
atomic bombing, 63
AT&T Bell Telephone Laboratories, 94
Australia, 228n88
aviation, Japanese, 33–38, 63–67

Avionics and Supersonic Aerodynamics (AVSA) research group: accomplishments of, 134; autonomy in, 84; camaraderie of, 79; challenges of, 69; costs of, 92; developments of, 136; funding for, 85; launch testing locations of, 110; local sourcing by, 93; origin of, 1–2, 7, 50–51; perks of, 84; photo of, 83; relationship with fishermen and, 111–12; structure of, 130

B-29 (bomber), 64, 69, 203
B50 (balloon), 145–46
B500 (balloon), 146
Baby (rocket), 93, 203
Baika (aircraft), *see Kawanishi Baika*
Bank of Japan, 167, 259n2
Basic Space Law, 4, 14, 75, 193.
Beggs, James, 162
Belgium, 228n88
Beliaev, Aleksander, 55
Beyond Gravity (company), 200
Biemiller, Andrew J., 157
blackboxing, 144, 156–58, 165
Bleriot, Louis, 33
BMW-1003a (engine), 42
Bor'ba v efre (Battle in the Ether) (Beliaev book), 55
Britain, rocketry program of, 52
BS-2A (satellites), 97
budget, 69-71, 87, 92, 95, 105, 148-149, 169, 171, 174-175, 179, 183-184, 186-187, 195
bullet train *see Shinkansen*
Buran space shuttle (USSR), 152

Callon, Michel, 9
Canada, 228n88
Cassini (probe), 133, 203
Castor II (rocket booster), 158
The Cheese and the Worms (Ginzburg book), 9
Chichibu Yoshida rocketry event, 66
children's literature, 59–60, 62–63, 66
China, 23, 61, 70, 126, 140, 155, 168
China–Japan space relations, 186–187, 234–236
Chisso (manufacturer), 117
Chūō Kōron, 58
coal industry, 32
Cold War, 14, 84
collective memory, 128
COMETS/Kakehashi (satellite), 176
commercialization of space, 41, 83, 131, 167, 219
Committee on Communications (Diet of Japan), 95
communications, Japanese reliance on, 97–98
Comprehensive Subcommittee (for ISAS reform), 183
Compton, Karl, 28
Crichton, Michael, 155
criticism, of rocketry, 70, 178-179, 182, 190
CS3 (satellite), 146
Culbertson, Frank, 163
Cumings, Bruce, 155
cyclotrons, 33
Cygnus-X-1 black hole, 145–46

Dainippon Chemicals, 44
d'Allest, Frederick, 154
dams, 117
Dare ni mo wakaru kagaku zenshū dai kyū ken (Science for Everyone Complete Series No. 9), 55
dark valley hypothesis, 225n51
DC-X Delta Clipper, 133
deaths, 102, 188, 233
decentralization, of space programs, 4, 33-34, 68-69, 234-236
Delta 3 (rocket) 179
demilitarization, 46–47, 54
Denmark, 228n88
Diet Budget Committee, 92, 182, 189
disaster management, 128, 137, 141, 154
disease, industrial 117
disillusion, rocketry in the Japanese consciousness and, 55–67
displacement, of locals, 117
dual-use, 4, 101
Dusinberre, Martin, 11, 112

Early Bird (satellite), 98
Easy Science workshops, 133
Eberstadt, George, 4, 5
ECS (satellite), 149
Eiffel, Gustave, 215n71
Eiffel-type wind tunnels, 215n71
Eighth Military Technical Laboratory, 220n145
electroencephalograph (EEG), 49
Endeavour (Space Shuttle), 127
Energia launch system (Soviet Union), 149
engineering, schools of, 34
Engineering Test Satellite (ETS) series, 146, 150, 154; ETS 1, 146; ETS-VI, 37, 146, 180, 184, 186, 190, 193; ETS-VIII, 89; ETS-VII, 193
engineer-managers, 32–33, 34, 81–82, 90, 197, 230n26
environmental impact, of rocketry, 52, 95, 137, 176, 245
Epsilon (rocket), 200
Europe–Japan space relations, 13, 161, 178–179, 203 234–236
European Space Agency (ESA), 13, 161, 203
exclusion zones (for launches), 112
explosives, development of, 47
Express (satellite), 180, 186
F4U (fighter), 64
F6F (fighter), 64
F11F (fighter), 67
Failure Knowledge Utilization Study Group, 188
firework rockets, 1, 58, 61
1st Military Laboratory, 220n145
fishermen: launch interruptions by, 118; liquid hydrogen testing and, 172; negotiations of, 119–21; power and influence of, 111–12; resistance of, 6, 116–18; schisms of, 119–21; settlement to, 119–21; Tanegashima and, 114–21; Uchinoura and, 114–21; unions of, 116–18
food, wartime consumption of, 38
forestry, 38
4FR-110 (rocket system), 42
France, 228n88
Frau im Mond (Woman in the Moon) (Lang film), 55
Fritzsche, Peter, 9–10
fuels, 43, 68. *See also specific fuels*
Fuji Seimitsu, 48, 93
Fujitsu, 99
Fujiyama, Aiichiro, 101
Fukamachi, Kiyotaka, 176–77
Fukuda, Kazuei, 117
Fukushima, Kimio, 164
Funaoka Naval Explosives Arsenal, 45
Fundamental Space Policy, 187
Funryū (surface-to-air missiles), 30
funshin (rocket artillery), 42
Furukawa, Hisao, 68

Gagarin, Yuri, 133
Galileo mission (United States), 149, 198
GDP, 222n180
Gekkan Yomiuri, 68
General Electric, 97
geostationary meteorological satellites (GMS), 146
geostationary orbit (GEO), 126
geostationary satellites, 145
geostrategy, 61, 70, 126, 142
Germany, 27–28, 31, 32, 65
Gerovitch, Slava, 128
Ginzburg, Carlo, 9
Glider Research Institute (Kyushu Imperial University), 38–39
Global Positioning Satellites (GPS), 2, 137, 141
Godai, Tomifumi, 10, 48, 147, 161–62, 169, 171, 178, 182–83
Goddard, Robert, 1, 35, 60, 62, 73
Great Tohoku Earthquake, 89
Griggs, David, 212n12
Gunji to Gijutsu (Military Affairs and Technologies) (magazine), 58, 62

H-1 (rocket): capabilities of, 105, 149; commercial use of, 154; cost of, 155; defined, 204; development of, 147, 158; domestic construction of, 158; licensing agreements and, 155; purpose of, 135–36
H-2A (rocket), 175–76, 181, 204
H-2B (rocket), 181, 204
H-2 Orbiting Plane (HOPE) project, 152, 153, 184, 191
H-2 (rocket): aftermath of, 188–90; ambitions of, 168–70; budget for, 174–75; challenges of, 133; collaboration of, 187; corporation investment into, 169; cost-cutting of, 175; cost of, 169, 186; defined, 204; as defining experience, 181; development of, 13, 170–72; expectations of, 168–70; explosion of, 4, 241n19; failures of, 162, 171–77; function of, 169–70; funding for, 169; interrogation regarding, 182–83; LE-7 engine in, 171–72 (*see also* LE-7 engine); mishaps of, 167, 168; models of, 170; origin of, 3, 168–69; professional frustration regarding, 178–80; public disappointment regarding, 178–80; purpose of, 135–36; quality control failures of, 189; rise and fall of, 168–81; significance of, 181; testing of, 171–72
H3/H-3 rocket, 181, 200–201, 204
Halley's Comet, 5, 87, 120
Hamada, Kōtarō, 119
Harada, Kazuo, 58
Harada, Mitsuo, 60
Haruka/MUSES-B (probe), 148
Harvey, Brian, 8
Hase, Hiroshi, 186
Hashimoto, Kazuhiko, 89
Hashimoto, Masako, 114
Hashimoto, Ryūtarō, 233n72
Hatsumei (Discovery), 58, 62–63
Hayabusa 2 mission, 199–200, 203
Hayabusa: Harukanaru kikan (Hayabusa: The Long Voyage Home) (film), 199
Hayabusa/MUSES-C (probe), 148
Hayabusa (probes), 2, 5, 10–11, 128–29, 198–99, 203
Hayakawa, Rieko, 141
Hendon, Hugh, Jr., 61
Hewitt, William R., 212n12
Hidehiko, Tamaki, 63
Hideo, Shima, 16, 48, 49, 88, 105, 125-127, 147, 158, 174, 190
Hideyoshi, Toyotomi, 1
Higuchi, Kiyoshi, 87, 127
Himawari (satellite), 105, 149
Hino, Kumao, 10, 16, 32–33, 35, 45–48, 49
Hino, Kumazō, 32, 33, 45
Hinotori (satellite), 134
Hiroshi, Saito, 178
Hiroshima, 50-51, 63, 131
Hirosige, Tetu, 81
HITAC 5202 (computer), 93
Hitachi, 2, 44, 93, 97, 98, 160, 198
Hiten/MUSES-A (probe), 86–87, 148, 149
Hodogaya Chemical, 37–38, 44
Hohmann, Walter, 60
Home, R. W., 19
Honeywell (manufacturer), 98
Hori, Shinzaburō, 58
Horikoshi, Jirō, 67
H series (rocket), 105. *See also specific types*
Hughes Aircraft Company, 98, 103, 177
Hughes Space and Communication, 170
HWK 109-509 (engine), 42
hydrogen peroxide (fuel), 37, 43–44, 68
Hyōdō, Tadashi, 33
Hyūga (battleship), 42

I-29 *Matsu* (submarine), 27–28
IBM, 99
Ichinen non Gakushū (magazine), 66

Iida, Takashi, 141
Illustrated London News (newspaper), 58
Imperial Japanese Army (IJA), 33, 35, 36, 39, 42, 69
Imperial Japanese Naval Aviation Headquarters, 27
Imperial Japanese Navy (IJN), 28, 30, 33, 35, 36, 38, 42–43, 44
India, 23, 99
indigenous launch capacity, quest for, 26–27, 115–116, 234
Indonesia, 5, 14, 92
industrialization, 32
Industrial Policy Association, 33
industrial waste, 110, 111, 117
INMARSAT (company), 170
Inoue, Ikutarō, 36–37
Institute of Space and Aeronautical Science (ISAS): bureaucratic victory of, 87; centralization of, 192; construction spree of, 145; costs of, 92; developments of, 147; failures of, 180; function of, 198; Hayabusa probes and, 198–99; independence of, 86, 87; institutional nature of, 148; investigation of, 75; local sourcing by, 93; name change of, 87–88; NASDA division with, 184; origin of, 4, 7, 86; outreach of, 109; productions of, 80; refusals of, 20; settlement by, 120; University of Tokyo and, 183
Intelsat I (satellite), 102
Intelsat II Lani Bird (satellite), 96
inter-continental ballistic missiles (ICBMs), 6, 65, 101, 102–3
interest rates, effect on Japanese economy, 167–68
International Academy of Astronautics, 139
International Geophysical Year (IGY), 50
International Institute of Space Law, 161
International Microgravity Laboratory Program, 151
International Space Station (ISS), 3, 5, 13, 151, 152, 154, 162–64, 204
International Telecommunications Satellite Consortium (INTELSAT), 103, 158
Inuzuka, Toyohiko, 40
Isao, Uchida, 184
Ise (battleship), 42
Ishihara, Shintarō, 125, 155
Ishikawajima Aircraft Manufacturing, 44
Ishikawajima-Harima Heavy Industries (IHI), 2, 97, 193
Isomura Manufacturing, 175
Italian Space Agency (ASI), 203
Italy, 228n88
Ito, Kenji, 67
Itokawa, Hideo: on American technology, 85–86; asteroid named after, 11, 198; challenges of, 82; characteristics of, 130; on Charles Lindbergh, 61; education of, 34–35, 129; expertise of, 69–70; fishermen and, 111–12; as founding father, 128–31; influence of, 10, 99–100, 198; leadership of, 1–2, 6, 54, 86; legacy of, 75; local sourcing by, 93; as mediator, 74; mission of, 101; motivations of, 128–29; myth of, 128–31; overview of, 48–51; personality of, 72, 73; photo of, 71, 83; popularity of, 73; popularizing postwar rocketry by, 69–75; as resource, 16, 17; rocket sales of, 92; as space program founder, 5; on Sputnik, 95; statue of, 72; on turbine technology, 63; at Uchinoura Launch Center, 115; vision of, 93–94; work of, 11; on World War II, 130–31; writings about, 129; writings of, 71, 130–31

ISRO (Indian Space Research Organization), 2, 14, 23, 85,
Iwatani Manufacturing Group, 171
Iwaya, Eiichi: death of, 69; developments of, 37, 42; education of, 49; humanizing wartime rocketry and, 67–69; military service of, 27–28; mission of, 101; photo of, 29; as resource, 16; wartime memoirs of, 54–55; writings of, 67–69

Japan: bright life in, 136–38; collaboration with the United States and, 85; dark valley of, 65; demilitarizing of, 46–47; economic decline of, 167–68, 174–75; economic growth of, 95, 125, 167; GDP of, 95; industrialization of, 32; infrastructure investment in, 110; interest rates in, 167–68; Lost Decades in, 140, 168; morale in, 61–62; national fertility in, 168; national pride of, 134–36; national security concerns of, 3; newspapers of, 16; political system of, 15; post-war innovations of, 211n4; self-defense forces of, 193; status insecurity of, 2–3; surrender of, 28, 81; tensions between United States and, 155–59; tensions with, 36; US trade deficit and, 155; war losses of, 27
Japan Aerospace Exploration Agency (JAXA): archive of, 15; budget of, 2; capabilities of, 135; employment at, 2; failures of, 200–201; foundation of, 233n72; function of, 193; growth of, 2; national security and, 14–15; origin of, 4, 12, 181, 194, 197; partnerships/contractors of, 2; road to, 190–94; space programs after, 185; success of, 198; support for, 201
Japan Astronaut Club, 194
Japan Business Federation, 6, 94, 98, 102, 103–5, 159, 204
Japan Communications Satellites Corporation, 89
Japanese Astronomical Society, 224n23
Japanese Broadcasting Corporation, 183
Japanese Experiment Module (JEM), 204
The Japanese Scientist (*Nihon kagakusha*) (journal), 139
Japanese Self-Defense Force (JSDF), 75, 137, 140, 193
Japan Federation for Commercial Satellites (JFCS), 126–27
Japan Hydrogen Fuel Corporation, 171
Japan Meteorological Agency, 183, 187
Japan National Railways, 65, 81–82, 110, 220n143
Japan Satellite Communications (JCSAT), 98
Japan Self Defense Forces, 75, 137, 140, 193,
Japan Socialist Party, 117, 163, 184
Japan Transport Association, 63
Jets (Frank Whittle book), 67
"Jettoki Shūsui ni tsuite," (Iwaya article), 69
J-I (rocket), 175
Johnson, Alexis, 103
Johnson, Lyndon B., 103
Johnson Memo, 103
Joint Intelligence Objectives Agency (JIOA), 205
Jumo-004b (engine), 42
Junichiro, Koizumi, 233n72

Kagaku no Shakaishi (A Social History of Science) (Hirosige), 81
Kagaku sen heiki (Nishizawa), 58–59
Kagaku Zasshi (Science Magazine), 62
Kagoshima Space Center, 114
Kaishima, Keitarō, 58, 59
Kakuda Space Center, 170, 172

Kallendar-Umezu, Paul, 140, 233n72, 238n77, 251n104
kamikaze. *See* Ōka (suicide air torpedo)
Kanadome, Mikiko, 16, 115
Kanaya, Arihiro, 172
Kaneshige, Kankuro, 86
Kappa 8-10 (rocket), 111
Kappa-8 (rocket), 79–80
Kappa class (rocket), 74, 112, 204
Kashima, Ibaraki, 97
Kashima Space Research Center, Ibaraki, 137
Kasumigaura research facility, 36
Katayama, Jin'ichi, 184
Kato, Yasuhiro, 165
Kawamura, Takeo, 233n72
Kawanishi Aircraft (Shin Meiwa), 47
Kawanishi Baika (aircraft), 40, 203
Kawanishi Maru Ka10 (engine), 40
Kawasaki conglomerate, 36, 47, 104
Kawasaki Heavy Industries, 104
Kawasaki Jet Ski, 97
Kawashima, Hidetaka, 175
KDD *see Kokusai Denshin Denwa*
Ki-43 Hayabusa *see Nakajima Ki-43 Hayabusa*
Ki-44 Shoki *see Nakajima Ki-44 Shoki*
Kibō module (International Space Station), 3, 5, 13, 162–64, 165, 204
Kikuna/WINDS (satellite), 89
Kimitsu heiki no zenbō (The Full Story of Secret Weapons) (Eiichi Iwaya book), 68
Kimiya, Kazuhiko, 182
Kimura, Hidemasa, 45, 65
Kinmen Island, 70
Kitazume, Susumu, 134, 141
Kiuchi, Shiro, 184
Kobe earthquake, 137
Kodama, Sueo, 117
Kodomo rigaku ebanashi (Science Picture-Stores for Children), 60
Koizumi, Kenkichirō, 82
Kōkū jōhō (Air News), 67
Kokusai Denshin Denwa (KDD), 93, 94, 96, 98–99, 183
Korea, invasion of, 1
Korolev, Sergei, 12, 73, 128
Kosmos-144 satellite (USSR), 103
Kumagai, Tsaburō, 159
Kumamoto (prefecture), 117
Kusayanagi, Daizō, 129, 246n21
Kyushu, Japan, 13, 112
Kyushu Denryoku, 93
Kyushu Institute of Technology, 192

Lambda 3-H (rocket), 80
Lambda-4S-3 (rocket), 118
Lambda 4S-5 (rocket), 192
Lambda class (rocket), 99, 204
laminar flow, 35
Lang, Fritz, 55
LE-5 (engine), 147, 176
LE-7 (engine), 167, 171, 173, 178
Lindbergh, Charles, 33, 34, 61
liquid fuel rocketry (LFR): advantages of, 99–102; anxieties of, 99–102; development of, 6, 88, 161–62; as dual use, 101; employment in, 102; function of, 94, 171, 207n4; H series of, 105; importance of, 106; importing, 85, 98, 142, 143–44, 165; internal and external pressure for importing, 102–4; N class of, 102, 105 (*see also* N-series rocket); origin of, 197; private sector in, 104–6; Q class of, 102 (*see also* Q rocket); research of, 101–2; United States deal regarding, 102–4
liquid hydrogen, 171, 172, 173
liquid oxygen, 165
Lockheed, 104
Logsdon, John, 157
Longacre, Andrew, 212n12
LOH *see liquid hydrogen*
LOX *see liquid oxygen*
Low, Morris, 19

low Earth orbit (LEO), 105
Lubrizol (company), 48
Lunar Cruiser, 165–66

M-3H (rocket), 147
M-3 (rocket), 180
M-3S (rocket), 147
M-4S (rocket), 147
Mabuchi, Chōko, 33, 219n142
MacArthur, Douglas, 46
Mainichi Shimbun (newspaper), 16, 70
Malina, Frank, 12
managed economy approach, 66
manager-specialists: aeronautical, 80–82; autonomy of, 84; influence of, 90; overview of, 79–80; public-private networks and, 89–90; recapturing status and entrenching influence by, 82–89; research agenda of, 85; in transwar period, 80–82
manganese, 39, 42
Marine Observation Satellite, 149
Mars exploration, 21-23, 133, 149, 152, 180
mass media, 16, 60, 96, 105, 178
Matogawa, Yasunori: on centralized space program, 190; education of, 49; on Itokawa, 129–30; Kimi Tanaka and, 114; leadership of, 10, 80; on the media, 178; promotion of space research and, 131–33; as resource, 16, 17; on rocketry, 137; on space dreams, 138; as space food judge, 136; on space program, 149; support from, 201; World War II and, 127; writings of, 132, 135, 138, 139
Matsu, Akira, 186
Matsuda, Kōyaku, 62
Matsushita Denki (company), 99
McDonnell Douglas, 104, 139–40
McDonnell Douglas DC-X Delta Clipper (spacecraft), 133
McDougall, Walter, 12, 55, 208n17
Me-163 Komet (aircraft), 37, 43, 204, 205
Me-262 (aircraft), 43
MeasuringWorth, 222n180
medium lift capability, 204
medium lift vehicle (MLV), 143–44
Meguro, Akira, 79, 80, 82, 89–90
Menzies, William Cameron, 55
MEXT *see Ministry of Education, Science, Sports and Culture of Japan*
MIC *see Ministry of Internal Affairs and Communication of Japan*
Michikawa Beach, 110–11
Midori (satellite), 137
Mie prefecture, 117
Miki, Tadanao, 35, 45, 65, 81–82
Miki, Yoshio, 117
militarism/militarization, 33, 54, 55–67, 119, 144, 212, 251n104
Mimura, Janis, 34, 66
Minamata, 109, 110
Minamitanechō, 116, 119, 120–21
Minerva lander, 198
Ministry of Education, Science, Sports and Culture of Japan (MEXT), 84, 187, 192, 204, 233n72
Ministry of General Machine Building (USSR), 12
Ministry of Internal Affairs and Communications of Japan (MIC), 192
Ministry of International Trade and Industry of Japan (MITI), 154, 187, 233n72
Ministry of Posts and Telecommunications of Japan, 96, 104, 183, 233n72
Ministry of Posts and Telecommunications Radio Research Laboratory *see Radio Research Laboratory.*
Ministry of Transportation of Japan, 187
Mir space station, 143
missiles, 4, 14, 29, 30, 39-41,

49, 63, 65, 69, 75, 101, 140-141, 168, 189, 193
MITI *see Ministry of International Trade and Industry of Japan (MITI)*
Mitsubishi A5M (aircraft), 67
Mitsubishi Aircraft Company, 37, 47
Mitsubishi Colt, 97
Mitsubishi (conglomerate), 44, 99, 139–40, 146, 170–71
Mitsubishi Electric Company (MELCO), 89–90, 98, 193–94
Mitsubishi Heavy Industries (MHI): aviation and, 36; contributions of, 165; cooperation of, 104; engine factory of, 146; JAXA and, 2; liquid hydrogen and, 171; master key technologies of, 147; problems of, 189; research and development (R&D) at, 51; rocketry and, 37, 93, 169; rocket testing ground of, 146; satellite communications and, 97
Mitsubishi A6M Zero (aircraft), 67
Mitsubishi J8M1/Ki-200 Shūsui (aircraft), 37, 40
Mitsubishi Shipbuilding, 37
Mitsui (company), 33
Mitsui Dye-Stuff, 44
Miyazaki prefecture, 116
Miyazaki Fishermen's Union, 108, 116, 119–20
Mizuno, Hideki, 134, 135, 139
Mizuno, Hiromi, 34, 65
mobilization for total war, 225–26n51
Momotarō no Umiwashi (Momotarō's Sea Eagles) (Seo film), 61
Momotarō: Umi no Shimpei (Momotarō: Sacred Sailors) (Seo film), 61
Moon, 2, 23, 126, 148, 151, 165–66
Moore, Aaron, 32, 65–66
Mori, Takeo, 27
Moridai, Yoshio, 108
Morita, Akio, 125
Morris-Suzuki, Tessa, 66
Motoki, Katsuhide, 199
MTSAT-1 (satellite), 177
Mukai, Chiaki, 48, 127, 136, 164
Murata, Kenrō, 234n7
Murata, Tsutomu, 36
Murayama, Ki'ichi, 183
Mu class (rocket), 86, 147, 204
Musashi (battleship), 27
Mu Space Engineering Spacecraft (MUSES) program, 148
M-V (rocket), 132–33, 180, 186, 193

N-1 (rocket), 105, 158, 204–5
N-2 (rocket), 147, 158, 205
Nagoya Aeronautical and Space Systems Facility, 37
Nagoya Aircraft Works, 146–47, 189
Nagoya Guidance & Propulsion Systems Works, 147, 172, 189
Nakahara, Nobuhei, 80–81, 82
Nakajima Aircraft, 36, 44, 47, 49, 204
Nakajima Ki-43 (aircraft), 11, 43 129, 198
Nakajima Ki-44 Shoki (aircraft), 49
Nakamura, Iwao, 234n7
Nakanishi, Fujio, 45
Nakasone, Yasuhiro, 95, 101, 162, 163
Nakazawa, Kisaburō, 234n7
Narita Airport, 109, 110
NASA: academics of, 84; budget of, 148; developments of, 203; financial support from, 161; Galileo mission of, 198; International Microgravity Laboratory Program of, 151; International Space Station (ISS) and, 162 (*see also* International Space Station (ISS)); Moon exploration of, 126, 165; warning from, 156
NASDA–ISAS rivalry, 32–34, 68–69
National Aerospace Laboratory, 3, 145,154
National Diet House of Representatives (NDHR), 15, 95, 164

nationalism, 11-12, 34, 54, 134–36
national prestige, 21, 24, 134-136, 152-154, 179
National Science Foundation, 150
national security, 3, 14–15, 194
National Space Activities Commission, 98, 102-3, 150, 162, 187, 189, 191, 192, 205
National Space Development Agency Law, 104
National Space Development Agency of Japan (NASDA): analyses of, 179; budget of, 105, 148, 169, 171, 174–75; closing of, 192; construction spree of, 145; criticism of, 179, 189; ESA agreement with, 161; failures of, 180, 189; goal of, 87; human-rated spaceflight and, 151; international projects of, 150; ISAS division with, 184; LFR development by, 161–62 (*see also* liquid fuel rocketry (LFR)); mission of, 88; origin of, 6, 7, 87; outreach of, 109; peaceful usage law of, 138–39; pressures of, 178; private sector and, 154–55; reliability of, 190; rise of, 104–6; satellites and, 97; settlement by, 120; space infrastructure of, 151; success of, 197; tensions with, 189
National Space Development Center, 86
National Space Development Promotion Committee, 94, 98, 104–5, 169
NavTech (US Naval Technical Mission to Japan), 29–30
NEC (Nippon Electric Company), 2, 97, 98, 99, 160, 189, 193
Netherlands, 228n88
New Horizons (probe), 205
newspapers, 16. *See also specific newspapers*
New Zealand, 99
Nihon hōsō kaisha (NHK), 60
Nihon Keizai Shimbun (newspaper), 16, 183
Nihon sōseiron (Itokawa book), 73, 74, 131
Niigata prefecture, 117
Niijima (island), 101, 109–12
Niijima Defense Agency site, 110
Nikai, Toshihiro, 177, 179
Nike (missile), 101
Nippon Chemical Synthesis (company), 44
Nippon Denki *see NEC*
Nippon Hōsō Kaisha (NHK), 95, 97, 105
Nippon Kayaku (company), 47
Nippon Telegraph and Telephone (NTT) Public Corporation, 89, 183
Nippon Television, 95
Nishiyama, Takashi, 81
Nishizaki, Kiku, 33, 219n142
Nishizawa, Yūshichi, 58–59
Nissan, 93, 104, 147, 148, 169, 192–93
Nixon, Richard, 155, 156
Nobuyoshi, Fugono, 10, 135, 188
Nomura, Masahiko, 220n145
Norakura shōi no roketto tokkantai (Second Lieutenant Norakura's Rocket Assault Team), 60, 62
North American Rockwell, 104
North Korea, 14, 23, 140, 141, 168
Nozomi (probe), 133, 180, 205
N-series (rockets), 88, 102, 105, 126, 147
nuclear power/nuclear weapons, 4, 11, 70, 109, 112, 140

Oberth, Hermann, 59, 60, 73
Obuchi, Keizō, 177
Office of Scientific Research and Development (United States), 84
Office of Strategic Services (OSS) (United States), 205

Ōfuna Naval Arsenal and Depot, 36, 42, 51
Ohara, Tsuneto, 63
Ohsumi (satellite), 80, 95, 115, 126, 134, 205
oil, substitutes for, 211n4
Ōka (suicide air torpedo), 28, 31, 40, 44, 64–65, 205
Okaeri, Hayabusa (Welcome Home, Hayabusa) (film), 199
Okazaki, Toshi, 177
Okinawa, 112
Okuda, Mariko, 172, 174
Okuda, Tadashi, 172, 174
Olympic Games, 96
Omi, Kōji, 195
Onoda, Junjiro, 84
Operation Meetinghouse, 38
Operation Paperclip, 31, 205
Order of the Rising Sun, 35
ortho tolyl urethane (OTU), 43, 44
O'Sullivan, John, 8, 16
Ōta, Atsuo, 152
outreach, 66, 115-117, 168, 189
Oya, Atsushi, 102
Oyama, Koichiro, 84
Ozawa, Ichiro, 179–80

pacifism, 64
Paine, Thomas O., 159
Pangborn, Clyde, 61
Pencil (rocket), 42, 51, 79, 112, 205
Philippines, 97–98
Phobos Mars lander (Soviet Union), 149
Pickering, Bill, 29
Pioneer (satellite), 50
platinum, 44
Plaza Accords, 160, 167, 175
postal rocket, 59
postwar managerialism, 66
Prince Motor Company, 93, 104, 169, 234n7
private sector, involvement in space sector, 93, 96, 97, 104–6, 154–55
protests, 6, 110, 118, 119, 121, 180
Proton (rocket), 176
public bads, 112, 120
public-private networks, in civilian space program, 89–90
public opinion, 55, 66, 78, 187-188

Qian Xuesen, 29, 73
Q (rocket), 88, 102
Quasi-Zenith Satellite System (QZSS), 2, 141, 193

R-7 Semyorka (rocket), 101
radar, 30, 135
Radio Regulatory Bureau, 119
Radio Research Laboratory, 96, 97, 104
Reagan, Ronald, 163
relocations, 117
research and development (R&D): concerns regarding, 119; costs of, 147–48; decentralization of, 101; growth of, 33, 51; innovation in, 42–44; material scarcity in, 39–40; private sector contributions in, 97; projects in, 39; of the Soviet Union, 94; by subcontractors, 43–44; success in, 42–44; variations of, 34
Research Commission on the Military Uses of Balloons, 34
Rikagaku Kenkyūjo (RIKEN), 33, 80
Robertson, Andrew, 64, 81
rocket cars, 59–60
rocket engines *see individual engines.*
rocket festivals, 66–67, 116, 137
rocket play, 111, 115
rocketry, Japanese: applied research for, 35–36; benefits of, 137; budget for, 95; centralization of, 233n72; cost of, 92; delusion and, 55–67; demilitarization process of, 54; designs of, 55; destruction to, 30; development of, 31–32; failures of, 38–40; humanizing wartime, 67–69; illustration of, 41; innovation in, 42–44; introduction to, 1–4; launching process in, 82; limitations of, 30–31, 38–40; militarism and, 55–67, 75, 140; peaceful, 72;

rocketry, Japanese (*cont.*)
prewar, 31, 33–38; processes of, 108–9; recapturing status and entrenching influence in, 82–89; reconfiguration of narratives of, 54; reputation of, 64–65; science fiction and, 58; specialists in, 45–51; success in, 42–44; surging of, 54; as symbol of power, 61–63; technologies for, 215n67; wartime, 38–40; as weapons, 3, 62, 141; writings regarding, 55–60, 62–63, 66
Rocketry Observation Cooperation Association, 110
Rocket System Corporation, 170, 175
Roketto (Itokawa book), 74
Roscosmos, 205
Rusk, Dean, 103
Russo-Japanese War, 34
ryūsei (wooden rockets), 66

S-520 (rocket), 146
Sagamihara, Japan, 121
Sagamihara Space Research Center and Campus, 136, 145, 165
Sagan, Carl, 196, 200
Saikin no hikōki to shōrai (The Present and Future of Flying Machines) (Kaishima), 55
Saito, Akio, 58, 62
Saito, Shigefumi, 135
Sakamoto, Hiroshi, 134
Sakidzu detarame nise bunka o hōmure (Bury the Bullshit and False Culture First) (Hori book), 58
Sakigake (probe), 87, 134, 197, 205
Sakin kodomo kagaku dokuhon (A Children's Reader on Recent Science), 59
Salyut space station, 149
Samuels, Richard, 2–3, 47
Sanboku Edogawa plant (Hodogaya Chemical) 44
Sanders, Murray, 212n12
Sarabhai, Vikram, 73
Sasakawa Peace Fund, 141
satellites: budget for, 96; capabilities of, 251n104; commercial use of, 95–99; communications of, 95–99; corporations and, 183; cost of, 98; death of industry of, 160–62; developments of, 94, 146; domestic communications, 98; economic tensions regarding, 160–62; engineering test, 146; exporting of, 146; function of, 23; geostationary, 145; importance of, 95–97; innovation in, 99; militarization and, 251n104; as motivators for space research, 95–99; origin of, 1–2; private sector companies and, 96; procedures of, 138; research and development (R&D) for, 97; restrictions regarding, 98–99; on rockets, 228n88; for surveillance, 141; for television, 95–96. *See also specific satellites*
Sato, Eisaku, 86, 99–100, 103, 104
Sato, Yasushi, 53–54, 84, 88, 89, 156
SCAPIN301 (directive), 46–47
schisms with fishermen, 119–21
Science and Technology Agency (STA), 15, 85, 89–90, 92, 101–2, 104, 112, 146, 190
Science and Technology Commission, 7
Science and Technology Committee, 182, 186
science fiction, rocketry and, 58
scientific nationalism, 34
Seisan kenkyū (Itokawa book), 73
Sekai no kūkō (World Aviation) (magazine), 68
Sekkeika no Yume ("Designers' Dreams") (magazine column), 55, 56–57
Senō, Tarō, 60
Seo, Mitsuyo, 61
Seventh Military Technical Laboratory, 220n145

Shenzhou I, 179
Shibato, Koji, 169
Shigemichi, Sonoyama, 159
Shigeru, Yoshida, 63–64
Shima, Hideo, 16, 48, 49, 88, 105, 125, 147
Shin Chūō Company, 93
Shin Meiwa (company), 47
Shinkansen (bullet train), 5, 19, 45, 65, 88, 97, 180,
Shinmura, Katsu, 163
shock wave theory, 47
Shōnen kurabu (magazine), 59
Showa Denko, 117
Shūkan shōkokumin (The Weekly Young Citizen), 62
Shūsui *see Mitsubishi J8M1/Ki-200 Shūsui*
Siddiqi, Asif, 9
Skipper, Howard E., 212n12
Skylab space station, 149
Smith, E. E. "Doc," 55
Sogame, Eiji, 177
solid-fuel rockets (SFRs), 85, 94, 96, 99, 207n4
Sora (Sky) (magazine), 55
sounding rocket, 205
Space Activities Agency, 187
Space Aeronautics Research Lab (Tokyo University), 48
Space and Technology Committee, 184
Space Applications Mission Directorate, 192
Space Battleship Yamato (2010 movie), 53–54, 126
Space Communications Corporation, 126
Space Development Promotion Headquarters, 183
space exploration: as defense industry, 12; global history of, 12–15; histories of the future of, 23–24; importance of, 23–24; justification of, 11; people's history of, 8–12. *See also specific aspects*
Space Flyer Unit, 187
space food competition, 136
Space Japan News (newspaper), 16
Space Japan Review (trade publication), 79, 127, 134, 137
space plane, 152-153, 184, 190,
space policy, 187; agitation toward, 94; Hideo Itokawa and, 128–31; influence to, 13, 98; introduction to, 125–28; peaceful purposes in, 14, 138–41; stakeholders in, 104; Yasunori Matogawa and, 131–33
space program, Japanese: accomplishments of, 126, 145–50; administrative structure of, 100; American influence on, 14; as bright life, 136–38; budget of, 195; challenges of, 165; challenges to peaceful usage of, 138–41; changes to, 127; civilian, 14, 89–90, 154; commercial benefit of, 154–55; contestation in, 7; decentralization of, 4; development of, 208n17; employee reductions in, 188; expansion of, 143, 145–50; failures of, 4, 197; focus of, 14; gender balance in, 127; geography of, 6; goal of, 13; growing ambitions of, 145–53; holistic portrait of, 4–8; ignorance regarding, 4–5; independence of, 153–64; infrastructure of, 5; justifications in, 127; local impact of, 13; medium-term plan of, 2; military involvement in, 193; national pride and, 134–36; opposition to, 109–10; overseas providers for, 14; overview of, 196–97; partnerships/contractors of, 5–6; peaceful purposes of, 14, 138–41; political concerns regarding, 183–88; principles of, 151; private enterprise in, 5–6; projects of, 3 (*see also specific projects*); psychological strain from, 79; public-private networks in, 89–90; reform of, 188–94; strategies of, 148–49; success of, 145–53, 197; summary of, 194–95

space research, Japanese: centralization of, 233n72; cost of, 184; everyday benefits of, 133–38; expenditures in, 127; involvement in, 3; outreach activities of, 127–28; overview of, 196–97; promotion of, 131–33; satellites as motivators for, 95–99
Space Shuttle, 105, 126, 127, 144, 159
Space Station Integration and Promotion Center, 163
Space Systems Laboratory (Kyushu Institute of Technology), 192
Space Transportation Mission Directorate, 192
Spain, 228n88
Special Committee for the Peaceful Usage of Space, 98
Sputnik, 50, 70, 95, 101, 106, 126
Stapledon, Olaf, 55
Stars and Stripes (newspaper), 85
Star Trek (TV show), 126
statutory corporations in Japan, 96
Stephenson, H. Kirk, 212n12
Strategic Defense Initiative, 163
Strategic Headquarters for Space Policy, 192
subcontractors, rocket manufacturing, 43–44
submarines, Japanese, 27, 28
Suisei (probe), 87, 205
Sumitomo Chemicals, 44
Super-30 clause 1, 170
Superbird (satellite), 126
Supreme Commander of the Allied Forces (SCAP), 46
surveillance satellites, 34, 141
Susumu, Nikaido, 101
Sweden, 228n88
Switzerland, 228n88
Syncom 2 (satellite), 102

Tachikawa Aircraft Company, 93
Taepodong (rocket), 14, 179
"Taiheiyō sensō mitsuwa kaitei ichi-man go-seri no misshi" (A Secret Tale of the Pacific War) (Iwaya), 68
Takashi, Nishiyama, 19–20
Takayama, Shōichi, 81
Takeo, Miki, 103
Takeo, Nagao, 220n145
Takimoto, Tomoyuki, 199
Tanaka, Kimi, 114
Tanaka, Makiko, 182
Tanegashima (island) 13, 179, 243n65
Tanegashima Launch Center: acceptance of, 11; arrival and welcome to, 114–16; development of, 106, 112; fishing unions and, 116–18; H-2 rocket testing at, 172; origin of, 87; photo of, 113; resistance to, 116–18; shutdowns of, 119; statistics of, 116
Tani, Ichirō, 35, 48
Tashiro, 146
Technical Department of the Imperial Japanese Army, 33
technical missions (Allied) 28–30
Technical Research Institute, 33
Technical Subcommittee for ISAS reform, 183
techno-fascism, 34, 66
technoimperialism, 66
technological expectation, prewar, 32–33
technonationalism, 3, 11, 34, 65
Telecommunications Advancement Agency of Japan, 233n72
telecommunications companies, space technological projects of, 94
televisions, 95–96, 126
Telstar 1 (satellite), 102
Tenma (satellite), 134
Tenmon Geppō (The Astronomical Herald) (journal), 224n23
textile industry, 32
Thailand, 97–98
Things to Come (Korda film), 55, 58
Thompson, E. P., 8

Thor-Delta 2914 (1974), 158
Thor-Delta launch system, 85, 88, 98, 103, 142, 143–44, 157, 165, 206
three-axis stabilization, 146
Tipton, Elise K., 225n51
Titan 4A (rocket), 179
Tōa Fuel Corporation, 82
Tokugawa, Yoshitoshi, 32, 33
Tokurō-2 (engine), 37
Tokuyama Dam, 194
Tokuyama facility of the Imperial Japanese Navy, 81
Tokyo, 38, 167
Tokyo Broadcasting System (TBS), 143, 151
Tokyo Imperial University, 34, 51
Tomifumi, Godai, 48, 148, 161–62, 169-172, 182-183,
Tomonaga, Shin'ichirō, 39
torpedoes, 37
Toshiba, 2, 97, 99, 160, 193
tourism, for rocket launches, 114–15
Trade Act of 1974, 160
transatlantic flights, 61
transfer orbit insertion, 146
Tropical Rainfall Measuring Mission, 163
TRW (company), 85, 98
Tsiolkovsky, Konstantin, 1, 55, 73, 138
Tsujimoto, Kiyomi, 187–88
Tsukuba Expo Center, 170
Tsukuba Space Center, 145, 163, 165
Tsuruta, Koichiro, 180
Type 96 (fighter), 67
Uchida, Isao, 177
Uchinourachō, Japan, 115, 119, 120
Uchinoura Ladies' Association, 83, 114
Uchinoura Launch Center: arrival and welcome to, 114–16; decamping to, 82, 84; development of, 106; fishing unions and, 108, 116–18; launcher at, 86; launch interruptions at, 118; origin of, 112; photos of, 115; probe launch at, 198; resistance to, 116–18; shutdowns of, 119; statistics of, 114
Uchū o sampo suru: roketto no zuihitsu (Esseys on Rockets) (Itokawa), 72
Ueda, Takeo, 140
Uesugi, Kuninori, 148
United Kingdom, 228n88
United Nations Transitional Authority in Cambodia (UNTAC), 140
United States: anti-Japanese sentiment in, 155; collaboration with Japan and, 85; government control in, 94; influence of, 142; Japanese negotiations with, 105; Japanese reliance on, 97–98; Japanese resistance to, 85; moneymaking goal of, 13; private sector involvement in, 94; research and development (R&D) in, 34; rocket mishaps in, 179; rocket purchases by, 103; rocketry cost in, 92; satellite industry concerns of, 160; satellite launches of, 102, 228n88; space program expansion and, 12; space program orchestration in, 94; telecommunications companies in, 94; tensions between Japan and, 155–59; trade deficit of, 155
University of Tokyo, 4, 7
Unmanned Space Experiment Free Flyer (USEF), 233n72
US Arms Control and Disarmament Agency, 102–3
US Army Scientific Intelligence Survey, 29
US Navy, 28–30
USS *Missouri,* 64
US-Japan relations, 34–35, 178–179, 234–236
USSR/Russia: Cold War and, 84; defense industry in, 12; failures of, 13; Japanese cooperation with, 103; launch market of, 176; moneymaking goal of, 13; research and development of, 94; rocketry research of, 70; satellite launches of, 228n88; space developments of, 23; space program expansion and, 12; space research of, 93–94; Sputnik of, 50, 70, 95, 101, 106, 126

USS *Sawfish,* 27
Usuda tracking center, 145

V-2 ballistic missile, 31, 206
Valier, Max, 60
Vaughn, Dianne, 188
Vietnam, 70
Vogel, Ezra, 136
von Braun, Wernher, 73
von Kármán, Theodore, 29
von Opel, Fritz, 59–60

Wada, Misao, 35
Wakayama University, 192
Walls, Gordon T., 212n12
Wall Street Journal (newspaper), 4, 74
Walter, Hellmuth, 42
Walther (engine), 43
War Decider/Kessen (weapon), 39
Warsaw Pact, 92–93, 156
Watanabe, Hirotaka, 101
Webb, James, 10, 86
Wege zur Raumschiffahrt (Ways to Space Travel) (Oberth), 59
Wells, H. G., 55, 58
West Germany, 228n88
Whittle, Frank, 67
World War I, 33
World War II, 1, 27, 38–39, 64, 196, 211n4
Wray, William, 138, 187
Wright, Wilbur, 33

Xuesen, Qian, 73

Yamamoto, Eisuke, 34
Yamamoto, Isoroku, 37
Yamana, Naruo, 82
Yamano, Masato, 176, 178
Yamanouchi, Shūichirō, 48, 190
Yamato (battleship), 27, 53, 54
Yamazaki, Naoko, 127
Yasunori, Matogawa, 30
Yasushigaya Chemicals, 44
yen, 160, 167, 174–75
YES 89 Space Hall (Yokohama Expo), 170
Yohkoh solar observatory, 87
Yokkaichi facility, 81
Yokohama Expo, 170
Yokosuka MXY-7 (suicide air torpedo), 205
Yomiuri Shimbun (newspaper), 16, 112, 130–31
Yoshio, Nishina, 61, 63
Youngs, Harry H., 212n12
Yugoslavia, 5, 14, 92–93, 99

Zambia, 99
Zuikaku (aircraft carrier), 42
Zwicky, Fritz, 29, 31, 45, 51

The authorized representative in the EU for product safety and compliance is:
Mare Nostrum Group
B.V Doelen 72
4831 GR Breda
The Netherlands

www.ingramcontent.com/pod-product-compliance
Lightning Source LLC
Chambersburg PA
CBHW030605230426
43661CB00053B/1846

* 9 7 8 1 5 0 3 6 4 4 7 8 6 *